Proceedings in Technology Transfer

The book series "Proceedings in Technology Transfer" provides a valuable platform for researchers, practitioners, and policymakers to share their experiences and insights related to technology transfer. Our aim is to facilitate knowledge exchange between academic researchers and industry professionals, accelerating the introduction of new technologies and driving economic growth.

Our series publishes books and conference proceedings that focus on original research findings, best practice recommendations, emerging trends, and technology evaluations. The series covers all topics which explore the transfer of technology in various industry sectors that require innovations and economic growth.

Join us in advancing the technology transfer and its impact on innovation. Publish your proceedings with us and offer insights and expertise to professionals in your field.

Mokhtar Awang
Editor

1st International Conference on Civil, Architecture, Environmental Engineering, and Technology

Springer

Editor
Mokhtar Awang
Mechanical Engineering Program, Faculty of Engineering
University of Technology Brunei
Bandar Seri Begawan, Brunei Darussalam

ISSN 2948-2321 ISSN 2948-233X (electronic)
Proceedings in Technology Transfer
ISBN 978-3-031-94346-1 ISBN 978-3-031-94347-8 (eBook)
https://doi.org/10.1007/978-3-031-94347-8

This Springer imprint is published by the registered company Springer Nature Switzerland AG
The registered company address is: Gewerbestrasse 11, 6330 Cham, Switzerland

Contents

Mitigation of Pier Structure Failure Due to Sediment

Muhammad Hawari Ikhwan(✉), Jojok Widodo Soetjipto, and Retno Utami Agung Wiyono

Program of Civil Engineering, Jember University, Jl Kalimantan 37, 68121 Jember, Indonesia
muh.hawari.ikhwan@gmail.com

Abstract. The pier structure on the Puger coast, Jember Regency, has the potential to influence the dynamics of sedimentation that can impact the coastal environment. This article aims to identify and evaluate mitigation strategies that can be applied to reduce the negative impacts of pier structures on sedimentation in the area. Through a comprehensive literature review and field data analysis, this study found that sedimentation can be controlled through breakwater structures, optimizing pier design, and applying environmentally friendly technologies. In addition, community participation in coastal management and regular monitoring are also essential elements in an effective mitigation strategy. The results of this study are expected to be a reference for the government and stakeholders in planning and implementing sustainable and environmentally friendly pier structures on the Puger coast and other areas.

Keywords: mitigation · pier structure · sedimentation

1 Introduction

Jember Regency is one area with abundant diversity and natural resources from the tourism sector in the southern coastal areas, such as Papuma Beach, Watu Ulo Beach, Payangan Beach, Puger Beach, and Bandealit Beach [1]. In addition to the growing potential in the tourism sector, many people use coastal areas to cultivate ponds, process fish, and buy and sell fishermen's catches. This is evidenced by the existence of a Fishing Port or Fish Landing Base (PPI) managed by the Coastal Fishing Port Management Unit (UPPPP) on the Puger Coast. The construction of a pier on Puger Beach, Jember Regency, aims to implement the importance of maintaining and cultivating marine products and providing an area for sellers of marine products worth visiting for both local and foreign tourists [2, 3].

In addition to abundant natural resources, Jember Regency is an area prone to disasters, one of which is significant climate change. This dramatically affects people's lifestyles, such as damage to coastal ecosystems, changes in sea waves, and rising sea temperatures. These conditions significantly affect people's lifestyles in the Puger Coast Area, Jember Regency [4]. Research by Dr. Gayatri Marliyani, S.T, M.Sc explains the

The original version of the chapter has been revised. The Author name has been changed to "Jojok Widodo Soetjipto" both in the Front Matter and also in the Chapter. A correction to this chapter can be found at https://doi.org/10.1007/978-3-031-94347-8_25

M. Awang (Ed.): ICCAEET 2024, PTT, pp. 1–5, 2025.
https://doi.org/10.1007/978-3-031-94347-8_1

potential for a tsunami to hit the southern coastal areas, including Jember Regency, East Java [5].

In addition, the sedimentation rate at PPI Puger is included in the category of reasonably rapid sedimentation [6], so it can damage building structures and infrastructure [7]. This research focuses on mitigation efforts due to sediment in the pier area at Puger Beach, Jember Regency so that the use of the pier can be maximized.

2 Method

Researchers use descriptive methods to collect data and focus on the methods and solutions. In the first step, researchers collect data by searching for articles on mitigating pier construction against sedimentation published in 2014 to 2024 and searched through several websites such as Science Direct, Springer, Google Scholar, Research Gate, and Publish or Perish. In this case, researchers took 25 articles with various methods for mitigating pier construction against sedimentation. The second step is to group the articles according to the methods used. The last step is to classify articles based on their solutions.

3 Classification of Sedimentation Control

Articles on sedimentation control can be classified into several main categories. First, some articles focus on concrete methods in physical form to reduce disaster risks, such as the use of technology and engineering structures. Second, there are articles that use environmentally friendly methods naturally or biologically. Third, there are articles that use a managerial approach or good management. Table 1 shows the classification of categories based on methods of mitigating excessive sediment.

Table 1. Classification of sediment control categories

Category	Approach Strategies
Physical Method	Retaining Walls are retaining walls built on riverbanks to prevent landslides and cliff erosion [8]
	Flow Deflectors are structures placed in river channels to direct water flow away from erosion-prone cliffs
	Sediment Retention Dam (checkdam) to hold sediment in river flow [1]
Natural Method	Reforestation of forests around river headwaters by planting trees and bushes in the river headwaters and along river banks to strengthen soil structure with a deep and extensive root system [9]

(*continued*)

Table 1. (*continued*)

Category	Approach Strategies
	Rainwater management can be achieved by directing rainwater flow to areas with dense vegetation or natural drainage systems that can absorb water and reduce erosion [10]
Managerial Method	Community Participation and Empowerment by encouraging active community participation in the planning and implementation of watershed management programs and encouraging cross-sector and cross-regional collaboration to address complex and multi-dimensional erosion problems [11]
	Institutions and Policies in the form of stakeholder involvement, including government, communities, and the private sector in watershed management and the development and enforcement of policies and regulations that support soil and water conservation and erosion management [12]

4 Mitigation in Sedimentation Control

4.1 Physical Approach

Riverbank retaining structures are used to prevent erosion and collapse of riverbanks. Different types of retaining structures can be used, depending on the river conditions and the intended use. Here are some types of river bank retaining structures [13]:

1. Gabions: Gabions are made of steel wire filled with rocks. They are often used because they are flexible, strong, and easy to adjust to the shape of the river bank. They are also effective in absorbing the energy of water flow.
2. Sheet pile: Retainers made of steel or concrete sheets embedded in the ground. Sheet piles are often used in areas with strong water flow or unstable soil.
3. Revetments: A layer of stone or concrete is installed along a river bank to protect it from erosion. Revetments can be made of stone (riprap), concrete, or a combination.
4. Retaining wall: Usually made of concrete or stone, it is used to hold back soil movement and maintain the stability of river banks.
5. Geotextile Structure: Utilizes geotextile material installed under the soil layer to prevent erosion and help stabilize the bank.
6. Riprap: A layer of large rocks is placed along a river bank to prevent erosion due to water flow. Riprap is often used in areas of rivers with fast-flowing currents.

Flow deflectors are structures placed in river channels to direct water flow away from eroding cliffs. Here are some types of flow deflectors that are often used: [13]

1. Groynes: Structures built perpendicular to the flow of the river. Groins work by reducing the speed of water flow near the cliff, thereby reducing erosion and helping to deposit sediment in the desired area.
2. Spur Dikes: Similar to groins, spur dikes are structures built perpendicular to the water flow or at a slight angle. They serve to change the direction of river flow and reduce the speed of water around eroding cliffs.

3. Vanes: Vanes are slanted structures installed in river channels to direct water flow away from eroding cliffs. They are usually made of wood, concrete, or steel.
4. Rock Deflectors: Made of large rocks placed in the river channel to change the direction of water flow. These structures are usually placed on the side of a river prone to erosion and help direct water flow away from that area.
5. Weirs: Low dams built across a river to raise the water level behind it and divert the flow. Weirs can be used to direct water flow away from vulnerable cliffs.

Flow deflectors effectively protect riverbanks from erosion by changing water flow patterns, but they must be carefully designed and installed to ensure long-term positive impacts.

4.2 Natural Approach

Reforestation around river headwaters by planting trees and shrubs is one effective way to strengthen soil structure and prevent erosion. The deep and extensive roots of trees and shrubs help hold soil, maintain moisture, and reduce surface water flow that can cause erosion. Some methods and types of reforestations that can be applied around river headwaters and along river banks include selecting plant species, gradual planting, and diverse planting patterns [9].

Rainwater management can be done by directing the flow of rainwater to areas with dense vegetation or natural drainage systems that can absorb water and reduce erosion. Some methods that can be used are Swale Making, Installing Rainwater Harvesting Systems, and Making Terracing (Multi-level).

4.3 Managerial Approach

Community participation and empowerment are key to successful watershed management and addressing complex erosion issues. This approach actively involves local communities in the planning and implementing watershed management programs, and encourages cross-sector and cross-regional collaboration. Increasing awareness and education, including in the planning process, empowerment through access to resources and training, and cross-sector and cross-regional collaboration.[14].

Institutions and Policies in watershed management refer to the involvement of various stakeholders, including government, communities, and the private sector, in managing water and land resources [14]. It also involves developing and enforcing policies and regulations that support soil and water conservation and erosion management with the implementation of Stakeholder Involvement from government, local communities, and the private sector.

5 Conclusion

In overcoming the failure of dock buildings due to sediment, a mitigation approach is attempted in controlling sedimentation, which includes three main approaches, namely physical, natural, and managerial. Physical approach with Retaining Walls consists of retaining walls built on riverbanks, and flow deflectors are structures placed in river channels to direct water flow away from erosion-prone cliffs to prevent landslides and cliff erosion. Natural approach with Reforestation of forests around river headwaters by planting trees and bushes in the river headwaters and along river banks to strengthen soil structure with a deep and extensive root system. Rainwater management can be achieved by directing rainwater flow to areas with dense vegetation or natural drainage systems that absorb water and reduce erosion. Managerial approach involving government, local communities, and private sectors to maintain all forms of development and planning to overcome pier structure failures.

References

1. Alfianto, A., Cecilia, S., Hidayah, A.N., Sukatja, C.B.: Perencanaan Sabo Untuk Mengendalikan Laju Sedimentasi Di Rawapening. J. SUMBER DAYA AIR. **17**(1), 1–12 (2021). https://doi.org/10.32679/jsda.v17i1.610
2. Alhasyi, G.P.A.: Karakteristik Sedimen Berdasarkan Analisis Ukuran Butir Di Pesisir Pantai Palanro Kabupaten Barru, Provinsi Sulawesi Selatan. Dintek. **17**(01), 9–19 (2024) Retrieved from https://www.jurnal.ummu.ac.id/index.php/dintek/article/view/1943
3. Apriyanti, D.: Dampak dan Upaya Penanggulangan Terjadinya Abrasi Menggunakan Citra Satelit Studi Kasus Di Wilayah Pesisir Tanjung Benoa Bali. J. Ilm. Geomatika. **1**(1), 39–47 (2021). https://doi.org/10.31315/imagi.v1i1.4732
4. Arif, A.M.S., Hamdi, N.F.: Pengendalian Abrasi Pantai Tumpaan, Amurang, Minahasa Selatan, Sulawesi Utara Dengan Menggunakan Pegar Geobag Rangka Bambu. Ris. Sains Dan Teknol. Kelaut. **5**(2), 138–142 (2022). https://doi.org/10.62012/sensistek.v5i2.24253
5. Arifin, F., Muhibuddin, A., Saleh, H.: Mitigasi Kawasan Rawan Banjir Pada Daerah Aliran Sungai Jeneberang Dengan Menggunakan Sistem Informasi Geografis. Urban Reg. Stud. J. **6**(1), 71–82 (2023). https://doi.org/10.35965/ursj.v6i1.3813
6. Pradanan, A.A.: Pola Distribusi Sedimen Di Perairan PPI PUGER Kab Jember. Jawa Timur. Universitas Hang Tuah (2018)
7. Aurel, T.: Analasis Mitigasi Bencana Abrasi Pada Wilayah Pantai Galesong Kabupaten Takalar. SENSISTEK. **6**, 241–254 (2023)
8. El-Mahdy, M.E.S., Saber, A., Moursy, F.E., Sharaky, A., Saleh, N.: Coastal erosion risk assessment and applied mitigation measures at Ezbet Elborg village, Egyptian delta. Ain Shams Eng. J. **13**(3), 101621 (2022)
9. Liu, T., Jiang, K., Tan, Z., He, Q., Zhang, H., Wang, C.: A method for performing reforestation to effectively recover soil water content in extremely degraded tropical rain forests. Front. Ecol. Evol. **9**, 643994 (2021). https://doi.org/10.3389/fevo.2021.643994
10. Firoozi, A.A., Firoozi, A.A.: Water erosion processes: Mechanisms, impact, and management strategies. Results Eng. **24**, 103237 (2024). https://doi.org/10.1016/j.rineng.2024.103237. https://www.sciencedirect.com/science/article/pii/S2590123024014919
11. Basuki, T., et al.: Improvement of integrated watershed management in Indonesia for mitigation and adaptation to climate change: A review. Sustain. For. **14**, 9997 (2022). https://doi.org/10.3390/su14169997
12. Pambudi, A.: Watershed management in Indonesia: A regulation, institution, and policy review. J. Perenc. Pembang. Indones. J. Dev. Plan. **3**, 185–202 (2019). https://doi.org/10.36574/jpp.v3i2.74
13. King, H.: 2015 State of Art: The Use of Groynes for Riverbank Erosion Protection (2015). https://doi.org/10.13140/RGs.2.12s773.3202
14. Narendra, B.H., et al.: A review on sustainability of watershed management in Indonesia. Sustain. For. **13**, 11125 (2021). https://doi.org/10.3390/su131911125

Penajam Tower Development Location Feasibility Study

Ahmad Imamuddin Hidayatullah(✉) and Alpian Nur

Department of Civil Engineering, University of 17 August 1945 Samarinda, Samarinda, Indonesia
ahmad.imamuddin.h@gmail.com

Abstract. The Penajam City Area, located near the National Capital Region, needs to strengthen its character with landmarks such as towers, monuments, or towers, as well as supporting functions such as commercial, tourism, and socio-cultural functions. The construction of the Penajam Tower will serve as a landmark building in the form of an observation tower with supporting facilities for tourism activities around the area. There are four alternative locations for the construction of the tower, including Water Front City, in front of the Regent's Office, in front of the Stadium, and behind the Stadium, with the feasibility determination based on six aspects such as Spatial, Technical, Transportation, Economic and Financial, Environmental, and Support. The analysis results indicate that location 2 is the best choice for the construction of the Penajam Tower as it meets the criteria with a higher weighted value compared to other locations.

Keywords: Feasibility Study · Penajam Tower · Location Alternative

1 Introduction

Over time, development and urban planning have become more advanced and changed. This has led to the emergence of a city identity crisis. Cities in Indonesia are experiencing an identity crisis, making cities in Indonesia look the same as each [1]. Seeing the existing problems, a monotonous city or illegibility city really needs design to create its own city identity, where identity formation can be done with the mental image of the people in the city, besides that a monotonous city makes people feel uncomfortable because they do not have a starting point/orientation. The starting point/orientation point can be assisted by the presence of landmarks [2].

The development of an urban area requires elements city builders who will produce the image of the city, where in the book Image of The City, Kevin Lynch [3] revealed that there are 5 elements that form the image physical city, namely: path, edge, district, nodes (nodes), and landmarks. These five elements can represent the taste of of an area and provide a strong image of the city. The existence of markers in a city will give it a strong character. For the city with the support of other elements.

The improvement of office building infrastructure is very necessary in line with the increasingly rapid socio-economic growth in almost all regions in Indonesia. So

M. Awang (Ed.): ICCAEET 2024, PTT, pp. 6–15, 2025.
https://doi.org/10.1007/978-3-031-94347-8_2

that the development of office building infrastructure is very important in supporting the achievement of economic growth rates. The development of office building infrastructure in the form of improving or renovating office buildings must be in accordance with the development of the need for additional economic services to the community.

Penajam City area as an area close to the National Capital area needs to strengthen its character that can provide a different nuance. This marker can be in various forms such as towers, monuments, sculptures or towers that need to be accompanied by supporting functions for the tower area such as commercial, tourism and socio-cultural functions. So it can be said that the Penajam Tower Development is a marker building in the form of a viewing tower building equipped with supporting facilities in the form of tourism activities around the area.

2 Research Methods

The location of this research is in Penajam District as shown in Fig. 1.

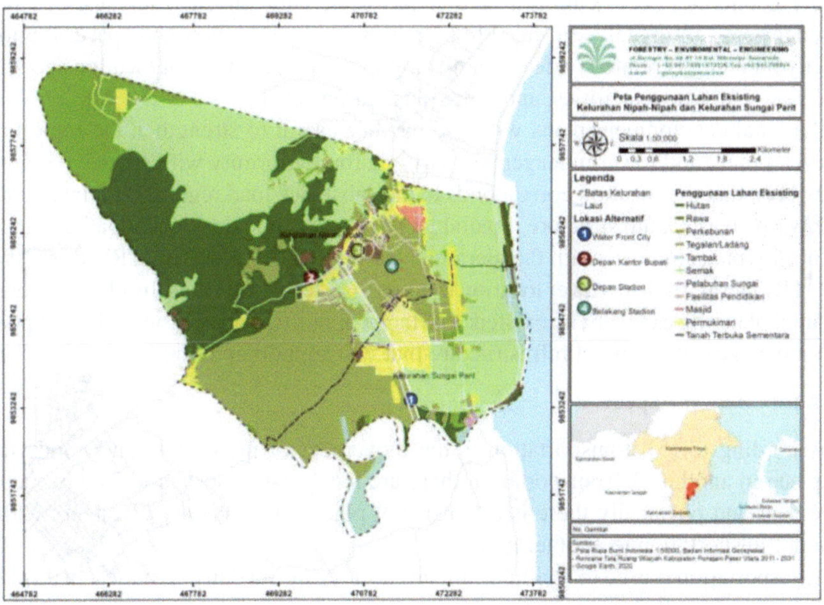

Fig. 1. Location of the Penajam Tower Development Plan

To determine alternative locations where towers will be built Sharpener with initial criteria:

- Located in a cultivation area and river border
- Government asset land
- Ease of Accessibility
- There is a Network of Infrastructure Facilities

- Stakeholder suggestions

Based on the initial criteria, the location plan is systematically determined as follows:

1. Location of Water Front City, Sungai Parit Subdistrict, Penajam District
2. Location in front of the Regent's Office, Nipah-Nipah Sub-district, Penajam District
3. Location in front of Nipah-Nipah Sub-district Stadium, Penajam District
4. Location Behind Nipah-Nipah Sub-district Stadium, Penajam District

Of the four alternative locations, the next step is to analyze the level of detailed feasibility of each location.

1. Determination of Location Eligibility Criteria

The criteria are determined based on several aspects, including:

- Spatial planning aspects.

Location and Space are the most fundamental parts of regional and city planning. Regional and city planning with various variations and all objects observed or planned, namely economic, socio-cultural, natural physical, built physical and various other related aspects, ultimately still need to be represented in location and space. Regional and city planning is related to decision making to determine the right location and space for various activities, facilities, and other planned objects [4].

Regional development plans will also be very useful to strengthen the importance of development and tower management. Criteria for conformity with spatial planning it can be seen that the location corresponds to the substance load regional spatial planning (RTRW) such as spatial structure, Spatial Patterns and Strategic Areas. Spatial planning conformity in accordance with the hierarchy starting from RTRWN, RTRWP, RTRWK, to RDTR. The alternative tower location is not in disaster-prone areas. In addition, related to spatial planning aspects The existence of the people's tower location is also supported by completeness of regional infrastructure that will support activities in the tower area.

- Transportation Aspects

According to [5], transportation is the movement of passengers and goods from one place to another. In transportation there are two most important elements, namely movement and physically there is a transfer of place for goods or passengers with or without transportation to another place.

Transportation is used to facilitate humans in carrying out daily activities. In this case, what is assessed in the transportation aspect is the impact on congestion, Accessibility & Traffic Conditions or existing traffic conditions in front of the location and Transportation Nodes (Intra-Intermodal), Distance between alternative tower locations and existing terminals.

- Technical Aspects

The technical approach in this case, is not comprehensive because it is still in the initial location search stage approaching conformity with several aspects: Mechanics and Soil Type, Construction Ease, Topography (Flat & Ramps), Area View of the Tower (2xh), Ideal Size Area (9 ha).

– Other Aspects

Socio-political factors are also an important concern, meaning that the support of the community and government in this matter the added value of determining is: Support Government, Community Support and Private Support (Table 1).

Table 1. Support Eligibility Assessment Matrix

No.	Variables	Assumptions	Criteria	Classification	Score
	Government Support	Government support is very important regarding readiness for implementation	Not Supported	Low	1
			Abstain	Currently	2
			Get Support	Tall	3
	Community Support	Support from the community around the location, has an impact on location supervision	Not supported	Low	1
			Abstain	Currently	2
			Get support	Tall	3
	Private Support	Private support in terms of development feasibility assessment	Not supported	Low	1
			Abstain	Currently	2
			Get Support	Tall	3

2. Weighting and Multi Criteria Analysis

Multicriteria analysis is a decision-making tool developed for complex multicriteria problems that include qualitative and/or quantitative aspects in the decision-making process [6].

Weighting is carried out by all relevant stakeholders such as regulator, operator, and user. Multicriteria analysis is performed by scoring each alternative port location based on established criteria previously (Fig. 2).

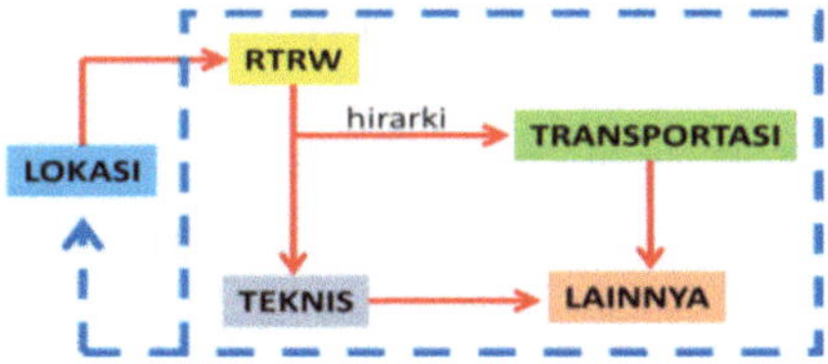

Fig. 2. Weighting Process

Based on the explanation above, the assessment is carried out by using a Likert scale for the assessment of each criteria that have been determined in the previous explanation. Table below explains the criteria with their values and weighting (Tables 2 and 3).

Table 2. Location Criteria Weighting Matrix

No.	Criteria	Description	Note	Score
1	Spatial Aspects			
	a. Compliance with RTRWN, RTRWP, RTRWK, RDTR	In accordance with the RTR and contained in the RTR as a whole	No	1
			Not enough	2
			In accordance	3
	b. Not Located in a Disaster Area	Not located in a disaster-prone area (far from the disaster location)	Near	1
			Currently	2
			Far	3
	c. Completeness of Regional Infrastructure Facilities	Available PLN Network, Telephone, Internet, Clean Water, Drainage, Garbage	< 50%	1
			> 50%	2
			complete	3
2	Technical Aspects			
	a. Soil mechanics and types	Efficiency of foundation and piling work implementation		
	b. Ease of construction			
	c. Topography (flat & sloping)			
	d. View of the area from the tower (2xh)	The position of the tower when it was built, had a good depth view		
	e. Ideal area (9 ha)			
	f. Topography (flat & sloping)	Available area in the flat – sloping category	< 30%	1
			30% - 60%	2
			> 60%	3
3	Transportation Aspects			
	a. Compliance with Tatranas, Tatrawil, Tatralok & RIPN	Appropriate and contained in the document as a whole	No	1
			Not enough	2
			In accordance	3
	b. Accessibility & Traffic Conditions	Congestion levels around the location area	Congested	1
			Currently	2
			Not enough	3

(*continued*)

Table 2. (*continued*)

No.	Criteria	Description	Note	Score
	c. Transportation Node (Intra-Intermodal)	Distance of location to transportation node (terminal)	> 1 km	1
			500 – 1 km	2
			< 500 m	3
4	Economics, Finance and Operations			
	a. Regional economic development	Impact on regional PAD and community welfare	A little	1
			Currently	2
			Lots	3
	b. Trading Area	Proximity to trading area locations	Far	1
			Currently	2
			Near	3
	c. Local Contribution (Socio-Cultural)	Can absorb local labor in each location	A little	1
			Currently	2
			Lots	3
	d. Financial Feasibility (IRR, NPV)	Project cannot be worked on	NPV < 0	1
		The project requires strong effort	NPV = 0	2
		Project can be done	NPV > 0	3
5	Environmental Aspects			
	a. Carrying Capacity and Capacity	DDDT from each location to the activities that will occur	Not enough	1
			Enough	2
			Capable	3
	b. Environmental Color (Flora & Fauna)	Minimal changes to the landscape & disturbance to flora/fauna	Tall	1
			Currently	2
			Low	3
6	Support Aspects			
	a. Provincial Government Support			
	b. District Government Support			
	c. Community Support			
	d. Private Support			

Table 3. Location Assesment Matrix

Recommendation	Grade	Score
Not feasible	Low	> 1
Eligible (with notes)	Currently	1 – 2
Worthy	Tall	2 – 3

3 Result and Discussion

The results of the feasibility analysis obtained a location ranking based on weighting, these ranking results may change based on other considerations due to changes in natural landscape conditions, a disaster occurred at that location, policy changes and so on. The current rankings are as follows:

1. Rank 1: location 2 in front of the Regent's Office, Kel. Nipah-Nipah District. Sharpener with a weight of 1.73
2. Rank 2: Location 3 In front of Nipah-Nipah Sub-district Stadium, Penajam District with a weight of 1.70
3. Rank 3: Location 4 Behind Nipah-Nipah Sub-district Stadium, Penajam District with a weight of 1.70 the same as location 3.
4. Rank 4: Location 1 Water Front City Sungai Parit Subdistrict Penajam District with a weight of 1.65.

The above ranking is a good description of the location among the locations. Others, but each has its own advantages and disadvantages, for example location 2 at rank 1 has a lack of topographic conditions so it is necessary engineering engineering related to the condition. Location 3 becomes the 2nd ranked location has a shortage of land area, so a strategy is needed to design tower building that adjusts the land area. For the location at Water Front City is land ownership so it requires financing and procurement documents land to be the location of Penajam Tower.

Based on the analysis results, at location 2 is a suitable location to build the Penajam Tower with several consideration. This ranking is the basis for selecting a location for construction of the Penajam Tower, but it is not an absolute necessity This location ranking, there are several considerations that can later change the ranking this, as over time there is change in funding capacity, changes policies and other conditions (Table 4).

Table 4. Analysis Result

No.	CRITERIA	Location 1: Water Front City		Location 2: In Front of Regent Office		Location 3: In Front of Stadium		Location 4: Behind Stadium	
		Score	Description	Score	Description	Score	Description	Score	Description
1	**Spatial Aspect**								
a	Compliance with RTRWN, RTRWP, RTRWK, RDTR	3.00	Residential Area	3.00	Residential Area	3.00	Residential Area	1.00	Green Open Space
b	Not in Disaster Area	3.00	No	3.00	No	3.00	No	3.00	No
c	Land Ownership Status	1.00	Medium	3.00	Government Land	3.00	Government Land	1.00	Government Land
d	Infrastructure Completeness	2.00	Medium	3.00	Complete	3.00	Complete	3.00	Complete
	Score 20% Total	**1.35**	2.25	**1.80**	3.00	**1.80**	3.00	**1.20**	2.00
2	**Technical Aspect**								
a	Soil Mechanics and Type	2.00	Medium	3.00	Good	3.00	Good	3.00	Good
b	Ease of Construction	2.00	Medium	2.00	Moderate	3.00	Easy	3.00	Easy
c	Topography (Flat & Sloping)	3.00	> 60% Flat	1.00	30%-60% Slightly Steep	3.00	> 60% Flat	3.00	> 60% Flat
d	View to Tower (2xH)	3.00	Good	3.00	Good	3.00	Good	3.00	Good
e	Ideal Area Size (9 ha)	3.00	Spacious	3.00	Spacious	3.00	Spacious	3.00	Spacious
	Score 20% Total	**3.90**	2.60	**3.60**	2.40	**4.50**	3.00	**4.50**	3.00
3	**Transportation Aspect**								
a	Traffic Impact	2.00	Moderate	1.00	Congested	1.00	Congested	3.00	Not Congested
b	Accessibility & Traffic Conditions	3.00	Smooth	2.00	Relatively Crowded	2.00	Relatively Crowded	3.00	Smooth
c	Transport Hub (Intra-Inter Modal)	1.00	Far from Terminal	1.00	Far from Terminal	1.00	Far from Terminal	1.00	Far from Terminal
	Score 20% Total	**0.90**	2.00	**0.60**	1.33	**0.60**	1.33	**1.05**	2.33
4	**Economic & Financial Aspect**								

(*continued*)

Table 4. (*continued*)

No.	CRITERIA	Location 1: Water Front City		Location 2: In Front of Regent Office		Location 3: In Front of Stadium		Location 4: Behind Stadium	
		Score	Description	Score	Description	Score	Description	Score	Description
a	Regional Economic Development	3.00	Moderate Growth	3.00	Moderate Growth	3.00	Moderate Growth	3.00	Moderate Growth
b	Proximity to Commercial Area	2.00	Fairly Close	2.00	Fairly Close	2.00	Fairly Close	2.00	Fairly Close
c	Contribution to Community	3.00	New Job Opportunities	3.00	New Job Opportunities	3.00	New Job Opportunities	3.00	New Job Opportunities
d	Financial Feasibility	3.00	NPV > 0	3.00	NPV > 0	3.00	NPV > 0	3.00	NPV > 0
	Score 10% Total	**2.20**	2.75	**2.20**	2.75	**2.20**	2.75	**2.20**	2.75
5	**Environmental Aspect**								
a	Land Carrying Capacity	3.00	High Land Support	3.00	High Land Support	2.00	High Land Support	3.00	High Land Support
b	Environmental Carrying Capacity	3.00	High Environmental Support	3.00	High Environmental Support	3.00	High Environmental Support	3.00	High Environmental Support
c	Environmental Conditions (Flora & Fauna)	2.00	Moderate Environmental Impact	3.00	Low Environmental Impact	3.00	Low Environmental Impact	2.00	Moderate Environmental Impact
	Score 10% Total	**0.80**	2.67	**0.90**	3.00	**0.80**	2.67	**0.80**	2.67
6	**Support Aspect**								
a	Provincial Government Support	2.00		2.00		2.00		2.00	
b	Regency Government Support	2.00		3.00		3.00		2.00	
c	Community Support	2.00		2.00		2.00		2.00	
d	Private Sector Support	2.00		2.00		2.00		2.00	
	Score 10% Total	**0.80**	2.00	**0.90**	2.25	**0.90**	2.25	**0.80**	2.00
	Grand Total	**9.95**	14.27	**10.00**	14.73	**10.80**	15.00	**10.55**	14.75
	Score	**1.66**	2.38	**1.67**	2.46	**1.80**	2.50	**1.76**	2.46

4 Conclusion

From the analysis results, a location ranking was obtained based on weighting which showed that location 2 in front of the Nipah-Nipah Village Regent's Office, Penajam District, had the highest ranking with a weight of 1.73. However, this ranking can change due to factors such as changes in natural conditions, disasters, policies, and others. Nevertheless, location 2 is considered feasible for the construction of the Penajam Tower with several considerations, such as topographic conditions that require engineering. This ranking is the basis for selecting a construction location, although it is not absolute because it can change over time and changes in certain conditions such as funding and policies.

References

1. Kompas: Ignoring the Potential of Cities in Indonesia Developing Generic. Retrieved March 15, 2024 , 2016 , from Ignoring Potential, Cities in Indonesia Develop Generic (kompas.com) (2016)
2. Ndolu, A.J., Sasongko, I., Reza, M.: Determination of Landmark Locations to Improve City Image in Kelapa Lima and Oebobo Districts, Kupang City, NTT Province . ITN (2018)
3. Lynch, K .: The Image of The City . New York (1960)
4. Suharto, B.B., Elysia, V.: The Importance of Location Analysis and Spatial Patterns in Regional and City Planning. Open University, Jakarta, pp. 1–321 (2014), ISBN 9789790118966
5. Hadihardaja, J.: Transportation System. Guru Darma University, Jakarta (1997)
6. Mendoza, G.A., Macoun, P.: Guide to Applying Multicriteria Analysis in Assessing Criteria and Indicators. SMK Grafika Mardi Yuana, Bogor (1999) ISBN 979–8764–41–2

Assessment of Damage Level and Estimation of Maintenance Costs for Architectural Components in the Rehabilitation Project for Building D of the Kutai Kartanegara Regent's Office Complex

Andi Aqsar Hamzah(✉), Wahyu Mahendra, and Eswan

Department of Civil Engineering, University of 17 August 1945 Samarinda, Jl. Ir. H. Juanda No. 80, Samarinda, East Kalimantan 75123, Indonesia
andiaqsar26@gmail.com

Abstract. Damage that occurs to a building must receive serious attention from office building managers. This is due to the potential for damage to the building that can cause safety risks and discomfort. State buildings should have a building life plan of up to 50 years. Therefore, maintenance and care of state buildings are carried out by considering the age of the building, depreciation, and building damage. Building D in the Kutai Kartanegara Regency Government Office Complex, which was built in 2004, is one example where there is some damage in the building caused by the conversion of several rooms, weather changes, and lack of maintenance. Building damage is feared to damage other components that can interfere with its function. This study was conducted to identify damage to architectural components, damage volume, and maintenance and care costs for Building D, Kutai Kartanegara Regency Regent Complex based on PU Regulation No. 22/PRT/M/2018, and to conduct an analysis by referring to the Indonesian National Standard (SNI) and the Construction Work Unit Price (HSPK) of Kutai Kartanegara Regency. The cost analysis focused on architectural components, namely walls, floors, and ceilings, as research objects. Based on the analysis, the percentage of damage to the architectural components of Building D of the Kutai Kartanegara Regent's Complex is 2.09% and is included in the type of minor damage. The budget plan for the maintenance work of Building D of the Kutai Kartanegara Regent's Office Complex on the architectural components is Rp. 914,215,152.83.

Keywords: Level of Building Damage · Volume of Building Damage · Cost Estimation

1 Introduction

A building is a facility for the building owner or user to carry out their daily activities optimally. The building is expected to be flexible, adapting to the changes that occur in the activities of the owner or user in the future [1]. Building maintenance and care are

M. Awang (Ed.): ICCAEET 2024, PTT, pp. 16–31, 2025.
https://doi.org/10.1007/978-3-031-94347-8_3

important issues that must be considered because they are related to the comfort and safety of every building user. Maintenance and care affect the planned life of the building so that it reaches the planned life [2–6]. The performance of a building can decline as the building ages. The decline in performance is generally caused by the influence of the surrounding environment, which leads to the degradation of building materials [7]. All potential damage must be identified from the operational phase. Therefore, early detection of damage is crucial, by responding to emerging signs to assess the possibility of other damages that could negatively affect the building [8]. This is observed in the damage that occurs in school buildings in general, whether it is minor, moderate, or severe damage, which are the three classifications of damage [9].

Based on the Regulation of the Minister of PUPR No. 22/PRT/M/2018 [10] concerning the maintenance and care of state buildings, the life of a building is the period of time a building can continue to fulfill the function and reliability of the building, in accordance with the requirements that have been set. The life of a building for state buildings (including state houses) is calculated at 50 years [11]. Maintenance and care of state buildings are carried out by considering the age of the building, depreciation, and damage to the building. The Kutai Kartanegara Regency Government Building Complex is one of the state asset government buildings built in 2004 and used for public services in Kutai Kartanegara Regency. Within a period of 20 years, an evaluation of the physical condition in August 2024 of Building D from the architectural components found several damages. The causes of damage include several converted rooms, changes in weather and lack of maintenance. This study aims to identify damage to architectural components of the building that require maintenance and care in Building D of the Kutai Kartanegara Regent's Office, to determine the volume of damage to the office building and to determine the costs of maintenance and care.

2 Literature Review

2.1 Definition

Building maintenance is an activity to maintain the reliability of a building and its infrastructure and facilities so that the building is always functional (*preventive maintenance*). Building maintenance is an activity to repair and/or replace parts of a building, components, building materials and/or infrastructure and facilities so that the building remains functional (*curative maintenance*).

2.2 Damage Level Assessment

Architectural Damage Grouping

The intensity of building damage can be classified into three levels of damage, namely: 1) Minor damage is damage mainly to non-structural components, such as roof coverings, ceilings, floor coverings, and filler walls. 2) Moderate damage is damage to some non-structural components, and/or structural components such as roof structures, floors, and others. 3) Severe damage is damage to most building components, both structural and non-structural, which if repaired can still function properly as they should. The maximum cost is 65% of the highest unit price for the construction of a new building in effect, for the same type/class and location.

Scope of Building Architectural Maintenance Activities

Maintenance work includes types of cleaning, tidying, inspection, testing, repair and/or replacement of building materials or equipment, and other similar activities based on building operation and maintenance guidelines. Maintaining the exit as a means of rescue (*egress*) for building owners and users properly and regularly. Maintaining the elements of the exterior of the building properly and regularly so that they remain neat and clean. Maintaining the elements of the interior and its equipment properly and regularly. Providing adequate and well-functioning maintenance systems and facilities, in the form of permanent equipment/equipment and/or work aids (*tools*). Carrying out the correct way of maintaining architectural ornaments and decorations by officers who have expertise and/or competence in their fields.

Maintenance of State Buildings or Structures

It is divided into three stages, namely: Analysis of the extent of damage, Analysis of the level of damage, and Estimation of the rough estimated price.

Extent and Volume of Damage

The area of damage is expressed in square meters (m2). To calculate the area of wall damage by first measuring the width and height of the damaged wall surface area. While to calculate the area of floor and ceiling damage is to measure the length and width of the damaged floor and ceiling surface area.

Architectural Damage Analysis

Analysis of the level of building damage including damage to architectural components can use the building damage level analysis table. Based on the regulations: 1) Regulation of the Minister of Public Works Number: 22/PRT/M/2018 concerning Technical Guidelines for the Construction of State Buildings. 2) Regulation of the Minister of Public Works Number: 16/PRT/M/2010 concerning the Building Damage Inspection Form.

Estimated Price Rough Estimate (Approximate Estimate)

The method used is the highest unit price. The amount of the required unit price is calculated based on the price of one highest ground floor per m2 of a multi-storey building multiplied by a certain coefficient or multiplier based on the number of floors in the building. Calculation using the highest unit price method is only used for buildings with a maximum of eight (8) floors. The value of the multiplier coefficient in determining the highest price of residential buildings according to the Regulation of the Minister of Public Works No. 22/PRT/M/2018 [10] can be seen in the Table 1.

3 Research Methods

The stages required to achieve the research objectives are as shown in the flow in Fig. 1.

Table 1. Highest Unit Price per m2

Number of Floors	Highest Unit Price per m2
2 Floors	1,090 highest standard price for high-rise buildings
3 Floors	1,120 highest price standard for high-rise buildings
4 Floors	1,135 highest price standard for high-rise buildings
5 Floors	1,162 highest price standard for high-rise buildings
6 Floors	1,197 highest price standard for high-rise buildings
7 Floors	1,236 highest price standard for high-rise buildings
8 Floors	1,265 highest price standard for high-rise buildings

3.1 Library and Literature Study

Literature study in this study was conducted to study building maintenance and care guidelines and maintenance and care cost estimation guidelines. In addition, studying several libraries that are sources of research for the basis of knowledge in conducting this study. This literature study will produce secondary data, namely data obtained or collected by people conducting research from existing sources. The secondary data in this study are as follows:

1. Floor plan and section drawings.
2. Regulation of the Minister of Public Works No. 24 / PRT / M / 2008 concerning Guidelines for Building Maintenance and Care.
3. Regulation of the Minister of Public Works No. 22 / PRT / M / 2018 concerning Technical Guidelines for the Construction of State Buildings.
4. Regulation of the Minister of Public Works No. 16 of 2010 concerning Technical Guidelines for Periodic Inspection of Buildings.
5. Government Regulation No. 36 of 2005 concerning Buildings.
6. SNI 2839: 2008 concerning Procedures for Calculating Unit Prices for Ceiling Work.
7. SNI 7395:2008 concerning Procedures for Calculating Unit Prices for Floor and Wall Covering Work.
8. Pt T-38-2000-C concerning Procedures for Calculating Unit Prices for Painting and Wall Finishing Work.
9. Construction Work Unit Price (HSPK) for Kutai Kartanegara Regency in 2024.
10. Regulation of the Governor of East Kalimantan No.24 of 2023 concerning the Standard Unit Price (SSH) for East Kalimantan Province in 2024.

3.2 Existing Condition Survey

Field survey (existing), in this stage prepares the location/object that will be used for research. Basically, it aims to obtain primary data, namely data obtained or collected directly in the field by the person conducting the research or the person concerned who needs it. The field *survey* conducted at Building D in the Kutai Kartanegara Regent's Office Complex, namely conducting visual observations and measurements (area and category of damage) of damage to floor, wall and ceiling components in the complex and finding unit prices and needs for maintenance.

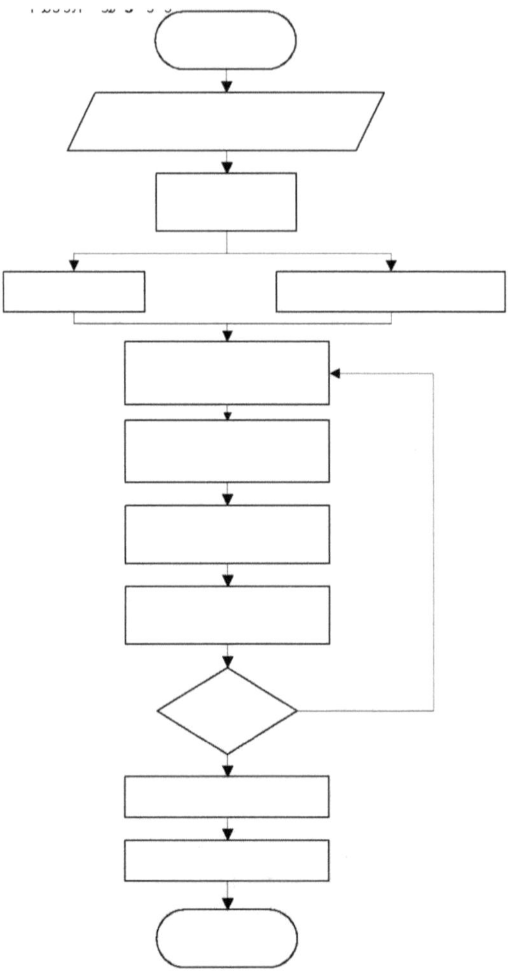

Fig. 1. Research Flow Diagram

3.3 Data Collection

The location used as the object of research is Building D in the Kutai Kartanegara Regent's Office Complex. Collecting primary and secondary data. The primary data in this study are:

1. Data on the volume of damage to wall, floor and ceiling components in Building D in the Kutai Kartanegara Regent's Office Complex and the type of damage (light, moderate, heavy and total).
2. Damage plan for walls, ceilings and floors of Building D.
3. Documentation.

While the secondary data in this study are:

1. Related regulations.
2. HSPK (Construction Work Unit Price) 2024.
3. Building plan drawings.

Direct inspection or survey is very important and must be done in planning architectural maintenance of buildings. In assessing the physical condition of a building component, a number of requirements or measurements are required that can be used as a reference for assessment and determining the condition of the building component. Assessment of the physical condition of building utilities is carried out by observing the utility components visually which is supported by documentary evidence when the survey is carried out.

3.4 Data Analysis Methods

The data analysis methods required are as follows:

1. Conducting a cost analysis of architectural maintenance and upkeep on Building D in the Kutai Kartanegara Regent's Office Complex using the Building Maintenance Cost Analysis Table. The analysis is carried out based on:
   (a) SNI 2839:2008 concerning Procedures for Calculating Unit Prices for Ceiling Work.
   (b) SNI 7395:2008 concerning Procedures for Calculating Unit Prices for Floor and Wall Covering Work.
   (c) Pt T-38-2000-C concerning Procedures for Calculating Unit Prices for Painting and Wall Finishing Work.
   (d) Construction Work Unit Price (HSPK) of Kutai Kartanegara Regency in 2024.
2. Calculating the project value of Building D according to PU Ministerial Regulation Number 45 of 2007, the unit price per m2 for multi-storey buildings, which is then multiplied by the building multiplier coefficient for the number of floors available. The calculation formula refers to Eq. 1.

$$Price\,per\,floor = SSH \times F \times L \tag{1}$$

   Where:
   SSH: East Kalimantan Provincial Government Unit Price Standards 2024
   F: Multiplier factor for multi-storey buildings
   L: Building area per floor
3. Conducting an analysis of the level of damage to Building D in terms of architectural damage using the Building Damage Level Analysis Table. The analysis was conducted based on the Regulation of the Minister of Public Works Number: 22/PRT/M/2018. The calculation formula refers to Eq. 2.

$$Percentage\ of\ Damage(\%) = \frac{Volume\ of\ Damage}{Total\ Volume} \times 100\% \tag{2}$$

4. Conducting a maintenance cost analysis of repainting walls using the Building Maintenance Cost Analysis Table. The analysis is carried out based on: a) Technical Instructions of Pt T 38-2000-C concerning Procedures for Calculating Unit Prices for Wall Painting and Finishing Work. b) Unit Price of Construction Work (HSPK) Kutai Kartanegara Regency in 2024.
5. Conduct a ceiling maintenance cost analysis using the Building Maintenance Cost Analysis Table. The analysis is conducted based on SNI 2839:2008 concerning Procedures for Calculating Unit Prices for Ceiling Work.

3.5 Research Location

The Kutai Kartanegara Regent's Office Complex is one of the government buildings of state assets used for public services located on Jalan AP Mangkunegoro, Tenggarong, Kutai Kartanegara Regency which is included in the office building complex of the Kutai Kartanegara Regency Regional Government Office, East Kalimantan Province. Building D is a public service building used as the office of the Regional Apparatus Organization (OPD) of the Investment and One-Stop Integrated Service Office (DPMPTSP), the Regional Finance and Asset Management Agency (BPKAD), the Office Empowerment Woman And Protection Children (DP3A), Food Security Service and Service Tourism, has 4 floors, the following is the division per floor (Fig. 2).

Floor 1: One Stop Investment and Integrated Services Service (Head of Office Room, Head of Subdivision Room, Head of Section Room, Staff Room, Lobby, Lactation Room, Toilet, Canteen)

2nd Floor : Regional Financial and Asset Management Agency (Head of Agency Room, Sub-Division Head Room, Section Head Room, Staff Room, Lobby, Meeting Room, Toilet, Hall)

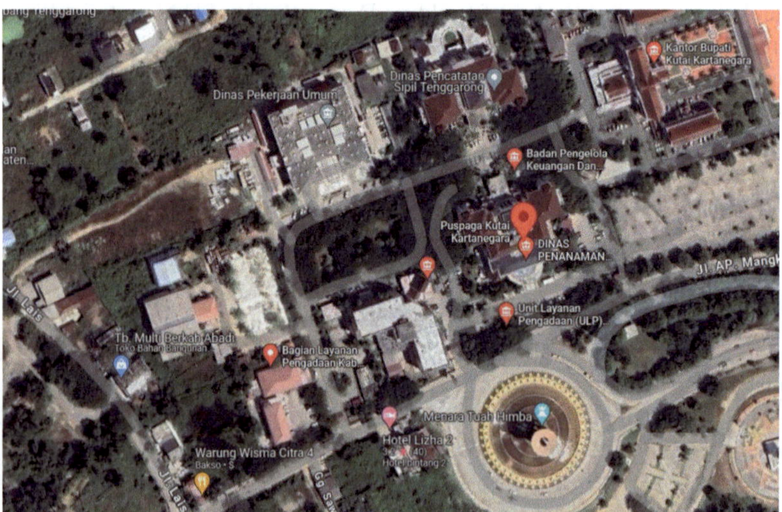

Fig. 2. Research Location

3rd floor: Empowerment Service Woman And Protection Children and Food Security Service (Head of Service Room, Sub-Division Head Room, Section Head Room, Staff Room, Lobby, Meeting Room, Toilet, Pantry)

4th floor:Tourism Office (Head of Department Room, Sub-Division Head Room, Head of Section Room, Staff Room, Lobby, Meeting Room, Toilet, Pantry, Archive Room)

Building Asset Price: Building Area × 1 m3 Building Rehabilitation (moderate) = 8380 m2 × Rp. 2,939, 760 = Rp24,629,309,280,-

4 Results and Discussion

4.1 Field Survey Results

At the field survey stage, this was carried out to determine the damage classification and the volume of damage can be seen in Table 2.

Table 2. Specifications of Building D, Kutai Kartanegara Regent's Office Complex

Maintenance Items	Specification
Interior Walls	Red Brick; Plaster; Cement; Wallpaper
Ceiling	Hollow Frame; Gypsum, Acoustic
Ceramic Floor	Granite Ceramic 60 × 60

4.2 Damage Map

In determining the damage classification, researchers refer to the Regulation of the Minister of Public Works Number: 16/PRT/M/2010 concerning Technical Guidelines for Periodic Inspection of Building Structures. Wall components are divided into two categories, namely interior walls and exterior walls. For roof coverings, there is one category, namely the ceiling. While for floor components, there is one category, namely ceramic floors. 1) Interior Wall Damage Plan. The interior walls of Building D in the Kutai Kartanegara Regent's Office Complex experienced damage in the form of hairline cracks and peeling paint. On the walls, the wallpaper was damaged, namely the wallpaper was peeling and moldy. Wall cracks occurred in several rooms. Peeling wallpaper occurred in almost all rooms. 2) Exterior Wall Damage Plan. The exterior walls of Building D in the Kutai Kartanegara Regent's Office Complex were in the form of wall painting. The walls were damaged, namely the wall paint was peeling and fading. 3) Ceiling Damage Plan. The ceiling in Building D, the Kutai Kartanegara Regent's Complex, experienced a lot of damage, namely leaks and seepage so that the paint peeled and changed color. Several ceilings on the outside of the room experienced leaks and bursts. 4) Floor Damage Map. The ceramic floor in Building D of the Kutai Kartanegara Regent Complex was damaged in the form of bubbling and cracked ceramics.

4.3 Damage Recapitulation

The damage recapitulation and damage analysis can be seen in Table 3 and Table 4.

Table 3. Damage Recapitulation

Floor	Component	Damage	Damage Volume (m^2)	Total of Volume (m^2)	Damage Percentage	Type of Damage
1st Floor	Wall	Peeling wallpaper	83.54	819.735	10.191	Low
		Peeling paint	156.54	819.736	19.096	Low
	Plafond	Detached and dirty	15.54	2373.627	0.655	Low
		Leaking	1073	2372.627	45.205	Moderate
	Floor	Damaged, cracked	121.6	2556.257	2.757	Low
		Detached	375.64	2556.257	14.695	Low
2nd Floor	Wall	Peeling wallpaper	85.64	227.08	37.714	Moderate
		Peeling paint	105.937	227.08	46.652	Moderate
	Plafond	Detached and dirty	19	1623.932	1.170	Low
		Leaking	529	1623.932	32.575	Moderate
	Floor	Damaged, cracked	78.597	1623.932	4.813	Low
		Detached	185.24	1623.932	2.164	Low
3rd Floor	Wall	Peeling wallpaper	5.54	529.994	1.054	Low
		Peeling paint	192.01	529.994	36.229	Moderate
	Plafond	Detached and dirty	15.51	1682	0.922	Low
		Leaking	268	1682	15.933	Low
	Floor	Damaged, cracked	78.597	1780	4.416	Low
		Detached	185.24	1780	10.407	Low

(continued)

Table 3. (*continued*)

Floor	Component	Damage	Damage Volume (m^2)	Total of Volume (m^2)	Damage Percentage	Type of Damage
4th Floor	Wall	Peeling wallpaper	45.54	665	6.848	Low
		Peeling paint	254	665	38.195	Moderate
	Plafond	Detached and dirty	25.34	1537	1.649	Low
		Leaking	187	1537	12.167	Low
	Floor	Damaged, cracked	106	1537	6.897	Low
		Detached	22.54	1537	1.466	Low

Table 4. Analysis of damage to standard work components

Floor	Building Component	Building Sub-Component	Damage (%)	Maximum Damage (%)	Damage Percentage (%)	Damage Value (%)
1	Foundation	Foundation	10	15	0	0
2	Structural	Column	27	30	0	0
		Beam				
		Slab				
		Ringbalk				
3	Roof	Roof Structure	8	100	0	0
		Roof Covering	2	100	0	0
4	Ceiling	Ceiling Structure	3.5	100	0	0
		Ceiling Covering	4.5	100	27.56	1.24
5	Wall	Brick	10	100	0	0
		Plastering	2	100	1.8	0.04
		Glass	1.5	100	0	0
		Acoustic	1.5	100	0	0
6	Floor	Floor Covering	10	100	0.01	0

(*continued*)

Table 4. (*continued*)

Floor	Building Component	Building Sub-Component	Damage (%)	Maximum Damage (%)	Damage Percentage (%)	Damage Value (%)
7	Utility	Electrical Installation	5	100	0	0
		Water Installation	1.5	100	0	0
		Waste Drainage	1.5	100	0	0
8	Finishing	Structure Finishing	1	100	0	0
		Finishing Plafond	2	100	1.18	0.02
		Finishing Wall	5	100	15.8	0.79
		Finishing Door	4	100	0	0
Total of Standar Work			**100**			**2.09**

Damage Classification Details:
Light < 30%
Medium > 30%–45%
Weight > 45%–65%
Special > 65%

Standard Work Unit Cost Needs Analysis

Analysis of Standard Work Unit Cost Needs includes: Building Level Coefficient (4 Floors)

Decree of the Minister of Settlement and Regional Infrastructure = 1,135

The Highest Unit Price Index for multi-storey buildings for moderate damage levels is adjusted to the highest Standard per meter for State Buildings (IHST) = Rp2,939,760,-

Analysis:
Treatment Type = Light Treatment
Standard Unit Cost of Work: 2.09% × 1.135 × Rp2,939,760 = Rp69,735.52
Demolition Cost: 10% × Rp69,735.52 = Rp6, 973.50
Standard Unit Cost of Work = Rp69,735.52 = Rp70,000,-
Fee application:
Standard work cost: = 8380 m2 × Rp70,000.00 = Rp586,600,000.00
Construction Costs
Construction Planning Cost (12.39%) = Rp72,679,740.00.
Construction Supervision Cost (7.72%) = Rp45,285,520.00.
Activity Management Fee (6.75%) = Rp.39,595,500.00.
Total Cost of Maintenance = **Rp744,160,760.00.**

In words: *Seven hundred and forty four thousand one hundred and sixty thousand seven hundred and sixty rupiah.*

(Source: PU Regulation No. 22/PRT/M/2018 [10])

Based on the analysis above, it can be seen that the percentage of damage to Building D of the Kutai Kartanegara Regent's Office in terms of damage to walls, ceilings and floors is 2.09%. The percentage of damage is categorized as Minor Damage, because the percentage of damage is less than 30%.

The analysis results obtained that the maximum maintenance cost for Building D of the Kutai Kartanegara Regent's Office, based on damage to walls, ceilings, and floors, was IDR 744,160,760.00. So based on Presidential Decree No. 16 of 2018 Article 1 paragraph 40 which states that construction work of less than IDR 200,000,000.00 will be carried out using the Tender (Auction) system. Unit price analysis for cracked wall maintenance work is guided by SNI 2837 - 2008 concerning Procedures for Calculating Unit Prices for Plastering Work, SNI 2839: 2008 concerning Procedures for Calculating Unit Prices for Ceiling Work, SNI 7395: 2008 concerning Procedures for Calculating Unit Prices for Floor and Wall Covering Work, Technical Instructions Pt - T - 38 - 2000 - C concerning Procedures for Calculating Unit Prices for Painting and Wall Finishing Work and Unit Prices for Construction Work (HSPK) Kutai Kartanegara Regency in 2024.

Initial Recapitulation of Treatment Costs

The maintenance costs for each maintenance job generated through the initial recapitulation calculation are not final. The initial recapitulation of maintenance costs is obtained by multiplying the Maintenance Work Unit Price (HSP) by the volume of damage can be seen in Eq. 3.

$$Initial\ Recapitulation = Damage\ Volume \times Maintenance\ HSP \quad (3)$$

Final Recapitulation of Treatment Costs

The final recapitulation is the final cost of each maintenance work item. The final recapitulation of maintenance costs is obtained by multiplying the floor factor by the Maintenance Cost can be seen in Eq. 4.

$$Final\ Recapitulation = Floor\ Factor \times Maintenance\ Cost \quad (4)$$

The result of final recapitulation of maintenance costs can be seen in Table 5.

Table 5. Final Recapitulation of Maintenance Costs

No	Work Description	Factor	Volume	Unit Price	Price	Total Price
	1st Floor					
A	Wall					
1	Maintenance of Damaged Walls	1.00	83.54	Rp311,216.00	Rp25,998,984.64	Rp25,998,984.64
2	Maintenance of Peeling Paint Walls	1.00	156.54	Rp99,466.00	Rp15,570,407.64	Rp15,570,407.64
B	Ceilings					
1	Maintenance of Peeling and Dirty Ceilings	1.00	15.54	Rp88,407.00	Rp1,373,844.78	Rp1,373,844.78
2	Maintenance of Leaking Ceilings	1.00	1073	Rp199,705.00	Rp214,283,465.00	Rp214,283,465.00
C	Floor					
1	Maintenance of Damaged/Cracked Floors	1.00	121.60	Rp496,000.00	Rp60,313,600.00	Rp60,313,600.00
2	Maintenance of Detached Floors	1.00	375.64	Rp183,612.00	Rp68,972,011.68	Rp68,972,011.68
	2nd Floor					
A	Wall					
1	Maintenance of Damaged Walls	1.09	85.64	Rp311,216.00	Rp26,652,538.24	Rp29,051,266.68
2	Maintenance of Peeling Paint Walls	1.09	105.937	Rp99,466.00	Rp10,537,129.64	Rp11,485,471.31
B	Ceilings					
1	Maintenance of Peeling and Dirty Ceilings	1.09	19	Rp88,407.00	Rp1,679,733.00	Rp1,830,908.97
2	Maintenance of Leaking Ceilings	1.09	529	Rp199,705.00	Rp105,643,945.00	Rp115,151,900.05
C	Floor					
1	Maintenance of Damaged/Cracked Floors	1.09	78.597	Rp496,000.00	Rp38,984,112.00	Rp42,492,682.08

(*continued*)

Table 5. (*continued*)

No	Work Description	Factor	Volume	Unit Price	Price	Total Price
2	Maintenance of Detached Floors	1.09	35.34	Rp183,612.00	Rp6,488,848.08	Rp7,072,844.41
	3rd Floor					
A	Wall					
1	Maintenance of Damaged Walls	1.12	5.54	Rp311,216.00	Rp1,724,136.64	Rp1,931,033.041
2	Maintenance of Peeling Paint Walls	1.12	192.01	Rp99,466.00	Rp19,098,466.66	Rp21,390,282.66
B	Ceilings					
1	Maintenance of Peeling and Dirty Ceilings	1.12	15.51	Rp88,407.00	Rp1,371,192.57	Rp1,535,735.68
2	Maintenance of Leaking Ceilings	1.12	268	Rp199,705.00	Rp53,520,940.00	Rp59,943,452.80
C	Floor					
1	Maintenance of Damaged/Cracked Floors	1.12	78.597	Rp496,000.00	Rp38,984,112.00	Rp43,662,205.44
2	Maintenance of Detached Floors	1.12	185.24	Rp183,612.00	Rp34,012,286.88	Rp38,093,761.31
	4th Floor					
A	Wall					
1	Maintenance of Damaged Walls	1.135	45.54	Rp311,216.00	Rp14,172,776.64	Rp16,086,101.49
2	Maintenance of Peeling Paint Walls	1.135	254	Rp99,466.00	Rp25,264,364.00	Rp28,675,053.14
B	Ceilings					
1	Maintenance of Peeling and Dirty Ceilings	1.135	25.34	Rp88,407.00	Rp2,240,233.38	Rp2,542,664.89
2	Maintenance of Leaking Ceilings	1.135	187	Rp199,705.00	Rp37,344,835.00	Rp42,386,387.73
C	Floor					

(*continued*)

Table 5. (*continued*)

No	Work Description	Factor	Volume	Unit Price	Price	Total Price
1	Maintenance of Damaged/Cracked Floors	1.135	106	Rp496,000.00	Rp52,576,000.00	Rp59,673,760.00
2	Maintenance of Detached Floors	1.135	22.54	Rp183,612.00	Rp4,138,614.48	Rp4,697,327.43
					Rp860,946,577.95	**Rp914,215,152.83**

5 Conclusions and Suggestions

5.1 Conclusions

From the results of the discussion on the analysis of maintenance and care costs for walls, floors and ceilings of Building D of the Kutai Kartanegara Regent Complex based on the regulations used, the following conclusions can be drawn:

1. Damage to architectural components on the walls of Building D, Kutai Kartanegara District Head Complex, includes dirty paint, peeling wallpaper, peeling and cracked paint.
2. Damage to architectural components on the floor of Building D, Kutai Kartanegara Regent Complex, includes damaged, broken and detached floors.
3. Damage to architectural components on the ceiling of Building D, Kutai Kartanegara District Head Complex, includes peeling paint and leaking ceilings.
4. Analysis of the percentage of damage to components of Building C of the Kudus Regent Complex based on PU Regulation No. 22/PRT/M/2018 is 2.09% and is included in the type of minor damage.
5. Damage analysis based on PU Regulation No. 22/PRT/M/2018 shows that the ceiling price for repairing Building D of the Kutai Kartanegara Regent Complex in terms of damage to walls, floors, and ceilings is IDR 744,160,760.

5.2 Suggestions

Based on these conclusions, the following suggestions can be made:

1. To minimize damage to larger architectural components, building owners/users are expected to immediately repair damage to walls, ceilings and floors in Building C, Kudus Regent Complex.
2. The owner/user carries out scheduled maintenance and upkeep on each component of the building periodically.
3. The results of the analysis of maintenance and care costs should be used as a basis for reference in maintenance and care, so that the maintenance results can be in accordance with the results of the analysis that has been carried out.

References

1. Fery, L.H., Trikomara, R., Taufik, H.: Estimasi Biaya Pemeliharaan Bangunan Berdasarkan PedomanPemeliharaan Dan Perawatan Bangunan Gedung (Permen Nomor: 24/Prt/M/2008)(Studi Kasus Bangunan Gedung Kantor Bupati Kampar). Jom Fteknik **3**(1), 1–12 (2016)
2. Kristiana, W., Nuswantoro, W., Yulfrida, D.A.: Manajemen Perawatan Dan Pemeliharaan Bangunan Kalimantan Tengah. J. Tek. **1**(1), 20–25 (2017)
3. Ali, A.M., Sangadji, F.A., Kempa, M.: Analisis Tingkat Kerusakan Bangunan Gedung Rusunawa Iain Ambon. Kokoh **22**(1), 1–14 (2024). https://doi.org/10.17509/k.v22i1.66208
4. Pramudya Restuning Kawedar, H., Trisiana, A., Arifin, S.: Analisis Tingkat Kerusakan Dan Estimasi Biaya Pemeliharaan Bangunan Gedung Rusunawa Putri Universitas Jember. J. Appl. Civ. Eng. Infrastruct. Technol. **3**(1), 1–6 (2022). https://doi.org/10.52158/jaceit.v3i1.148
5. Ratnaningsih, A., Firnanda, K.W., Trisiana, A.: Analisis Operasional Dan Pemeliharaan Gedung Berdasarkan Tingkat Kerusakan Dan Anggaran Biaya Dengan Metode Activity Based Costing Analysis (ABC). Teras J. J. Tek. Sipil **13**(2), 353 (2023). https://doi.org/10.29103/tj.v13i2.886
6. Devina, P.I.: Rudi Waluyo, and Veronika Happy Puspasari, "Analisis Estimasi Biaya Perawatan Bangunan Gedung Di Universitas Palangka Raya." J. Tek. Sipil **17**(2), 95–102 (2023). https://doi.org/10.24002/jts.v17i2.6407
7. Hidayatullah, I.: Analisis Tingkat Kerusakan Pada Bangunan Gedung Sekolah Dasar Negeri Di Kota Sukabumi. J. Ilm. Wahana Pendidik. **10**(1), 467–477 (2024)
8. Herzanita, A., Arini, R.N., Andreas, A.: Analisis Identifikasi Kondisi Struktur Bangunan (Studi Kasus: Gedung Magister Manajemen, Fakultas Ekonomi Dan Bisnis, Universitas Pancasila). Janata **3**(2), 63–72 (2023)
9. Iriana, R.T., Riana, A.: Analisa Tingkat Kerusakan Dan Estimasi Biaya Perbaikan Bangunan Gedung Sekolah. J. Tek. Sipil, no. November, 1–14 (2012)
10. PERMEN. PermenPUPR22–2018, vol. 3, no. 2 (2018)
11. Kusumastuti, D.R., Setiawan, D.B., Rahmi, D.R., Rochimawati, M., Supriyo, S.: Analisis Tingkat Kerusakan Dan Estimasi Biaya Pemeliharaan Komponen Arsitektural Pada Gedung C Di Komplek Gedung Kantor Bupati Kudus. Bangun Rekaprima **8**(1), 9 (2022). https://doi.org/10.32497/bangunrekaprima.v8i1.3556

Investigation Survey and Design of Bontang River

Aulia Muttaqin(✉)

University of 17 August 1945, Samarinda, East Kalimantan, Indonesia
auliamuttaqin19@gmail.com

Abstract. In the Bontang watershed there are 8 flood-prone areas, among others: Telihan area, Kanaan area, Disnaker Resident area, Bethlehem & Bontang Permai Resident area, Imam Bonjol area, Jl. A. Yani, Awang Long and Sendawar area, Rawa Indah area and Jl. KS. Tubun. Bontang City experiences flooding problems, although many flood control efforts have been made from planning to construction. For this reason, it is necessary to conduct an evaluation and integrated management in this area. To secure the results of the development and development of Bontang City in accordance with the direction of regional development and regional development plans, it is necessary to conduct a detailed study of regional planning, especially in terms of flood control. To obtain a flood control building construction that meets the criteria in accordance with current conditions, both in terms of environmental, technical, economic, and social and cultural aspects, it is necessary to carry out the Bontang River SID (Survey, investigation and Design). In connection with the above, in Fiscal Year 2023, the East Kalimantan Provincial Government through the Public Works, Spatial Planning and Public Housing Office intends to carry out Bontang River SID Work in order to realize and realize the construction of flood control buildings so that the utilization and utilization of water resources in the Bontang City area can take place harmoniously and optimally, in accordance with the needs and carrying capacity of the environment.

Keywords: Flood · Flood Control · Flood Control Building

1 Introduction

The topography of Bontang City is generally undulating hills and plain areas with many swamps. As with other cities in East Kalimantan, flooding is one of the most important problems in Bontang City. Flood-prone areas in Bontang City are located in 4 watersheds, namely the Bontang watershed, Guntung watershed, Tanjung Limau watershed and Belimbing watershed [1]. The following are the causes of flooding based on the existing watersheds. In the Bontang watershed, there are 8 flood-prone areas, among others: Telihan area, Kanaan area, Perum Disnaker area, Bethlehem & Bontang Permai Resident area, Imam Bonjol area, Jl. A. Yani, Awang Long and Sendawar area, Rawa Indah area and Jl. KS. Tubun. Bontang City experiences flooding problems, although many flood control efforts have been made from planning to construction. For this reason, it is necessary to conduct an evaluation and integrated management in this area. To secure

M. Awang (Ed.): ICCAEET 2024, PTT, pp. 32–40, 2025.
https://doi.org/10.1007/978-3-031-94347-8_4

the results of the development and development of Bontang City in accordance with the direction of regional development and regional development plans, it is necessary to conduct a detailed study of regional planning, especially in terms of flood control [2]. To obtain a flood control building construction that meets the criteria in accordance with current conditions, both in terms of environmental, technical, economic, and social and cultural aspects, it is necessary to carry out the Bontang River SID. In connection with the above, in Fiscal Year 2023, the East Kalimantan Provincial Government through the Public Works, Spatial Planning and Public Housing Office intends to carry out Bontang River SID Work in order to realize and realize the construction of flood control buildings so that the utilization and utilization of water resources in the Bontang City area can take place harmoniously and optimally, in accordance with the needs and carrying capacity of the environment.

2 Research Methods

2.1 Socio-Economique and Cultural Survey

The survey was conducted using a qualitative method because it focuses more on the substance of the meaning of the phenomenon under study. Qualitative analysis emphasizes the power of words and sentences used. Therefore, Basri et al. [3] concluded that the focus of qualitative research is on the process and the meaning of the results. Qualitative research attention is more focused on human elements, objects, and institutions, as well as the relationship or interaction between these elements, in an effort to understand an event, behavior, or phenomenon [4]. With this qualitative method, we try to understand the attitudes and perceptions of the community towards the business plan and/or activities that will be carried out. Community attitudes and perceptions are subjective, individual, and related to value content. The use of this method is influenced by and represents a paradigm that reflects a point of view on reality [5]. Meanwhile, according to McCusker et al. [6], the choice of using qualitative methods in terms of research objectives is to understand how a community or individuals accept certain issues. The implementation of the social and economic condition survey work was carried out by conducting primary and secondary data collection. Primary data collection activities are carried out by conducting: Observation, Interviews, Questionnaires to the communities affected by flooding.

2.2 Topographic Survey

This topographic mapping survey activity is intended to obtain a description of the topographic conditions at the work site, for which measurements are taken with measuring instruments used including waterpasses and Total Stations so as to produce measurement data, and then this measurement data is analyzed so as to produce positions in the form of coordinates and elevations of points that can represent the state of the field on the map to be used for the purposes of the design process and subsequent analyses. Based on observations in the field orientation, the survey team should make a work plan and work map that contains the following: A network of primary polygon points, and their binders, Network of secondary polygon points (if required), Position of Benchmark (BM) & Control Point (CP) and other stakes, Situation measurement path plan, Measurement schedule.

2.3 Geological Investigation and Soil Mechanics

The scope of the Bontang River SID field survey activities includes Soil boring there are 3 boring points, 2 (two) points are located on the existing bridge, boring B-2 (15 m) left and boring B-3 (15 m) right while 1 boring point B-1 (15 m) on the existing right upstream of the bridge and Cone Penetration Test work as many as 5 (five) Cone Penetration Test points, which are scattered upstream and downstream of the bridge plan as follows (Tables 1, 2 and 3).

Table 1. Table Boring locations and coordinates

CODE	DEPTH (M)	BORING POINT	COORDINATES	
			X	Y
B - 1	15	Existing Tiles Upstream Right of Bridge	550,342	014428
B - 2	15	Existing Tiles Left Downstream of Bridge	550,336	014441
B - 3	15	Existing Tiles Right Downstream of Bridge	550,342	014428

Table 2. Table Cone Penetration Test locations and coordinates

CODE	CONE PENETRATION TEST POINT	COORDINATES	
		X	**Y**
S - 1	G. Telihan Village, Bontang Barat District	551,022	014449
S - 2	Kanaan Village, Bontang Barat District	548,917	014963
S - 3	G. Telihan Village, Bontang Barat District	549,845	014594
S - 4	Kanaan Village, Bontang Barat District	551,022	014449
S - 5	G. Elai Village, Bontang Barat District	552,734	014766

2.4 Rainfall Analysis

Based on the results of statistical analysis of rain data and data selection and data correction, it is known that the rain data of Kanaan Rain Station, Santan and Sei Siring Station have climatic similarities with Global Precipitation Measurement (GPM) data and can be used for GPM data correction, meaning that the data is feasible or can be used for further analysis. The further analysis required is the analysis of rainfall forecasting, then continued with the screening of selected stations.

Table 3. Selected Rainfall Plan of Bontang Watershed

Periodicites	GPM Grid 1 Corr	GPM Grid 2 Corr
2	99.58	98.26
5	132.09	127.30
10	156.86	149.44
20	183.39	173.14
25	192.42	181.21
50	222.27	207.88
100	255.18	237.29
1000	392.06	359.58
R100/R2	2.56	2.41

Rainfall forecasting is the largest annual rainfall with a certain chance that may occur in an area, or rain with a certain return period possibility. The method for calculating the amount of rainfall forecasting used in this activity is the Generalized Extreme Value (GEV), which is based on statistical parameters of annual maximum daily rainfall data.

The reasonableness check of the calculated design rainfall can be done by comparing the 100-year return period (R100) to the 2-year return period rainfall (R2). The R100/R2 ratio should be above 1.7 and less than 3.2. From the results of the design rainfall calculation above, it was obtained that in Grid 1 Corrected R100/R2 was 2.56, and Grid 2 Corrected was 2.41. So, the design rainfall calculation can be accepted.

2.5 Flood Discharge Analysis

Existing Condition Plan Flood Discharge. The existing conditions of the Bontang River scheme can be explained as follows:

1. In the upstream section of the river, the Sukarahmat Dam Sub-watershed is formed;
2. In the next section of the river is made Sub-watershed 1;
3. In the next section of the river there is a building retarding basing Lake Kanaan, so this section as Sub-watershed 2;
4. In the next river section, Sub-watershed 3 is formed; In this study, the plan flood discharge analysis was conducted with the help of HEC-HMS software. This analysis was conducted on 2 (two) conditions, namely existing conditions and planned conditions (conditions after the construction of the Sukarahmat Dam, Telihan Polder, and Kanaan Depression).
5. The downstream river segment is made Sub-watershed 4 to the estuary.

Flood Discharge Plan. The plan condition, the Bontang River scheme can be explained as follows:

1. In the upstream section of the river there is a Sukarahmat Dam plan, so the analysis in this section considers the outflow from the Sukarahmat Dam reservoir;

2. The next river section is made Sub-watershed 1, in this section there is Telihan Polder and the planned Bontang Dam;
3. The next river segment is made Sub-watershed 2, in this river segment there is a retarding basing building of Lake Kanaan;
4. The next river segment is made Sub-watershed 3, in this river segment there is a retarding basing of the Kanaan Lake Depression;
5. The downstream river segment is made Sub-watershed 4 to the estuary.

2.6 Hydrological Analysis

Data Used. The data required for river hydraulics analysis include the following:

1. Hydrological Data. Plan flood discharge data with a return period of 2, 5, 10, 25, 50, and 100 years.
2. Topographic Measurement Data, which includes: a) Topographic map scale 1: 50,000 to make river schematization and river area map; b) Situation map of the river trajectory at a scale of 1: 2.000; c) Longitudinal cut of the river at a scale of 1: 2.000 H; 1: 200 V; d) Cross section of the river scale 1: 200.
3. River Data. Data of the river channel and riverbank area to estimate the flow roughness coefficient.
4. Data of existing river buildings, including Data on river crossing gate buildings
5. Tidal height data. Tidal observation data is accessed through the Geospatial Information Agency website. The data was then processed with the Least Square Method. Tidal observations were made at the mouth of the Bontang River.

HEC-RAS Hydraulics Model. In actual conditions, the river cross section is not prismatic and even very diverse in shape. Flow profile analysis, in addition to using the steady flow approach as described above, will also be analyzed using the unsteady flow approach. Analysis of the water level profile with an unsteady flow approach is because one of the things that affects the water level is the amount of flow that is not constant in the time dimension. The calculation of the water level profile with unsteady conditions uses the help of the HEC RAS program package (Hydraulic Engineering Center - River Analysis System from the US Army Corps of Engineers) version 4.0, March 2001, Public Domain edition.

Model Input. The data inputs performed for the river hydraulic model include:

1. River geometry data, including: a) River scheme coordinates (X, Y); b) Transverse river profile (distance and elevation); c) The distance of each transverse profile (left cliff and right cliff); d) The position limit of the river trough (left cliff and right cliff); e) Manning's roughness coefficient (n) on riverbeds, left and right riverbanks; f) Plan the position of the river embankment (if any); g) Position plan and cross-sectional dimensions plan for river normalization (if any).

2. Building data in the river, such as:a) Culvert building data (if any); b) Weir building data (if any); c) River door data (if any).

Plan discharge data at the points reviewed with a return period of 2, 5, 10, 25, 50 and 100 years.

HEC-RAS Modeling Results. The results of the hydraulics analysis obtained are flood water level profile data for each series of analysis for each condition of the planned flood discharge along the river section under review. From this water level profile, locations that experience overflow will be known.

The results of the hydraulics analysis obtained are flood water level profile data in each analysis series for each planned flood discharge condition along the river sections under review. From this water level profile, locations that experience overflow will be observed.

Bontang River Normalization Plan Condition. The results of the The design of the normalization plan is as follows: Dredging Plan: 1 m; Normalization Width Plan: 12 m; Normalization Length: 12.92 km (Patok CT.0 - CT. 119); The following is a picture of the location of the normalization plan and the results of the 25 years return period design simulation (Figs. 1, 2 and 3).

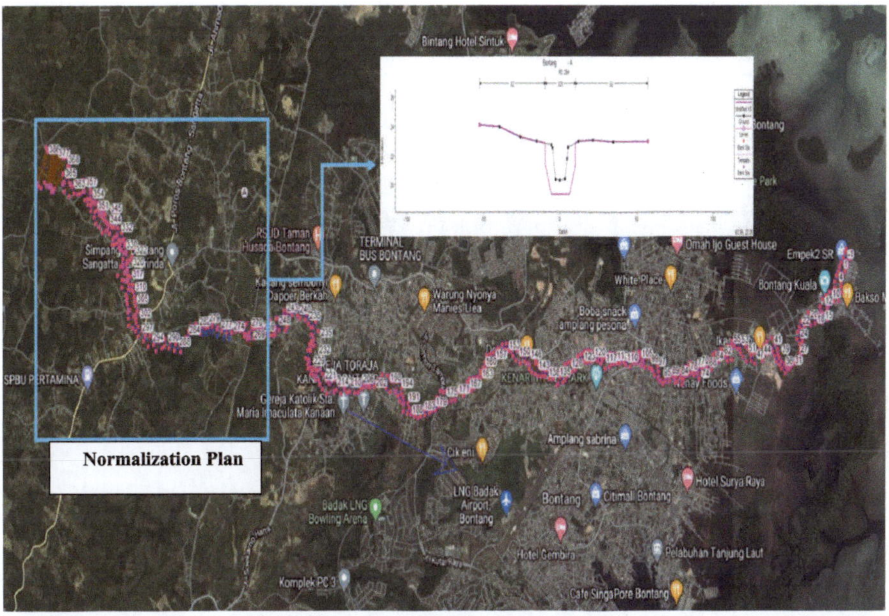

Fig. 1. Bontang River Normalization Plan

3 Results and Discussion

Flood control with a management approach is a flood management method that prioritizes persuasive approaches to flood-affected communities, or with local government policies, rather than technical planning. This method is used in flooded areas or areas in residential areas where the level of non-technical difficulties is very high (population relocation/land acquisition plan) [7]. Based on the results of direct observations in the field and the development program plan of the Bontang City government as well as time and cost

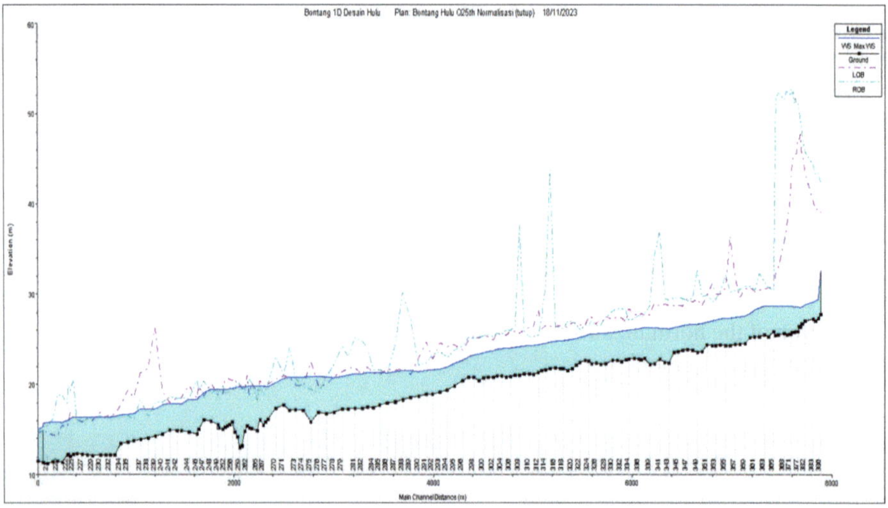

Fig. 2. Longitudinal Water Level Profile Normalized Design Condition with Q25years Return Period in Bontang River (Upstream - Middle)

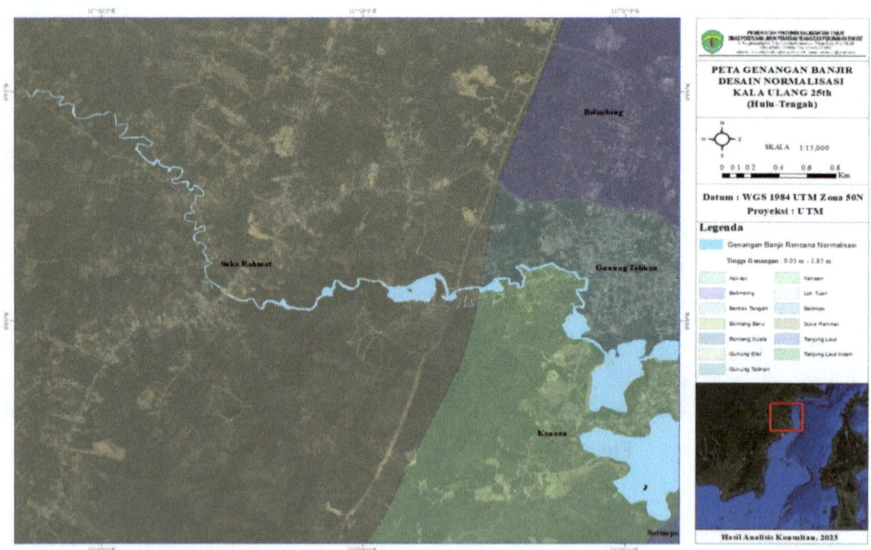

Fig. 3. Normalized Design Normalization Design Flood Inundation Map with Q25th Return Period in Bontang River (Upper-Middle)

limitations, the flood control plan in the Bontang watershed is made on a priority scale based on the orientation of the watershed identified as a flood area that needs to be addressed immediately. Based on the results of field identification, the flood problems that often occur in this watershed area are due to the overflow of the Bontang River. And almost along the riverbanks and floodbanks have been utilized as dense residential areas.

So that technical handling sometimes experiences obstacles in implementation because it often clashes with community interests. Hence the need for flood management while taking into account the function and behavior of the river [8]. Providing an understanding to the community that the river is one part of living life on the riverbanks needs to be emphasized. "Learning to live with the river" is a motto that can be used as a reference. Supporting aspects in the selection of priority watersheds that need to be addressed immediately are as follows:

1. Area affected by flooding; 2. Number of people affected by flooding; 3. Location where the flood occurred;4. Frequency of flood occurrence; 5. Cumulative impact of flooding on damage to public facilities and obstacles to community activities; 6. The level of difficulty faced in planning flood control in the area concerned.

Based on the factors determining the priority scale, for flood management plans in the city of Bontang in general and in the Bontang watershed in particular. Some recommendations for management aspect activities related to flood control plans in Bontang City are: 1. Mapping flood areas and establishing these flood areas as prohibited areas for settlement or cultivated land; 2. Provide recommendations for the height of the floor of the house in accordance with the existing flood peil. This activity is carried out for residential areas in flood areas with a relatively small risk and frequency of flood events; 3. Flood proofing of existing buildings in the are; 4. Conducting conservation activities in water catchment areas (making infiltration wells, reforestation, land use utilization that pays attention to water resources); 5. Develop a city spatial plan based on the development of water resources; 6. Flood forecasting and early warning system.

4 Conclusions

Geological and soil mechanics investigation activities include soil boring there are 3 boring points, 2 (two) points are located on the existing bridge, boring B-2 (15 m) left and boring B-3 (15 m) right while 1 boring point B-1 (15 m) on the Existing Right upstream from the Bridge and Cone Penetration Test work as many as 5 (five) Cone Penetration Test points, which are scattered upstream and downstream of the planned Bridge 3. The causes of flooding in Bontang City are:

a. The characteristics of the Bontang watershed area are flood plain areas with a slope percentage of 0–12%; b. Flooding occurs due to the overflow of the Bontang River, especially during rain events that coincide with sea tides; c. Sedimentation in the river channel;d. Land use change reduces the water catchment area. Flood control infrastructure are not yet comprehensive according to the needs of the water system.

Management of floodplain areas: a. Land use regulation in the river basin; b. Community counseling on flood problems.; c. Developing city spatial plans based on the development of water resources.; d. Establishing a flood hazard warning system.

References

1. Nandiasa, J.E.: Analisis Pengendalian Banjir Kota Bontang Kalimantan Timur. Rekayasa Sipil. **9**(1), 1 (2020). https://doi.org/10.22441/jrs.2020.v09.i1.01

2. Saragih, D.: Problems and Solutions for the Flood Control Program in Medan City and its Surroundings. J. Civ. Eng. Forum. **9**(January), 47–58 (2022). https://doi.org/10.22146/jcef.4784
3. Basri, H.: Using qualitative research in accounting and management studies: Not a new agenda. J. US-China Public Adm. **11**(10), 831–838 (2014). https://doi.org/10.17265/1548-6591/2014.10.003
4. Abd Majid, A., Mohamed, Z., Majid, A.., Norsiah: Tapping new possibilities in accounting research, qualitative research in accounting - Malaysian cases (2010)
5. Sale, J.E.M., Lohfeld, L.H., Brazil, K.: Revisiting the quantitative-qualitative debate: Implications for mixed-methods research. Qual. Quant. **36**(1), 43–53 (2002). https://doi.org/10.1023/A:1014301607592
6. McCusker, K., Gunaydin, S.: Research using qualitative, quantitative or mixed methods and choice based on the research. Perfusion. **30**(7), 537–542 (2015). https://doi.org/10.1177/0267659114559116
7. Wang, L., et al.: A review of the flood management: from flood control to flood resilience. Heliyon. **8**(11), e11763 (2022). https://doi.org/10.1016/j.heliyon.2022.e11763
8. Shah, M.A.R., Rahman, A., Chowdhury, S.: Challenges for achieving sustainable flood risk management. J. Flood Risk Manag. **11** (2015). https://doi.org/10.1111/jfr3.12211

Analysis of Time and Cost Performance of the Construction of the Madrasah Darussalam Samarinda Building with the Earned Value Method Approach

Naufal Rhiswanda Jasman(✉)

Department of Civil Engineering, University of 17 August 1945 Samarinda, Jl. Ir. H. Juanda No. 80, Samarinda, East Kalimantan 75123, Indonesia
naufalr66@gmail.com

Abstract. Planning, cost and time control are part of overall construction project management. In addition to quality assessment, the presentation of a project can also be assessed in terms of cost and time. Possible deviations from the plan in the aspects of cost and time used for a construction work should be measured on an ongoing basis. Significant cost and time deviations indicate poor project management. As for the Performance Analysis of Time and Cost with the Earned Value Analysis Method on the Construction of Madrasah Darussalam Samarinda, this uses the Earned Value Analysis Method with data collection in the field on the Madrasah Darussalam Samarinda Building Construction Project. In this study, Performance Analysis of Time and Cost with Earned Value Analysis Method on the Construction of Madrasah Darussalam Samarinda Building, the SPI value is 0.853 while the CPI is 1.006 which means that the project is delayed in terms of schedule but incurs less actual costs from the work that has been carried out. For the ETC value of Rp. 13,380,535,283 and the EAC value of Rp. 30,483,085,771 while the ECD value is 294 days for the project completion time, it is recommended that related parties control performance progress in the following weeks and find solutions so that the project is completed on time with the remaining costs.

Keywords: Cost · Schedule · Earned Value Analysis

1 Introduction

Planning, cost and time control are part of overall construction project management. In addition to quality assessment, the presentation of a project can also be assessed in terms of cost and time [1]. Possible deviations from the plan in terms of cost and time spent on a construction job should be measured on an ongoing basis. Significant cost and time deviations indicate poor project management [2]. Therefore, project control is necessary to prevent the execution of project work from deviating from the plan, so that it is expected that the elements and objectives of the control will not change significantly, even though the project work is carried out according to the plan [2–5].

M. Awang (Ed.): ICCAEET 2024, PTT, pp. 41–54, 2025.
https://doi.org/10.1007/978-3-031-94347-8_5

The concept of "Earned Value" is one of the tools used in project management that integrates cost and time [6–9]. The earned value concept presents three dimensions, namely the physical completion of the project (the percent complete) which reflects the planned absorption of costs (budgeted cost), the actual costs that have been incurred or called actual cost and what is obtained from the costs that have been incurred or called earned value [10]. From these three dimensions, the earned value concept can be linked between cost and time performance derived from the calculation of the variance of cost and time [11]. In this study, an analysis of the time and cost performance of the construction of the Darussalam Samarinda Madrasah building will be carried out using the Earned Value method.

2 Literature Review

The project control method used is the Integrated Cost and Schedule Control Method (Earned Value). This method examines the tendency of the Schedule Variant and Cost Variant at a period of time during the project [6–9]. The project cost components can be explained, namely:

1.Direct costs, which consist of material costs, labor costs, sub-contractor costs and equipment costs.

2.Indirect costs consisting of office and field overheads.

The picture above is the project cost components. Meanwhile, time control is about scheduling. Scheduling is made to describe planning on a time scale. Scheduling determines when activities are started, postponed and completed. So that financing and usage will be adjusted according to the needs to be determined.

The Earned Value method is a control method used to control project cost and schedule in an integrated manner. This method provides information on the status of project performance at a reporting period and provides information on the predicted cost and time to complete all work based on performance indicators at the time of reporting. Variance Analysis Method is a method to control the cost and schedule of a construction project activity. In this method, identification is done by comparing the amount of costs actually incurred against the budget. Variance analysis is carried out by collecting information on the latest status of project progress at the time of reporting by counting the number of units of work that have been completed and then comparing with planning or looking at resource usage records.

Another way to demonstrate the existence of variances is by using graphs. The "S" graph will illustrate the progress of the volume of work completed throughout the project cycle. When the graph is compared to the work completed throughout the project cycle. When the graph is compared using a similar graph prepared on the basis of the basic plan, it will be immediately apparent if there are any deviations. The use of the Earned Value Concept in project performance assessment can be explained through Fig. 1. Some terms related to this assessment are Cost Variance, Schedule Variance, Cost Performance index, Schedule Performance index, Estimation at Completion Date.

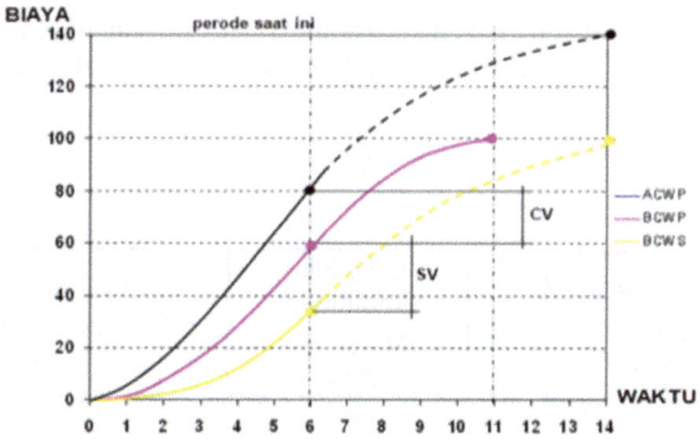

Fig. 1. Earned Value S Curve Graph (source: Soaemardi et al.)

3 Method

The technique in analyzing data is a sequence of steps carried out systematically and logically according to the theoretical basis of the problem so that an accurate analysis is obtained to achieve the goal. Stages in this study are as follows:

1. Before conducting research, it is necessary to conduct a literature study to deepen the knowledge related to the research topic. Then determine the formulation of the problem until the compilation of data.
2. Calculating ACWP (Actual Cost of Work Performance), BCWP (Budgeted Cost of Work Performance), BCWS (Budgeted Cost of Work Schedule). ACWP (Actual Cost of Work Performance) is calculated from total direct costs. BCWP (Budgeted Cost of Work Performance) is calculated from the actual weight of all work against the contract value. BCWS (Budgeted Cost of Work Schedule) is calculated from the weight of the work against the cost budget plan.
3. Calculate SV (Scheduling Variance), CV (Cost Variance), CPI (Cost Performance Index), SPI (Scheduling Performance Index), ETC (estimated cost for remaining work). SV (Scheduling Variance) is calculated from the difference in BCWS (Budgeted Cost of Work Schedule). CV (Cost Variance) is calculated from the difference between BCWP (Budgeted Cost of Work Performance) and ACWP (Actual Cost of Work Performance). CPI (Cost Performance Index) is calculated from the comparison of BCWP (Budgeted Cost of Work Performance) and ACWP (Actual Cost of Work Performance). CPI (Cost Performance Index). SPI (Scheduling Performance Index) is calculated from BCWP (Budgeted Cost of Work Performance)/BCWS (Budgeted Cost of Work Schedule). ETC (estimated cost for remaining work) is calculated from the difference between the budget and BCWP (Budgeted Cost of Work Performance) divided by CPI (Cost Performance Index). Calculating EAC (estimated total project cost), EAC (estimated total project cost) is calculated using the ACWP (Actual Cost of Work Performance) formula. CPI (Cost Performance Index) + ETC. (estimated cost for remaining work)

4. Discussion and conclusion. This discussion explains the calculations that have been carried out. Conclusions are also called decision making. At this stage, the data that has been analyzed is made a conclusion related to the research objectives. The data processing process can be seen in Fig. 2.

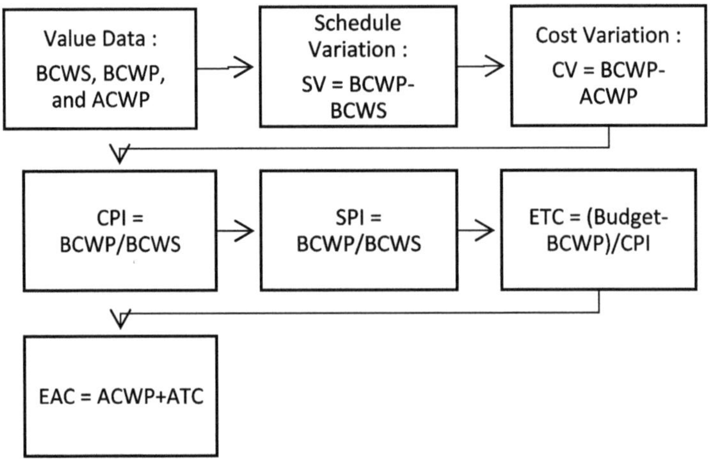

Fig. 2. Data Processing

4 Result and Discussion

4.1 Earned Value Analysis

Calculation of Planned Value BCWS (Budgeted Cost of Work Schedule) period week 25. The calculation of planned value in the seventh month uses data on the weight of the plan in the week under review, which can be observed on the S-curve and the amount of the cost budget plan for a job. The BCWS (Budgeted Cost of Work Performance) calculation formula in the 25th week period of the seventh month of review is:

Accumulated weight of roof structure work plan = 1.23%
Total weight of roof structure work = 13.30%
Roof structure work budget = Rp. 4,077,848,177, then:
BCWS of roof structure work =1.23%/13.20%x 4,077,848,177
= Rp. 377,231,970.

For the calculation of the BCWS (Budgeted Cost of Work Schedule) value in the week before and the following week, the same method as the calculation can be seen in Table 1.

4.2 Earned Value BCWP (Budgeted Cost of Work Performance) Week 25 Period

The calculation of Earned Value in the seventh month of this review can be calculated using data on the percentage of work realization in the week under review which can

Table 1. BCWS (Budgeted Cost of Work Performance) Calculation Results; 7th month period

Number	Job Description	Budget Plan (Rp)	Percentage of Work Plan (%)	Percentage of Total (%)	BCWS
1	Structure Roof	4.077.848.717	1,23	13,30	377.231.970
2	Couple Wall, Partition (1st Floor)	605.755.465	0,33	1,98	100.959.243
3	Finishing (1st Floor)	1.004.173.002	0,27	3,27	83.681.083
4	Couple Wall, Partition	3.071.301.786	0,90	10,01	276.023.392
5	Finishing	34.242.041	0,02	0,11	5.707.007
		Total			843.602.694

be observed in the weekly project report and the cost budget for a job. The BCWP calculation formula for the seventh month of the review period in Eq. 1.

$$BCWP\ per\ task = \frac{Actual\ Job\ Weight}{Total\ Job\ Weight}\ x\ Job\ Budget \tag{1}$$

Accumulated weight of 1st floor structural work realization = 0.478%.

Total weight of 1st floor structural work = 14.387%.

1st floor structural work budget = Rp. 4,412,491,841, Thus, the calculation is shown in Table 2.

Table 2. BCWP Calculation Results (Budgeted Cost of Work Performance) 7th period

Number	Job Description	Budget Plan (Rp)	Percentage of Work Plan (%)	Percentage of Total (%)	BCWS
1	Structure	4.412.491.841	0,478	14,387	146.713.360
2	Structure Roof	4.077.848.177	1039	13,296	318.508.436
3	Couple Wall, Partition	605.755.465	0,197	1975	60.467.066
4	Couple Wall, Partition	3.071.301.786	0,072	10,014	21.938.385
		Total			547.627.246

4.3 ACWP (Actual Cost of Work Performance) for the 7th Month Period

The Actual Cost value is obtained from the total costs incurred during the period under review, obtained by calculating all labor costs, materials imported and tools used in each week. An example of the calculation of Actual Cost at week 25 in month 7 of the greening period is as follows:

- Labor wage = IDR 13,622,750.
- Material cost = Rp. 532,376,333.
- Equipment cost = Rp. 532,909.

Total cost = Rp. 546,531,992.

4.4 Variance Calculation and Project Performance Index

From the value of BCWS (Budgeted Cost of Work Schedule), BCWP (Budgeted Cost of Work Performance), and ACWP (Actual Cost of Work Performance), the value of time variance / SV (Scheduling Variance), cost variance / CV (Cost Variance), schedule performance index / SPI (Scheduling Performance Index) and cost performance index / CI (Cost Performance Index) is obtained. The following is an example of calculating time variance / SV (Scheduling Variance), cost variance / CV (Cost Variance), schedule performance index / SPI (Scheduling Performance Index) and cost performance index / CPI (Cost Performance Index) is shown in Table 3.

4.5 SV (Scheduling Variance)

The SV value can be calculated by subtracting the BCWP (Budgeted Cost of Work Performance) value from the BCWS (Budgeted Cost of Work Schedule) value. Example of calculating the SV value in week 25:

SV = BCWP – BCWS.

= Rp. 17,208,447,335 – Rp. 20,163,118,635.

= – 2.954.671.300.

The calculation of the SV (Scheduling Variance) can be seen in Table 4.

4.6 CV (Cost Variance)

The calculation of the CV (Cost Variance) value is obtained from the difference between the BCWP (Budgeted Cost of Work Performance) value and the ACWP (Actual Cost of Work Performance) value. Example of CV value calculation in week 25:

CV = BCWP - ACWP

= Rp.17,208,447,335 - Rp. 17,102,550,488

= Rp. 105,896,847

For the calculation of the CV (Cost Variance) can be seen in Table 5.

Table 3. Value of BCWS (Budgeted Cost of Work Scheduling), BCWP (Budgeted Cost of Work Performance) and ACWP (Actual Cost of Work Performance)

Week	BCWS (Budgeted Cost of work Scheduling)	BCWP (Budgeted Cost of Work Performance)	ACWP (Actual Cost of Work Performance)
1	75.157.792	38.584.080	38.198.239
2	639.956.800	52.608.045	52.081.964
3	1.204.755.808	88.019.075	87.138.884
4	1.769.554.816	105.144.940	104.093.491
5	2.259.196.031	292.423.730	290.810.444
6	2.748.837.247	658.024.779	655.497.491
7	3.238.625.422	1.091.067.838	1.084.210.119
8	3.728.266.637	1.760.683.260	1.747.129.386
9	4.844.412.225	2.142.873.630	2.128.173.186
10	5.988.434.061	3.559.201.055	3.540.959.792
11	7.132.455.897	4.726.166.940	4.696.256.018
12	8.503.372.079	6.248.482.257	6.203.348.182
13	9.248.635.238	8.031.302.631	7.980.820.095
14	9.965.273.699	9.290.257.518	9.236.627.595
15	11.685.648.502	9.787.800.001	9.733.921.306
16	13.092.771.119	11.259.344.038	11.198.107.623
17	14.261.229.909	12.565.541.840	12.501.693.030
18	14.960.489.170	13.628.046.027	13.553.572.175
19	15.659.748.431	14.770.158.251	14.684.263.277
20	16.944.855.524	16.338.107.923	16.236.533.452
21	18.393.898.656	16.554.749.465	16.451.008.578
22	19.319.515.940	16.660.820.088	16.556.018.496
23	20.163.118.635	17.208.447.335	17.102.550.488
24	75.157.792	38.584.080	38.198.239
25	639.956.800	52.608.045	52.081.964

4.7 SPI (Scheduling Performance Index)

The SPI (Scheduling Performance Index) value is obtained from the comparison between the value of physically completed work / BCWP (Budgeted Cost of Work Performance) with the planned cost expenditure incurred based on the work plan / BCWS (Budgeted Cost of Work Schedule). An example of calculating the SPI (Scheduling Performance Index) value in week 25:

SPI = BCWP / BCWS

Table 4. SV (Scheduling Variance) values up to week 25

Week	BCWP (Budgeted Cost of Work Performance)	BCWS (Budgeted Cost of work Scheduling)	SV (Scheduling Variance)
1	38.584.080	75.157.792	−36.573.713
2	52.608.045	639.956.800	−587.348.755
3	88.019.075	1.204.755.808	−1.116.736.733
4	105.144.940	1.769.554.816	−1.664.409.876
5	292.423.730	2.259.196.031	−1.966.772.302
6	658.024.779	2.748.837.247	−2.090.812.468
7	1.091.067.838	3.238.478.462	−2.147.410.625
8	1.760.683.260	3.728.266.637	−1.967.583.378
9	2.142.873.630	4.844.412.225	−2.701.538.595
10	3.559.201.055	5.988.434.061	−2.429.233.006
11	4.726.166.940	7.132.455.897	−2.406.288.957
12	6.248.482.257	8.503.372.079	−2.254.889.822
13	8.031.302.631	9.248.635.238	−1.217.332.608
14	9.290.257.518	9.965.273.699	−675.016.181
15	9.787.800.001	11.685.648.502	−1.897.848.502
16	11.259.344.038	13.092.771.119	−1.833.427.082
17	12.565.541.840	14.261.229.909	−1.695.688.069
18	13.628.046.027	14.960.489.170	−1.332.443.143
19	14.770.158.251	15.659.748.431	−889.590.180
20	16.338.107.923	16.944.855.524	−606.747.601
21	16.554.749.465	18.393.898.656	−1.839.149.191
22	16.660.820.088	19.319.515.940	−2.658.695.852
23	17.208.447.335	20.163.118.635	−2.954.671.300
24	38.584.080	75.157.792	−36.573.713
25	52.608.045	639.956.800	−587.348.755

= Rp. 17,208,447,335 / Rp. 20,163,118,635.
= 0,853
For the calculation of the SPI (Scheduling Performance Index) can be seen in Table 6.

Table 5. CV (Cost Variance) values up to week 25.

Week	BCWP (Budgeted Cost of Work Performance)	ACWP (Actual Cost of Work Performance)	CV (Cost Variance)
1	38.584.080	38.198.239	385.841
2	52.608.045	52.081.964	526.080
3	88.019.075	87.138.884	880.191
4	105.144.940	104.093.491	1.051.449
5	292.423.730	290.810.444	1.613.286
6	658.024.779	655.497.491	2.527.288
7	1.091.067.838	1.084.210.119	6.857.719
8	1.760.683.260	1.747.129.386	13.553.873
9	2.142.873.630	2.128.173.186	14.700.444
10	3.559.201.055	3.540.959.792	18.241.263
11	4.726.166.940	4.696.256.018	29.910.922
12	6.248.482.257	6.203.348.182	45.134.075
13	8.031.302.631	7.980.820.095	50.482.536
14	9.290.257.518	9.236.627.595	53.629.923
15	9.787.800.001	9.733.921.306	53.878.694
16	11.259.344.038	11.198.107.623	61.236.415
17	12.565.541.840	12.501.693.030	63.848.810
18	13.628.046.027	13.553.572.175	74.473.852
19	14.770.158.251	14.684.263.277	85.894.974
20	16.338.107.923	16.236.533.452	101.574.471
21	16.554.749.465	16.451.008.578	103.740.887
22	16.660.820.088	16.556.018.496	104.801.593
23	17.208.447.335	17.102.550.488	105.896.847

4.8 CPI (Cost Performance Index)

The CPI (Cost Performance Index) value can be calculated by comparing the value of physically completed work / BCWP with the costs that have been incurred in the same period / ACWP. Example of calculating CPI (Cost Performance Index) in week 25:

CPI = BCWP / ACWP

= Rp. 17,280,891,079 / Rp. 17,174,664,284

= 1006

The calculation of the CPI (Cost Performance Index) can be seen in Table 7.

Table 6. SPI (Scheduling Performance Index) values up to week 25

Week	BCWP (Budgeted Cost of Work Performance)	BCWS (Budgeted Cost of work Scheduling)	SPI (Scheduling Performance Index)
1	38.584.080	75.157.792	0,513
2	52.608.045	639.956.800	0,082
3	88.019.075	1.204.755.808	0,073
4	105.144.940	1.769.554.816	0,059
5	292.423.730	2.259.196.031	0,129
6	658.024.779	2.748.837.247	0,239
7	1.091.067.838	3.238.478.462	0,337
8	1.760.683.260	3.728.266.637	0,472
9	2.142.873.630	4.844.412.225	0,442
10	3.559.201.055	5.988.434.061	0,594
11	4.726.166.940	7.132.455.897	0,663
12	6.248.482.257	8.503.372.079	0,735
13	8.031.302.631	9.248.635.238	0,868
14	9.290.257.518	9.965.273.699	0,932
15	9.787.800.001	11.685.648.502	0,838
16	11.259.344.038	13.092.771.119	0,860
17	12.565.541.840	14.261.229.909	0,881
18	13.628.046.027	14.960.489.170	0,911
19	14.770.158.251	15.659.748.431	0,943
20	16.338.107.923	16.944.855.524	0,964
21	16.554.749.465	18.393.898.656	0,900
22	16.660.820.088	19.319.515.940	0,862
23	17.208.447.335	20.163.118.635	0,853

Earned Value Analysis Results Up to the Last Week of Reporting From the previous calculation analysis, the value of time and cost indicators and the value of project performance analysis up to the last week of reporting week 25 are obtained. The following are the results of the Earned Value analysis up to week 25 can be seen in Table 8.

While the graph of the relationship results of BCWS (Budgeted Cost of Work Schedule), BCWP (Budgeted Cost of Work Performance) and.

ACWP (Actual Cost of Work Performance) up to the 25th week of reporting can be seen in Fig. 3.

From the BCWS (Budgeted Cost of Work Schedule), BCWP (Budgeted Cost of Work Performance) and ACWP (Actual Cost of Work Performance) relationship graph

Table 7. CPI (Cost Performance Index) values up to week 25

Week	BCWP (Budgeted Cost of Work Performance)	ACWP (Actual Cost of Work Performance)	CPI (Cost Performance Index)
1	38.584.080	38.198.239	1010
2	52.608.045	52.081.964	1010
3	88.019.075	87.138.884	1010
4	105.144.940	104.093.491	1010
5	292.423.730	290.810.444	1006
6	658.024.779	655.497.491	1004
7	1.091.067.838	1.084.210.119	1006
8	1.760.683.260	1.747.129.386	1008
9	2.142.873.630	2.128.173.186	1007
10	3.559.201.055	3.540.959.792	1005
11	4.726.166.940	4.696.256.018	1006
12	6.248.482.257	6.203.348.182	1007
13	8.031.302.631	7.980.820.095	1006
14	9.290.257.518	9.236.627.595	1006
15	9.787.800.001	9.733.921.306	1006
16	11.259.344.038	11.198.107.623	1005
17	12.565.541.840	12.501.693.030	1005
18	13.628.046.027	13.553.572.175	1005
19	14.770.158.251	14.684.263.277	1006
20	16.338.107.923	16.236.533.452	1006
21	16.554.749.465	16.451.008.578	1006
22	16.660.820.088	16.556.018.496	1006
23	17.208.447.335	17.102.550.488	1006

in week 25 is located below the BCWS (Budgeted Cost of Work Schedule) curve, this means that the work done is slower than the planned schedule. Meanwhile, the ACWP (Actual Cost of Work Performance) curve is also below the BCWP (Budgeted Cost of Work Performance) curve, which illustrates that the costs incurred are lower than the budgeted costs for the work that has been carried out.

Project condition information until the last week of reporting.

At the end of the 25th review week, the work performance was 56.12%, whereas it was planned to be 65.743%. With a planned budget of.

Rp. 20,163,118,635, from these achievements the BCWP (Budgeted Cost of Work Performance) value is Rp. 17,208,447,335, with an actual budget spent of

Table 8. Earned Value analysis results up to week 25

% Plan	% Realization	Time and Cost Indicators		
		BCWS	BCWP	ACWP
65.743	56.12	20.163.118.635	17.208.447.335	17.102.550.488
Performance Analysis				
	SV	CV	SPI	CPI
−2.954.671.300		**105.896.847**	**0,853**	**1006**

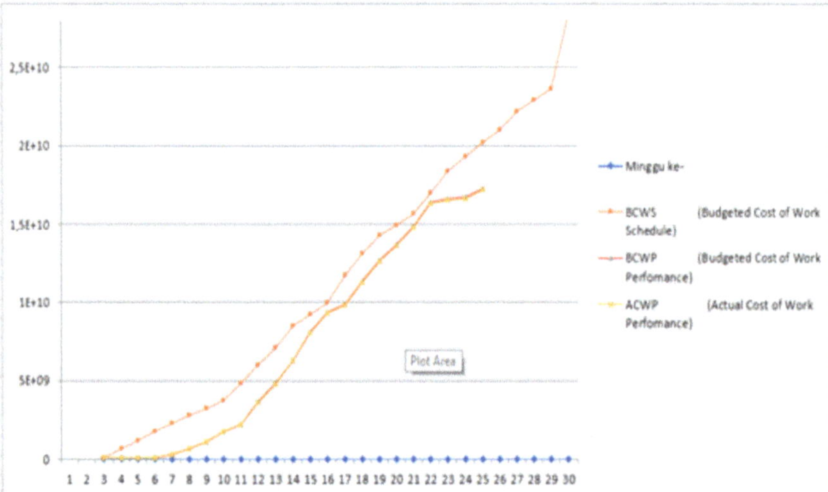

Fig. 3. Earned Value S Curve Graph (source: Soaemardi et al.)

Rp. 17,102,550,488. Of the three performance indicator values obtained, the SV (Scheduling Variance) value is −2,954,671,300, a negative value indicates that the project implementation time is delayed from the planning schedule. While the CV (Cost Variance) value is Rp. 105,896,847, a positive value indicates a lower cost for work that has been carried out that week.

The SV (Scheduling Variance) and CV (Cost Variance) values show that the project conditions until week 25 are delayed from the planned schedule but are experiencing cost savings from those budgeted. For the SPI (Scheduling Performance Index) value until week 25 there is 0.853 this figure shows the performance of the project which is delayed from the schedule. While the CPI (Cost Performance Index) value is 1.006 which shows lower cost expenditures than the budget for the work that has been carried out.

4.9 Estimated Project Completion Time

The project was planned to last 280 days. Reporting is done at the end of the 25th week on July 23, 2018, which is on the 175th day of the project. From the results of the project analysis, the schedule performance index SPI (Scheduling Performance Index) value is 0.857, then the analysis to estimate the final time (ECD) if known:

Time remaining = 280 – 175 = 105 days

Time traveled = 175 days

SPI = 0.857

ECD = (Remaining time/SPI) + time done

= (105/0,857) + 175

= 297.52 = 297 days

It is estimated that the project will be completed in 297 days. If performance is like this, it will be delayed from the original planning schedule.

Estimated Project Completion Costs and Total Project Costs.

Because the percentage of work up to week 25 has reached above 50%, the assumption used to predict the budget for the remaining work uses the formula:

ETC = (Budget – BCWP)/CPI

= (Rp.30,669,265,830 – Rp. 17,208,447,335)/1.006.

= Rp. 13,380,535,283.

From the ETC (estimate for the remaining work), the total project cost estimate is obtained:

EAC = ETC + ACWP

= Rp.13.380.535.283 + Rp. 17.102.550.488

= Rp. 30,483,085,771.

Cost value has decreased from the budgeted cost for the work performed.

5 Conclusions

The schedule variance (SV) value is almost the same as the schedule performance index (SPI). From the review until the 25th week, the value of SV (Scheduling Variance) is −2,954,671,300 minus value illustrates the magnitude of project delays expressed in cost. The schedule variance (CV) value is almost the same as the cost performance index (CPI). The difference between the two is the depiction of greater or lesser actual costs for work that has been completed. For CV, the depiction of the budget overrun or underrun is expressed in terms of cost while the CPI (Cost Performance Index) depicts the budget overrun or underrun as an index value. From the review until the last week of reporting, namely in week 25, the CV (Cost Variance) value is 105,896,847 which means that there is an improvement or the contractor is making a profit from the work that has been carried out.

References

1. Wicaksono, R.M.B.: Analisis kinerja biaya dan waktu menggunakan metode earned value pada Proyek Pembangunan Infrastruktur Laboratorium Tradisional Food GMP Facility (Paket 3). Akselerasi J. Ilm. Tek. Sipil. **3**(1), 41–49 (2021). https://doi.org/10.37058/aks.v3i1.3558

2. Bahri, M., Sembiring, K.: Analisis Kinerja Waktu Menggunakan metode earned value pada Proyek Pembangunan Gedung Kejaksaan Tinggi DKI Jakarta. J. Tek. **15**(2), 107–116 (2023). https://doi.org/10.30736/jt.v15i2.1086
3. Priyo, M., Indraga, K.F.: Analisis Kinerja Biaya Dan Jadwal Terpadu Dengan Konsep earned value method (Studi Kasus: Proyek Pembangunan Gedung). J. Ilm. Semesta Tek. **18**(2), 106–121 (2015)
4. Nurafandi, M.R., Rusdianto, Y., Andardi, F.R.: Jurnal Teknik Sipil Unaya Analisa Estimasi Biaya dan Waktu Menggunakan. J. Tek. Sipil Unaya. **9**(1), 55–63 (2023)
5. Putra, R.G., Fatmawati, W., Mas'idah, E.: Analisa Waktu Dan Biaya Proyek Konstruksi Pembangunan Gedung Gudang Dan Kantor PT ABC Semarang Dengan Earned Value Analysis. Konf. Ilm. Mhs. Unissula. **3**, 101–126 (2020)
6. Aditama, R., Witjaksana, B.: Analisis Biaya dan Waktu Menggunakan Metode EVM (Earned Value Method) Pada Proyek Konstruksi (Studi Kasus Pada Proyek Pembangunan Gedung Laundry RSUD Sidaorjo). Junal Univ. 17 Agustus 1945 Surabaya. (45), 1–15 (2021)
7. Indramanik, I.B.G., Astariani, N.K., Sudiarsana, I.W.: Analisa Kinerja Biaya Dan Waktu Menggunakan Metode Konsep Nilai Hasil (Earned Value Concept) (Studi Kasus: Proyek Pembangunan Gedung Ruang Kelas Baru Madrasah Tsanawiyah Negeri (MTsN), Amlapura, Kabupaten Karangasem). J. Tek. Gradien. **14**(02), 37 (2022)
8. Sunatha, I.G.N., Ritaka Wangsa, A.A.R., Praganingrum, T.I., Loviani, P.M.: Kinerja Biaya Dan Waktu Proyek Dengan Metode earned value management. J. Tek. Gradien. **15**(02), 39–47 (2023). https://doi.org/10.47329/teknik_gradien.v15i02.1085
9. Ramadhan, A.H., et al.: Menggunakan Metode earned value Pada. **04**(01), 69–77 (2024)
10. Momao, Y., Triwuryanto, Maulana, R.: Analisis Kinerja, Waktu Dan Biaya Pelaksanaan Proyek Pembangunan hotel Dengan Metode earned value. Equilib. **01**(01), 11–19 (2020)
11. Jaya, A., Putra, D., Nusa, A.B.: Pembangunan Gedung Sekolah Minggu Resort Aek Kanopan Kabupaten Labuhan Batu Utara Dengan Menggunakan Metode earned value. J. Tek. Sipil. **2**(2) (2023)

Jack-Up Rig Leg Penetration Evaluation for Wellhead Platform Development in Offshore Southeast Asia

Ervandi Ary Rahmadhani(✉)

Master of Engineering Program, University of 17 August 1945, Samarinda, East Kalimantan, Indonesia
ervandiary@gmail.com

Abstract. This research involves a comprehensive geotechnical investigation at the jack-up rig location on the Natuna offshore platform, Southeast Asia. The primary objective of this investigation is to provide detailed geotechnical data to support engineering and design work for the Natuna Block development project. The collected data is used to ensure that the soil conditions at the site meet the technical requirements for rig operations, offering more accurate guidance in determining foundation parameters. At the planned location for the jack-up rig installation, the geotechnical investigation included two composite boreholes and piezocone penetration tests (PCPT) with a maximum depth of 36 m at two points: WHP-P4S and WHP-P4P. Additionally, a sampling borehole with a maximum depth of 36 m was conducted at WHP-P3B-BH, along with an extra PCPT borehole of the same depth at WHP-P3B-CPT. Each borehole and penetration test were designed to identify the geotechnical characteristics of the site, serving as a basis for planning and executing the rig operations. This evaluation provides a detailed assessment of the leg penetration of the jack-up rig "X" at the drilling location in the Natuna Sea. The study covers several key aspects, including: Evaluation of piezocone penetration test (PCPT) results and geotechnical data analysis at each leg location of the rig, an in-depth analysis of the jack-up rig spudcan penetration, a crucial element in assessing the stability and safety of rig operations, recommendations based on the data and analysis results regarding the rig leg penetration, along with further discussion on the implications of these findings for rig operations at this site. Through a systematic and data-driven approach, this study provides the necessary technical guidance to ensure that the rig foundation design meets the expected safety and performance standards, thus supporting smooth operations on the Natuna offshore platform.

Keywords: Jack-up Rig Leg Penetration · Geotechnical Investigation · Piezocone Penetration Test (PCPT)

1 Introduction

This geotechnical investigation at the jack-up rig location for the offshore development project in Natuna, Southeast Asia, details the investigation process, which involved drilling, sampling, in-situ Piezocone Penetrometer Tests (PCPT), and laboratory testing

M. Awang (Ed.): ICCAEET 2024, PTT, pp. 55–73, 2025.
https://doi.org/10.1007/978-3-031-94347-8_6

in the Natuna Sea [1]. Conducted using the specialized vessel M.V. Mariner, the investigation included activities both offshore and onshore to gather essential data for the project's foundation design.

The primary objective of the investigation was to conduct a soil survey beneath the seabed to sufficient depths to determine soil layers and mechanical properties [2]. This data will inform the technical parameters for pile foundation design, pile drivability studies, mudmat design, and jack-up analysis. The scope of work at the jack-up location "X" involved continuous PCPT and sample drilling up to 36 m at proposed foreleg, port leg, and starboard leg locations, along with logging data from testing, measuring water depth, and laboratory analysis of soil samples [3].

Fig. 1. General Location Map

The location coordinates for the boreholes are illustrated in the accompanying figures and tables. The actual coordinates at the planned jack-up location "X" were determined based on survey data, and additional details about the borehole positions and water depth measurements are provided [4]. Water depth was monitored before and during drilling using a combination of electronic sensors, drill strings, and echo sounders, with periodic measurements taken every 10 min. For accurate bathymetry data, further assessment is recommended from relevant geophysical report [5] (Figs. 1 and 2).

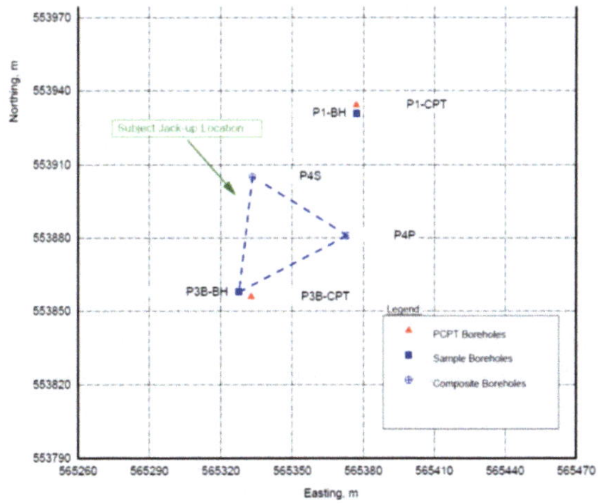

Fig. 2. Proposed Drilling Plan

The parameters used for the coordinates mentioned above are as follows:

- Datum: WGS 1972 BE
- Projection: UTM Zone 48 North
- Central Meridian: 105° E
- Latitude of Origin: 0° N

2 Water Depth Measurement

The summary of water depth measurements taken at the beginning of each drilling operation, conducted with electronic base sensors, can be found in Table 1, 2. Water depth was measured at the start of the drilling activities using a combination of electronic base sensors, drill string, and echo sounder. Additionally, periodic measurements were made every 10 min to monitor changes in water depth during the drilling and sampling process. Data from the echo sounder, along with the corrected water depths, can be found in the Appendix below.

The information provided here regarding water depth is for reference only. For accurate bathymetric data, a more detailed assessment of water depth, or jacket elevation design (Fig. 3).

Table 1. Coordinates

Borehole Designation	Final Depth (m)	UTM Coordinates - Easting (m)	UTM Coordinates - Northing (m)	Geographical Coordinates - Latitude (N)	Geographical Coordinates - Longitude (E)
P3B-PCPT	35.8	565,332.73	553,855.86	05° 00′ 37.86″	105° 35′ 21.68″
P3B-BH	35.8	565,327.50	553,857.85	05° 00′ 37.92″	105° 35′ 21.51″
Borehole Designation	**Final Depth (m)**	**UTM Coordinates - Easting (m)**	**UTM Coordinates - Northing (m)**	**Geographical Coordinates - Latitude (N)**	**Geographical Coordinates - Longitude (E)**
Starboard Leg					
P4S	37.9	565,333.29	553,904.98	05° 00′ 39.45″	105° 35′ 21.70″
Port Leg					
P4P	36.3	565,372.06	553,880.85	05° 00′ 38.67″	105° 35′ 22.96″

Table 2. Water Depth Measurements

Borehole Designation	Measured Water Depths - Metres	Measured Water Depths - Time, Date
Bow Leg		
P3B-CPT	82.50	1900 h
P3B-BH	82.20	2330 h
Borehole Designation	**Measured Water Depths - Metres**	**Measured Water Depths - Time, Date**
Starboard Leg		
P4S	82.20	2330 h
Port Leg		
P4P	82.20	1610 h

3 Interpretation of Investigation Results

3.1 Soil Stratigraphy

The soil stratigraphy obtained from field investigations and laboratory tests at the planned Jack-up evaluation site is presented in the borehole logs found in the Illustration section. This stratigraphy is based on the classification of soil samples obtained from drilling, in-situ test data, and observations made during drilling activities. The primary stratigraphic interpretation for each leg location from the sample drilling and PCPT can be found in Tables 3–5.

Details regarding the soil, including stratigraphy, strength, color variations, and inclusions, are recorded in the borehole logs found in Plates (2a & 2b), (3a & 3b), and (4a & 4b) for the rig legs at the bow, starboard, and port positions. An overview of the soil

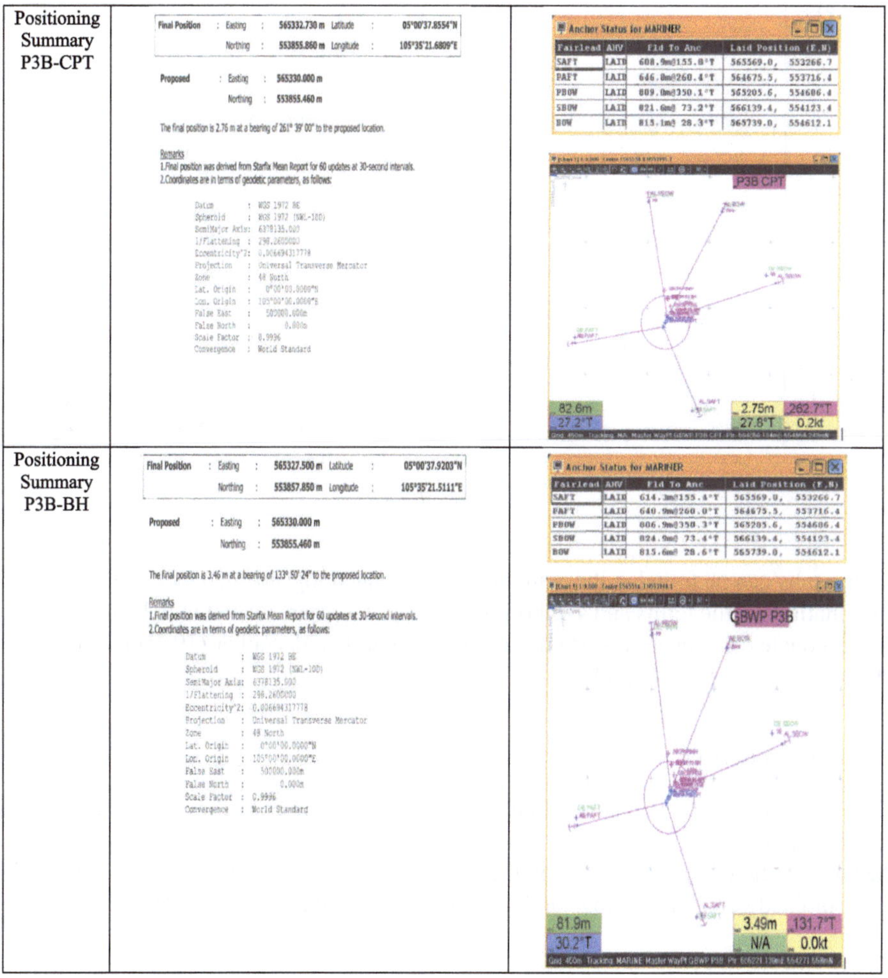

Fig. 3. Positioning Summary

Positioning Summary P3B-P4P	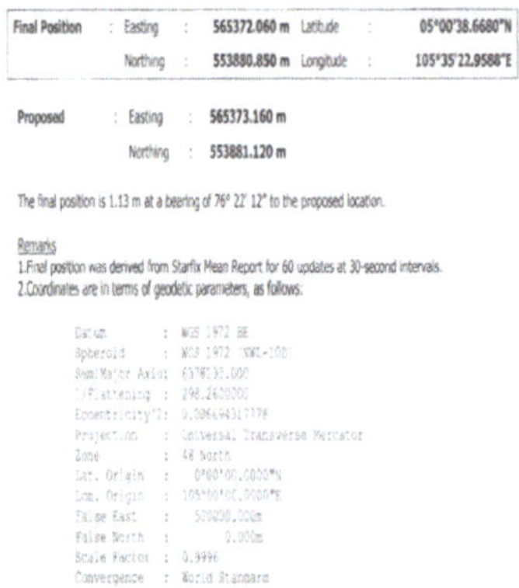	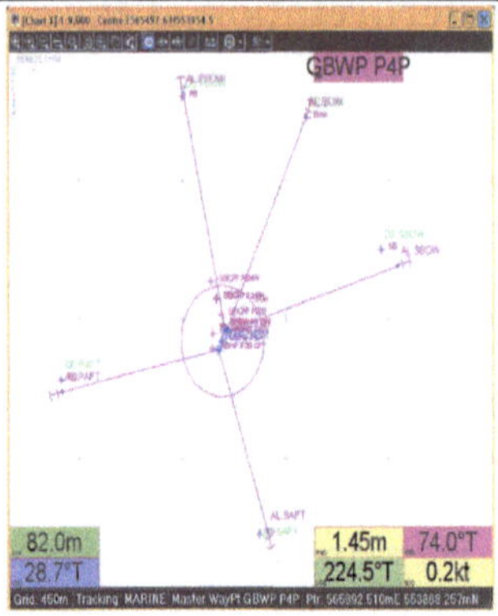
Positioning Summary P3B-P4S	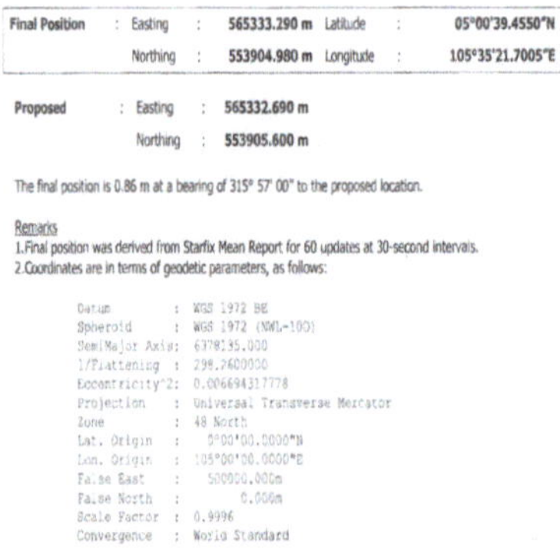	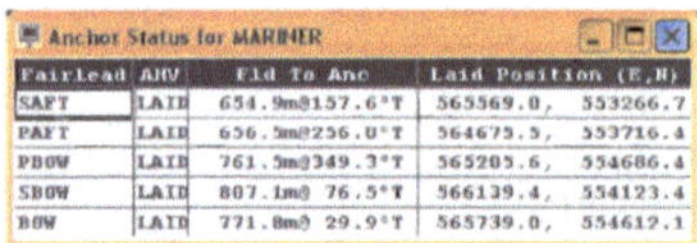 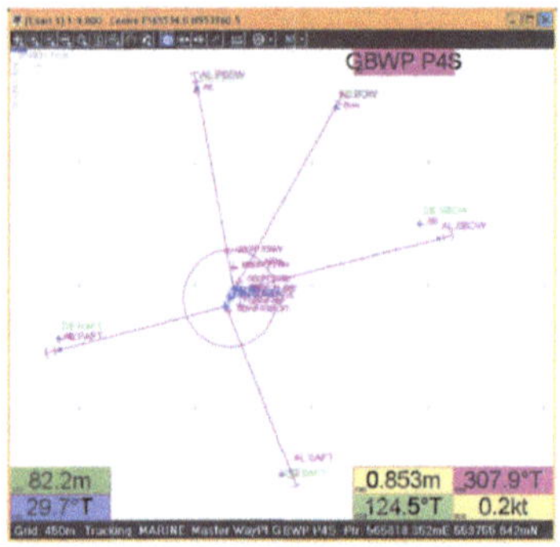

Image 1 (Positioning Summary P3B-P4P, report):

Final Position : Easting : 565372.060 m Latitude : 05°00'38.6680"N
Northing : 553880.850 m Longitude : 105°35'22.9588"E

Proposed : Easting : 565373.160 m
Northing : 553881.120 m

The final position is 1.13 m at a bearing of 76° 22' 12" to the proposed location.

Remarks
1.Final position was derived from Starfix Mean Report for 60 updates at 30-second intervals.
2.Coordinates are in terms of geodetic parameters, as follows:

Datum : WGS 1972 BE
Spheroid : WGS 1972 (NWL-10D)
SemiMajor Axis: 6378135.000
1/Flattening : 298.2600000
Eccentricity^2: 0.00669431778
Projection : Universal Transverse Mercator
Zone : 48 North
Lat. Origin : 0°00'00.0000"N
Lon. Origin : 105°00'00.0000"E
False East : 500000.000m
False North : 0.000m
Scale Factor : 0.9996
Convergence : World Standard

Image 2 (Positioning Summary P3B-P4P, chart):

Image 3 (Positioning Summary P3B-P4S, report):

Final Position : Easting : 565333.290 m Latitude : 05°00'39.4550"N
Northing : 553904.980 m Longitude : 105°35'21.7005"E

Proposed : Easting : 565332.690 m
Northing : 553905.600 m

The final position is 0.86 m at a bearing of 315° 57' 00" to the proposed location.

Remarks
1.Final position was derived from Starfix Mean Report for 60 updates at 30-second intervals.
2.Coordinates are in terms of geodetic parameters, as follows:

Datum : WGS 1972 BE
Spheroid : WGS 1972 (NWL-10D)
SemiMajor Axis: 6378135.000
1/Flattening : 298.2600000
Eccentricity^2: 0.006694317778
Projection : Universal Transverse Mercator
Zone : 48 North
Lat. Origin : 0°00'00.0000"N
Lon. Origin : 105°00'00.0000"E
False East : 500000.000m
False North : 0.000m
Scale Factor : 0.9996
Convergence : World Standard

Image 4 (Anchor Status for MARINER):

Fairlead	AHV	Fld To Anc	Laid Position (E,N)	
SAFT	LAID	654.9m@157.6°T	565569.0,	553266.7
PAFT	LAID	656.5m@256.0°T	564675.5,	553716.4
PBOW	LAID	761.5m@349.3°T	565205.6,	554686.4
SBOW	LAID	807.1m@ 76.5°T	566139.4,	554123.4
BOW	LAID	771.8m@ 29.9°T	565739.0,	554612.1

Image 5 (Positioning Summary P3B-P4S, chart):

Fig. 3. (*continued*)

stratigraphy can be found in the combined geotechnical logs located in Plates 5a and 5b. Explanations of the terms and symbols used in these logs are available in Plate 6a, while the carbonate classification system according to Clark and Walker (1977) is detailed in Plate 6b (Figs. 4, 5, 6, 7 and 8).

3.2 Soil Properties

The soil at this location consists of layers of clay extending from the current seabed to a depth of approximately 35.8 to 37.9 m. Generally, these clay layers are consistent across the three studied locations, although there are variations in layer thickness as detailed in the combined geotechnical logs on Plates 5a and 5b (Tables 6, 7, 8 and 9).

Table 3. Soil Stratigraphy P3B BH/PCPT Location (Bow Leg)

Stratum	Penetration, m		Thickness, m	Description
	From	To		
1	0	1.1	1.1	Very soft calcareous CLAY
2	1.1	9	7.9	Very soft to firm calcareous CLAY
3	9	12.2	3.2	Stiff to hard CLAY
4	12.2	17.9	5.7	Stiff to very stiff lean CLAY
5	17.9	22.9	5	Stiff to very stiff CLAY
6	22.9	30.2	7.3	Stiff to very stiff lean CLAY
7	30.2	35.8	5.6	Stiff to very stiff CLAY

Table 4. Soil Stratigraphy – P4S (Starboard Leg)

Stratum	Penetration, m		Thickness, m	Description
	From	To		
1	0	0.8	0.8	Very soft calcareous CLAY
2	0.8	9.6	8.8	Soft to stiff calcareous CLAY
3	9.6	13.6	4	Stiff to hard CLAY
4	13.6	16.8	3.2	Stiff to very stiff lean CLAY
5	16.8	28.9	12.1	Stiff to very stiff CLAY
6	28.9	37.9	9	Stiff to very stiff CLAY

Table 5. Soil Stratigraphy - P4P Location (Port Leg)

Stratum	Penetration, m		Thickness, m	Description
	From	To		
1	0	1.1	1.1	Very soft calcareous CLAY
2	1.1	10.2	9.1	Soft to stiff calcareous CLAY
3	10.2	15.4	5.2	Stiff to hard CLAY
4	15.4	25.4	10	Stiff to very stiff CLAY
5	25.4	33.4	8	Stiff to very stiff CLAY
6	33.4	36.3	2.9	Stiff to very stiff CLAY

Table 6. Rig Information

Plan area	152.6 m^2
Equivalent diameter	13.9 m
Elevation of footing tip below the major cross-section of the spudcan	2.4 m
Volume of tank footing	404.5 m^3
Lightship weight	38.5 MN
Maximum preload	72.2 MN
Available leg length	131.3 m

Table 7. Plates (7a & 7b)

Ground Unit	Depth (m)	Ground Model	ϕ' (deg)	$c_{u,top}$ (kPa)	$c_{u,bot}$ (kPa)	γ' (kN/m^3)
	0					
1	1.1	Clay		2	4	4.6
2	1.9	Clay		10	32	7
3	2.9	Clay		32	42	7
4	7.6	Clay		42	42	7
5	7.9	Clay		160	160	9.8
6	9	Clay		42	42	7
7	12.2	Clay		173	116	9
8	14	Clay		116	116	8.9
9	17.9	Clay		116	131	8.9
10	22.9	Clay		116	105	8
11	30.2	Clay		105	105	8.5
12	35.8	Clay		105	105	7.8

- Squeezing Factor : 1.50
- Squeezing Depth Factor : 4.00
- Undrained Shear Strength Depth Factor: 4.00
- Location: GBWP-P38-BH/PCPT (Bow Leg) - High Estimate

Groun d Unit	Dept h (m)	Groun d Model	ϕ' (deg)	$c_{u,top}$ (kPa)	$c_{u,bot}$ (kPa)	γ' (kN/m^3)
	0					
1	1.1	Clay		2	4	4.6
2	1.9	Clay		10	28	7
3	2.9	Clay		28	38	7
4	7.6	Clay		38	38	7
5	7.9	Clay		40	40	9.8
6	9	Clay		38	38	7
7	12.2	Clay		157	104	9
8	14	Clay		104	104	8.9
9	17.9	Clay		104	119	8.9
10	22.9	Clay		104	95	8
11	30.2	Clay		95	95	8.5
12	35.8	Clay		95	95	7.8

- Squeezing Factor : 1.50
- Squeezing Depth Factor : 4.00
- Undrained Shear Strength Depth Factor: 4.00
- Location: GBWP-P38-BH/PCPT (Bow Leg) - Low Estimate

Table 8. Plates (9a & 9b)

Ground Unit	Depth (m)	Ground Model	ϕ' (deg)	$c_{u,top}$ (kPa)	$c_{u,bot}$ (kPa)	γ' (kN/m^3)
	0					
1	0.8	Clay		1	3	5.6
2	2.8	Clay		16	40	7.0
3	6.5	Clay		40	40	7.0
4	8.4	Clay		40	63	7.0
5	9.6	Clay		63	32	7.0
6	13.6	Clay		168	105	9.1
7	16.8	Clay		105	147	9.2
8	20.0	Clay		147	110	8.4
9	25.4	Clay		110	110	8.4
10	26.3	Clay		110	158	8.4
11	28.9	Clay		158	105	8.4
12	37.9	Clay		105	110	7.8

- Squeezing Factor : 1.50
- Squeezing Depth Factor : 4.00
- Undrained Shear Strength Depth Factor: 4.00
- Location: GBWP-P4S (Starboard Leg) - High Estimate

Ground Unit	Depth (m)	Ground Model	ϕ' (deg)	$c_{u,top}$ (kPa)	$c_{u,bot}$ (kPa)	γ' (kN/m^3)
	0					
1	0.8	Clay		1	3	5.6
2	2.8	Clay		14	36	7.0
3	6.5	Clay		36	36	7.0
4	8.4	Clay		36	57	7.0
5	9.6	Clay		57	28	7.0
6	13.6	Clay		152	95	9.1
7	16.8	Clay		95	133	9.2
8	20.0	Clay		133	100	8.4
9	25.4	Clay		100	100	8.4
10	26.3	Clay		100	142	8.4
11	28.9	Clay		142	95	8.4
12	37.9	Clay		95	100	7.8

- Squeezing Factor : 1.50
- Squeezing Depth Factor : 4.00
- Undrained Shear Strength Depth Factor: 4.00

Location: GBWP-P4S (Starboard Leg) - Low Estimate

Table 9. Plates (9a & 9b)

Ground Unit	Depth (m)	Ground Model	ϕ' (deg)	$c_{u,top}$ (kPa)	$c_{u,bot}$ (kPa)	γ' (kN/m³)
	0					
1	1.1	Clay		1	2	5.0
2	3.0	Clay		18	36	7.0
3	10.2	Clay		36	42	7.0
4	15.4	Clay		178	84	9.0
5	20.4	Clay		110	136	8.2
6	23.2	Clay		136	105	8.2
7	25.4	Clay		105	105	8.2
8	33.4	Clay		110	110	8.7
9	36.3	Clay		110	110	7.6

- Squeezing Factor : 1.50
- Squeezing Depth Factor : 4.00
- Undrained Shear Strength Depth Factor: 4.00

Location: GBWP-P4P (Port Leg) - High Estimate

Ground Unit	Depth (m)	Ground Model	ϕ' (deg)	$c_{u,top}$ (kPa)	$c_{u,bot}$ (kPa)	γ' (kN/m³)
	0					
1	1.1	Clay		1	2	5.0
2	3.0	Clay		16	32	7.0
3	10.2	Clay		32	38	7.0
4	15.4	Clay		162	76	9.0
5	20.4	Clay		100	124	8.2
6	23.2	Clay		124	95	8.2
7	25.4	Clay		95	95	8.2
8	33.4	Clay		100	100	8.7
9	36.3	Clay		100	100	7.6

- Squeezing Factor : 1.50
- Squeezing Depth Factor : 4.00
- Undrained Shear Strength Depth Factor: 4.00

Location: GBWP-P4P (Port Leg) – Low Estimate

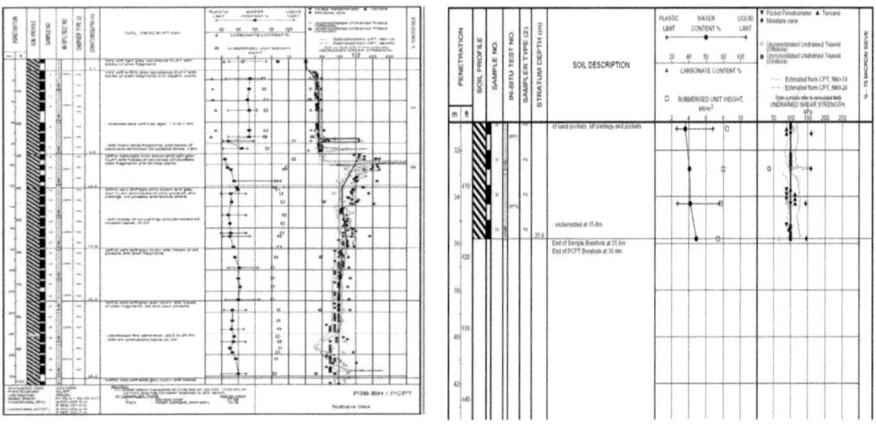

Fig. 4. Plates (2a & 2b)

At the study site for the jack-up rig, the clay is generally firm to very firm, except for a very soft surface clay layer. A thin layer of this clay is present at the P3B and P4S locations at depths ranging from 12.2 to 17.9 m. This clay exhibits high plasticity, with a Liquid Limit (wL) between 51 and 104% and (cu/σ′vo)nc values between 0.22 and 0.38. The thin clay layer has a wL between 48 and 49% and an (cu/σ′vo)nc value of approximately 0.21.

The clay is calcareous (containing calcium carbonate) from the mudline to a depth of approximately 9.0 to 10.2 m, with carbonate content ranging from 7 to 33%. The term "calcareous" refers to soil with a carbonate content between 10 and 50% based on the carbonate classification system of Clark and Walker (1977). Inclusions found include coral and shell fragments, pockets/voids of silt and sand, hardened silt pockets, mica spots, as well as organic and iron stains at various depths.

3.3 Interpretation of Soil Boring Data

The strength data recorded in the borehole logs and PCPT results, along with other soil test results, are used to establish the stratigraphy and soil strength parameters [6].

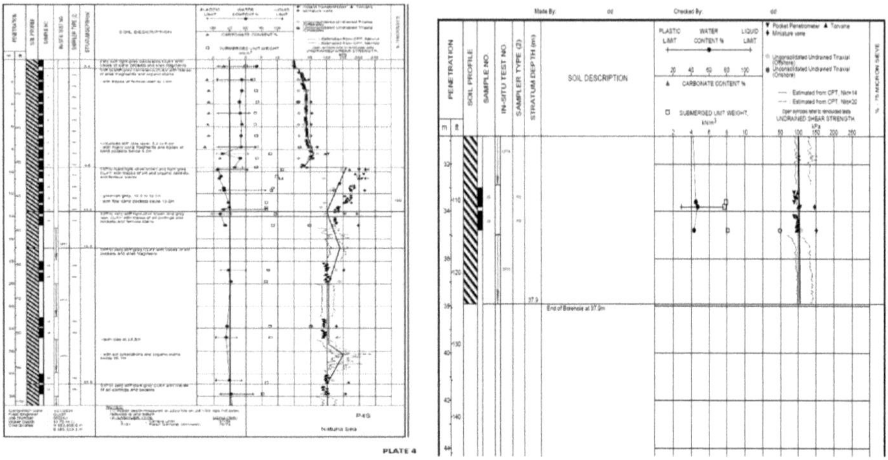

Fig. 5. Plates (3a & 3b)

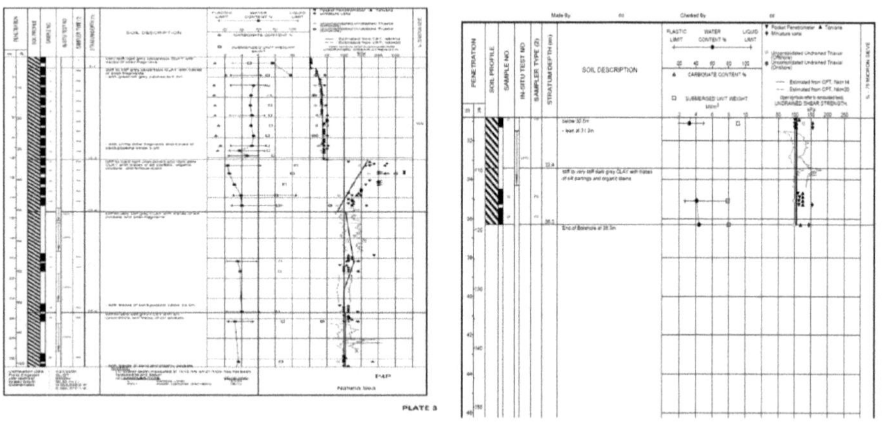

Fig. 6. Plates (4a & 4b)

The unit weight (UW) of the soil can also be calculated based on its moisture content, assuming the soil is fully saturated. The procedure for calculating unit weight is described below. In-situ unit weight can be estimated by calculating the value based on the moisture content and soil density, assuming the soil is in a saturated condition, using the following formula:

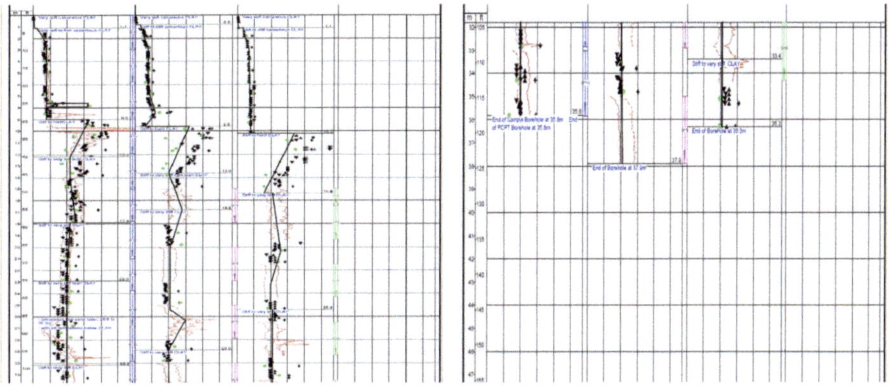

Fig. 7. Plates (5a & 5b)

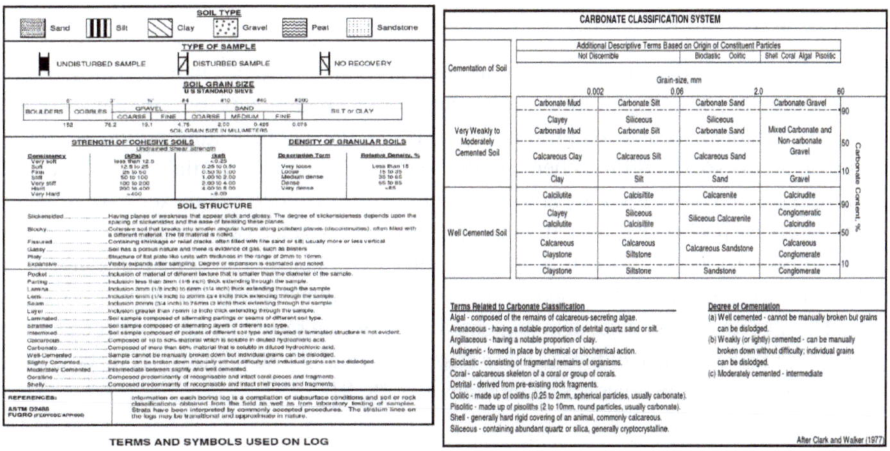

Fig. 8. Plates (6a & 6b)

$$\gamma = 9.81\left\{\frac{G_s(1+0.01w)}{(1+0.01wG_s)}\right\}$$

where, γ = unit weight, kN/m^3;
w = water content, percent; and
G_s = specific gravity of solids.

For cohesive soils, G_s, can be computed as follows:

$$G_s = 0.00072\,(w_L) + 2.667$$

where, w_L = liquid limit, percent.

Additionally, moisture content and Atterberg Limits are used to provide further information for determining soil stratigraphy. From spesific information on moisture content are available, while Atterberg Limits data is presented, Plasticity and Liquidity Indexes can be found.

The calculated (cu/σ'vo)nc values are also considered in the development of strength parameters and can be obtained These values are calculated using Skempton's (1957) formula:

$$(c_u/\sigma'_{vo})_{nc} = 0.11+0.0037I_p$$

where	σ'_{vo}	=	effective overburden pressures, kPa; and
	I_p	=	plasticity index

3.4 Interpretation of Cone Penetrometer Data

The results from the penetrometer tests are used to Assess material characteristics, establish stratigraphy and soil conditions, select appropriate parameters for granular soils and estimate undrained shear strength for clay soils [7]. Piezocone testing provides continuous data on the soil conditions beneath the seabed, with results available in this report. Data from the PCPT are used to identify strength patterns and to avoid excessive reliance on strength test results that may be influenced by disturbance or premature failure due to structural features such as slick surfaces, blocks, or slabs. The cone resistance data collected during the tests are used to assess the undrained shear strength of cohesive soils. The sleeve friction data, also presented as a friction ratio (the ratio between sleeve friction and tip resistance expressed as a percentage), along with measured excess pore pressure, are used to evaluate soil characteristics [8, 9].

3.5 Correlation Between Soil Boring and PCPT Data

The measurement of cone tip resistance in cohesive soils, along with shear strength data from laboratory tests, is recorded in the borehole logs. The ratio of sleeve friction to cone tip resistance, along with excess pore pressure readings, generally supports the classification of clay and sand/silt soils and aids in identifying soil layers and stratigraphy [10].

3.6 Estimate of Shear Strength in Cohesive Soils

The results from cone penetration tests can be used to estimate shear strength in cohesive materials. Subsequently, the PCPT data are correlated with the undrained shear strength obtained from laboratory tests using the following equation [11]:

$$c_u = \frac{q_t - \sigma_{vo}}{N_{kt}} = \frac{q_{net}}{N_{kt}}$$

where,

c_u	=	undrained shear strength;
q_t	=	corrected PCPT tip resistance;
q_{net}	=	net cone resistance;
σ_{vo}	=	total overburden pressure, (including hydrostatic); and
N_{kt}	=	cone factor.

According to Schmertmann (1975), the value of Nkt is influenced by various factors, including the method used to determine undrained shear strength, in-situ soil stress conditions, stress history, soil structure, soil sensitivity, soil plasticity characteristics, type of cone penetrometer used, PCPT operational mode and penetration rate. In various studies, the value of Nkt has been reported to range from 4 to 30. At the proposed Jack-up study site, Nkt values between 14 and 20 were applied to determine the undrained shear strength from the PCPT data. The obtained shear strength data and conclusions are presented in the borehole logs and undrained shear strength plates available in the illustrations.

4 Leg Penetration Analyses

4.1 Description of Rig

The jack-up rig's legs will penetrate the soil beneath the seabed until the soil's bearing capacity matches or exceeds the pressure applied by the legs [12]. This study aims to provide information on the extent of penetration achievable by the legs, enabling the Operator to determine whether the length of the jack-up rig's legs is sufficient to ensure operational safety at the site. Additionally, the study evaluates the potential for sudden deep penetration that may occur due to the transition from harder soil layers to softer underlying layers. The planned MODU is a jack-up drilling rig with three legs, each functioning independently and equipped with spudcans mounted on the structural framework [13]. The dimensions of the support legs and related load information for this rig, according to the data provided by design, are summarized as follows:

The weight of the empty vessel refers to the vertical load applied to the foundation when the jack-up rig has lowered its legs to the seabed and raised its hull to a state of being free from air, without adding water ballast.

4.2 Design Parameter

Design parameters, such as shear strength and unit weight, are determined based on the analysis of laboratory shear strength test data as well as PCPT data. In developing the shear strength profile for the jack-up rig analysis, the measured shear strength trends are compared and evaluated with data from torvane, pocket penetrometer, laboratory mini vane, and PCPT data [14]. Unconsolidated undrained (UU) shear strength data are considered along with other laboratory shear strength test data, PCPT data, and observed shear strength trends to determine the design parameters. For clay, we apply a high estimate of +5% and a low estimate of −5% to the specified design shear strength profile [15]. At the P3B BH/PCPT location, very stiff clay was detected at depths of 7.6 to 7.9 m, which was not observed at P3P and P3S locations. Therefore, higher shear strength from the UU data is used as the high estimate, while lower shear strength from the PCPT data is used as the low estimate. The design parameters used in the analysis can be found in the illustrations on Plates (7a & 7b), (9a & 9b), and (11a & 11b) for the P3B BH/PCPT (fore leg), P4P (port leg), and P4S (aft leg) locations. The design shear strength profiles are also graphically displayed in the boring logs (Figs. 9 and 10).

5 Results and Analysis

The results of the analysis are presented in Plates 8, 10, and 12, showing the soil resistance curves against spudcan penetration below the seabed for each leg. A summary of these analysis results can be found in the following table:

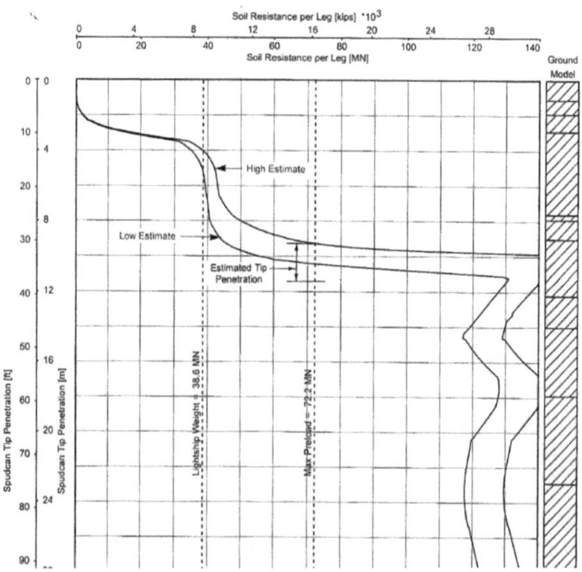

Fig. 9. Plate 8

The projected final depth for each leg penetration, as shown in the table above, is calculated from the current seabed level. As previously mentioned, very stiff clay was found at depths of 7.6 to 7.9 m at the front leg location and has been incorporated into the High Estimate parameters. Therefore, the anticipated penetration range for the front leg is greater compared to the port and stern leg locations (Fig. 11).

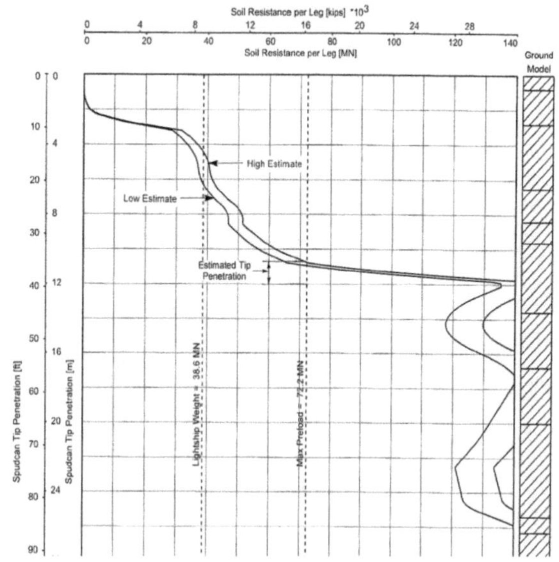

Fig. 10. Plate 10

Table 10. Expected Leg Penetrations

Spudcan Leg	Maximum Preload (MN)	Expected Final Penetration (m)	Possible Rapid Penetration during Application of Lightship Weight or Initial Stage of Preloading (m)	Potential Punch-through If yes, magnitude (m)
Bow Leg	72.2	9.2 – 11.4	4.0 (4.0 to 8.0)	Not Indicated
Starboard Leg	72.2	10.7 – 12.0	2.0 (4.4 to 6.4)	Not Indicated
Port Leg	72.2	11.5 – 12.6	3.6 (5.6 to 9.2)	Not Indicated

5.1 P3B BH/PCPT Location (Bow Leg)

The spudcan leg tip of designate rig is estimated to stabilize at a depth between 9.2 and 11.4 m below the seabed, with a maximum load of up to 72.2 MN. During loading approaching lightship or in the early preloading phase (with loads between 38 and 42 MN), there is a possibility of rapid penetration of approximately 4.0 m (from a depth of around 4.0 to 8.0 m). After this, the leg will penetrate gradually until it reaches the maximum preloading of 72.2 MN. If the leg is at a depth of around 4.0 m at lightship, careful attention is needed during the early preloading phase. No potential punch-through was found during the final penetration under maximum preloading conditions.

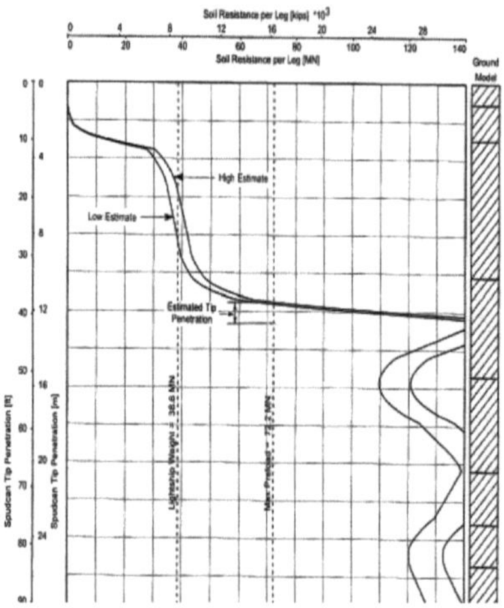

Fig. 11. Plate 12

5.2 P4S Location (Starboard Leg)

The final penetration depth of the starboard leg of designate rig is estimated to be between 10.7 and 12.0 m below the seabed, with a maximum preloading load of 72.2 MN. Similar to the bow leg, there is an indication of a potential rapid penetration of approximately 2.0 m (from a depth of about 4.4 to 6.4 m) during lightship loading and/or the early preloading phase, with loads ranging from 36 to 40 MN. After this, the leg will penetrate steadily until it reaches the final depth of between 10.7 and 12.0 m with a maximum preloading load of 72.2 MN. If the leg penetration reaches around 4.4 m at lightship load, extra caution should be applied during the early preloading phase. There are no indications of potential foundation failure or "punch through" during the preloading process or at the estimated final penetration under maximum preloading load.

5.3 P4P Location (Port Leg)

At the port leg location, the spudcan tip of designate rig is estimated to penetrate to a final depth of approximately 11.5 to 12.6 m below the seabed with a maximum preloading load of 72.2 MN. As with other locations, there is an indication of a potential rapid penetration of about 3.6 m (from a depth of approximately 5.6 to 9.2 m) during lightship loading and/or the early preloading phase, with reaction loads ranging from 35 to 40 MN. After this phase, the leg will continue to penetrate steadily until it reaches the final depth of 11.5 to 12.6 m with a maximum preloading load of 72.2 MN. If the leg penetration reaches around 5.6 m during lightship loading, extra caution is required during the early preloading phase. No indications of potential "punch-through" foundation failure were found at the final penetration under maximum preloading load.

6 Summary and Recommendations

6.1 Summary

In the geotechnical survey for the Natuna Southeast Asia Offshore Development project Indonesia, the investigation covered two platform locations: Wellhead Platform "A" and Wellhead Platform "B," as well as a jack-up rig location and the pipeline route. The purpose of this survey was to provide the geotechnical data necessary to support engineering work for the Natuna Sea Development Project. The geotechnical investigation was carried out using a specialized drilling vessel. The survey included three locations with three sample borings, three PCPT borings, and two composite sample-PCPT borings with target depths ranging from 36 to 150 m below the seabed at each of the studied platforms, as well as three gravity core drops along the pipeline route.

This report presents engineering analyses related to the installation of the jack-up rig, including soil parameter evaluation and the development of penetration curves for the jack-up designate rig at Wellhead Platform "A." The main findings and conclusions of this study are summarized as follows:

Coordinates of the soil boring locations and the recorded water depths are provided in Tables 11 and 12.

Table 11. Coordinates

Borehole Designation	Final Depth (m)	UTM Coordinates		Geographical Coordinates	
		Easting (m)	Northing (m)	Latitude (N)	Longitude (E)
Bow Leg					
P3B-PCPT	35.8	565,332.73	553,855.86	05° 00′ 37.86"	105° 35′ 21.68"
P3B-BH	35.8	565,327.50	553,857.85	05° 00′ 37.92"	105° 35′ 21.51"
Starboard Leg					
P4S	37.9	565,333.29	553,904.98	05° 00′ 39.45"	105° 35′ 21.70"
Port Leg					
P4P	36.3	565,372.06	553,880.85	05° 00′ 38.67"	105° 35′ 22.96"

The soil at the site consists of clay layers extending from the seabed to a depth of approximately 35.8 to 37.9 m. Overall, these layers are uniform and consistent across the three investigated locations, although there are variations in thickness, as summarized in Tables 2 to 4.

Table 12. Water Depth Measurement

Borehole Designation	Measured Water Depths	
	Metres	Time
Bow Leg		
P3B-CPT	82.5	1900 h
P3B-BH	82.2	2330 h
Starboard Leg		
P4S	82.2	2330 h
Port Leg		
P4P	82.2	1610 h

Estimation of the penetration of the spudcan tip of desinate rig under a maximum load of 72.2 MN will reach depths of 9.2 to 11.4 m at the bow leg location, 10.7 to 12.0 m at the starboard leg location, and 11.5 to 12.6 m at the port leg location, as summarized in Table 10. At the final penetration under maximum load, there is no potential for "punch-through." However, the bow leg may experience minor, non-continuous movement (around 1 m) when encountering coral reefs at depths greater than 8 m.

6.2 Recommendations

Spudcan on jack-up rigs often experience lateral movement during the installation and preloading processes [16]. If not properly managed, this can lead to the spread of the rig legs. The horizontal displacement of the spudcan, or slippage, is typically influenced by the following factors: (1) Seabed slope, (2) Existing spudcan footprints, (3) Buried channels (spudcan near the edge of a channel), (4) Variations in lateral soil strength, and (5) Harder materials (such as solid clay, sand, or silt) within generally soft to firm clay layers.

To mitigate the risk of rapid penetration at this site, we recommend preloading each rig leg separately. The preloading process and spudcan penetration monitoring should be conducted carefully, with the vessel's hull maintained at a safe height.

These predictions are specific to rig "West Callisto" at the GBWP Jackup site. The study assumes that the soil stratigraphy revealed through drilling is applicable only to the respective leg locations. Considerations regarding lateral variations in soil conditions, seabed slope, geological conditions, or other factors that may affect foundation behavior are not within the scope of this study.

Variations in subsurface conditions across the site may result in differences in foundation penetration compared to our analytical estimates, as well as uneven penetration among the rig legs. Additionally, discrepancies between observed and estimated penetration may indicate that (1) soil conditions or preloading differ from the assumptions of this study, and/or (2) measured water depth or water-to-hull distance differs from actual measurements.

References

1. Jerman, J.: Evaluation of a rigid element for offshore applications (2014)
2. Zhao, J., Zhu, Q., Huang, W., Zhang, Y., Liu, L.: Spudcan deep penetration in multi-layered soils incorporating sand relative density. J. Mar. Sci. Eng. **11**, 33 (2022). https://doi.org/10.3390/jmse11010033
3. Jun, M.J., Kim, Y.H., Hossain, M.S., Cassidy, M.J., Hu, Y., Park, S.G.: Global jack-up rig behaviour next to a footprint. Mar. Struct. **64**, 421–441 (2019). https://doi.org/10.1016/j.marstruc.2018.12.002
4. GEO-2. Hong Kong, Geoguide 2- Guide to Site Investigation. Geotech. Eng. Off., p. 349 (2017)
5. Kim, H., Aoki, S., Kim, K., Kim, J., Shin, B.-S., Lee, K.: Bathymetric survey for seabed topography using multibeam echo sounder in Wando. In: Kim, H., Aoki, S., Kim, K., Kim, J., Shin, B.-S., Lee, K. (eds.) Bathymetric Survey for Seabed Topography using Multibeam Echo Sounder in Wando, Korea. J. Coast. Res. vol. 95, p. J. Coast. Res. vol. 95, p. 527 (2020). https://doi.org/10.2112/SI95-102.1
6. Urmi, Z., Ansary, M.: Interpretation of soil stratigraphy and geotechnical parameters from CPTu at Bhola, Bangladesh, pp. 643–648 (2018)
7. Campanella, R., Robertson, P., Gillespie, D.: Cone penetration testing in deltaic soils. Can. Geotech. J. **20**, 23–35 (2011). https://doi.org/10.1139/t83-003
8. Hossain, M.I.: Evaluation of Undrained Shear Strength and Soil Classification from Cone Penetration Test (2018)
9. Jaksa, M., Kaggwa, W., Brooker, P.: An Improved Technique for Evaluating the CPT Friction Ratio (2000)
10. Alshibli, K., Okeil, A., Alramahi, B., Zhang, Z.J.: Reliability analysis of CPT measurements for calculating undrained shear strength. Geotech. Test. J. **34**, 721–729 (2011). https://doi.org/10.1520/GTJ103771
11. Zein, A.: Estimation of undrained shear strength of fine grained soils from cone penetration resistance. Int. J. Geo-Eng. **8** (2017). https://doi.org/10.1186/s40703-017-0046-y
12. Kee, R., Ims, B.W.: Geotechnical Hazards Associated with Leg Penetration of Jack-up Rigs BT - Seabed Mechanics, pp. 169–174 (1984)
13. Li, Y., et al.: Causes of post-installation penetration of jack-up spudcan foundations in clays. PLoS ONE **13**, e0206626 (2018). https://doi.org/10.1371/journal.pone.0206626
14. Eslami, A., Mohammadi, A.: Drained soil shear strength parameters from CPTu data for marine deposits by analytical model. Ships Offshore Struct. **11**, 1–13 (2016). https://doi.org/10.1080/17445302.2015.1131082
15. Lemos, S., Pires, P.: The undrained strength of soft clays determined from unconventional and conventional tests. Soils Rocks **40**, 291–301 (2017). https://doi.org/10.28927/SR.403291
16. Stewart, D., Finnie, M.: Spudcan-footprint interaction during jack-up workovers. Proc. Int. Offshore Polar Eng. Conf. **1** (2001)

Literature Review: Green Concrete Materials as a Sustainable Innovation

Annisa' Carina[1(✉)], Machmud Sugandi[2], Rina Riefqi Mariana[2], and Eko Daniyanto[1]

[1] Universitas Islam Darul 'Ulum, Sukodadi, Jawa Timur, Indonesia
annisacarina@unisda.co.id

[2] Universitas Negeri Malang, Malang, Jawa Timur, Indonesia

Abstract. Concrete is a building material in the construction industry with constituent components consisting of fine aggregate, coarse aggregate, water, and cement. Concrete materials continue to evolve in the global construction industry, with various technological innovations and new materials keeping them relevant and essential. The concrete manufacturing industry has a significant impact on the environment, society, and the economy. Some of the impacts of the concrete manufacturing industry include carbon emissions, exploitation of natural resources, waste, and pollution. Therefore, literature review research was carried out with the aim of creating concrete constituent materials to reduce the carbon footprint and increase sustainability in the construction industry. The method used in this research was SLR (*Systematic Literature Review*), from literature searches, filtering, and quality assessment, 25 articles were found. Based on the literature obtained, the use of green concrete materials includes rice husk ash, fly ash originating from city dust, recycled concrete aggregate, ceramic powder waste, granite powder waste, steel powder, plastic fiber with the type of plastic commonly used being polyethylene. Terephthalate (PET), polyethylene (PE), and polypropylene (PP), glass fiber with types of glass waste used such as bottles, windows, car glass, and glass from industrial waste, rubber fiber from used tires as a substitute material. These materials are supported by many universities or research institutions in America, China, Thailand, India, Germany, and several countries on the European continent which are conducting studies on the influence of the materials that make up green concrete on its mechanical properties and durability.

Keywords: Green Concrete · Material · Literature Review

1 Introduction

Concrete is a building material in the construction industry with constituent components consisting of fine aggregate (such as sand), coarse aggregate (such as gravel or crushed stone), and cement as a binding agent mixed with water as an ingredient that starts the chemical process (hydration) in the cement, which makes the mixture harden and strengthen [1, 2]. Concrete is used in various construction applications, including building foundations, bridges, roads, tunnels, and various other structures because of its extraordinary strength and long durability [3].

M. Awang (Ed.): ICCAEET 2024, PTT, pp. 74–86, 2025.
https://doi.org/10.1007/978-3-031-94347-8_7

Concrete has the privilege of being able to work together or known as a composite system so that it can improve the quality of concrete. Composite concrete is enhanced with additional materials or reinforcements to provide certain properties that are better than conventional concrete [3]. The most used composite concrete in the construction industry is reinforced concrete, namely concrete reinforced with steel rods or steel mesh to increase its tensile strength [4].

As the construction industry develops, concrete has undergone significant developments since it was first used in ancient times. Around 3000 BC, the Egyptians used a mixture of mud and straw to build bricks [5]. They also used gypsum and lime as binding materials. The Romans developed hydraulic concrete around 200 BC by mixing lime and volcanic ash (pozzolana), which produced water-resistant concrete [6]. Famous examples are the Pantheon and the Colosseum.

20th century France, developed prestressed concrete in 1928, which increased the load capacity and structural flexibility of concrete [7]. Concrete is the main material in the construction of skyscrapers, large bridges, dams, and other infrastructure. As technology advances, various research, and development of high-performance concrete with a compressive strength of more than 50 MPa is carried out [8].

Concrete materials continue to evolve in the global construction industry, with various technological innovations and new materials keeping them relevant and essential. The concrete manufacturing industry has a significant impact on the environment, society, and the economy. Some of the impacts of the concrete manufacturing industry include, (a) carbon emissions, where the component that makes up concrete is cement. The cement manufacturing process requires high temperatures which are usually achieved by burning fossil fuels. This process is estimated to be responsible for around 8% of total global CO2 emissions [9]; (b) exploitation of natural resources. Another cement constituent material is aggregate, aggregate is obtained through a mining process. Mining sand, gravel, and rock for concrete aggregate can damage local ecosystems and cause soil erosion. Apart from aggregate, the process of making concrete requires large amounts of water, which can have an impact on the availability of clean water [10, 11]; (c) waste and pollution. Cement and concrete factories can produce dust and other air pollutants, which contribute to public health problems and air pollution. In addition, concrete production produces solid waste, including construction and demolition waste, which must be managed properly [12].

In his research, Emrén [13], found that the rate of change in CO2 concentration is controlled by global temperature. Thus, global warming has a negative impact on the sustainability of the earth. To reduce this impact, various efforts can be made, one of which is carrying out a more efficient production process by using technology that reduces energy consumption and emissions. The green concrete production process is designed to reduce the negative environmental impacts of traditional concrete production and use. The main goal of green concrete is to reduce carbon emissions, use recycled materials, and save natural resources [14]. With ongoing efforts in technological innovation and environmentally friendly practices, the green concrete industry is becoming an increasingly important solution in global efforts to reduce the carbon footprint and increase sustainability in the construction industry.

2 Research Methods

Research methods are ways of working to answer *research needs.* The method used in this research is SLR (*Systematic Literature Review*), which is a research method that combines the results of several independent *studies* but with interrelated topics to develop theoretical and practical aspects as the basis for research hypotheses [15]. The data used comes from textbooks, journals, scientific articles. The steps for a literature review according to Yu Xiao and Maria Watson are as follows (Fig. 1):

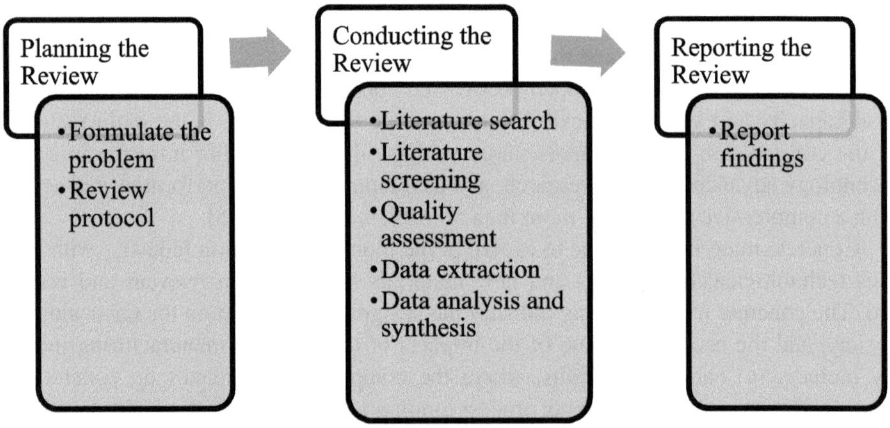

Fig. 1. Systematic Literature Review Method [15]

2.1 Formulate the Problem

The problems in this research are:

a. What impacts do the materials that make up concrete have on the environment?
b. What environmentally friendly materials can be used in preparing green concrete?

2.2 Review Protocol

The review protocol describes all elements of the review, the review protocol in this study is:

a. Knowing the impact that the materials that make up concrete have on the environment,
b. Formulate environmentally friendly concrete building blocks

2.3 Literature Search

The process of finding material for review in this research uses key words as in Table 1.

Table 1. Key words for literature search

Research questions	Keywords
What impacts do the materials that make up concrete have on the environment?	Concrete constituent materials, impact of the use of concrete constituent materials
What environmentally friendly materials can be used in preparing green concrete?	Green concrete, environmentally friendly materials that make up concrete, the strength of green concrete

2.4 Quality Filtering

Next, filter each article to determine whether the article is suitable and can be used for data extraction and analysis. The following are the criteria used in this research,

a. Research published in the last five years
b. Research published in international journals
c. Research is concerned with research questions

2.5 Quality Assessment

Quality assessment of each article was carried out using the checklist method as in Table 2.

Table 2. Quality Assessment Checklist Method

Sections	Checklist items	√
Title	The research title contains the words green concrete, environmentally friendly concrete, substitute materials for concrete	
Introduction	Rational and objective in nature which supports the research title	
Method	Using testing methods from concrete laboratories	
Results	Producing concrete with substitute materials that make up concrete that are environmentally friendly and have compressive strength test results that comply with standards	
Conclusion	Produce research on concrete constituent materials with environmentally friendly substitute materials	

2.6 Data Extraction

The total final paper obtained will go through a data extraction process where the data needed to answer the research question will be described using the following data extraction template:

a. Publication details: Title, author, year
b. Research problem and objectives: Green concrete
c. Concrete testing: Strength of green concrete

2.7 Data Analysis and Synthesis

Summarize and integrate data into literature review results.

2.8 Report Findings

Reports findings from literature searches, screening, and quality assessments.

Using the basic literature review according to Yu Xiao and Maria Watson, researchers began collecting data, resulting in the following flow diagram.

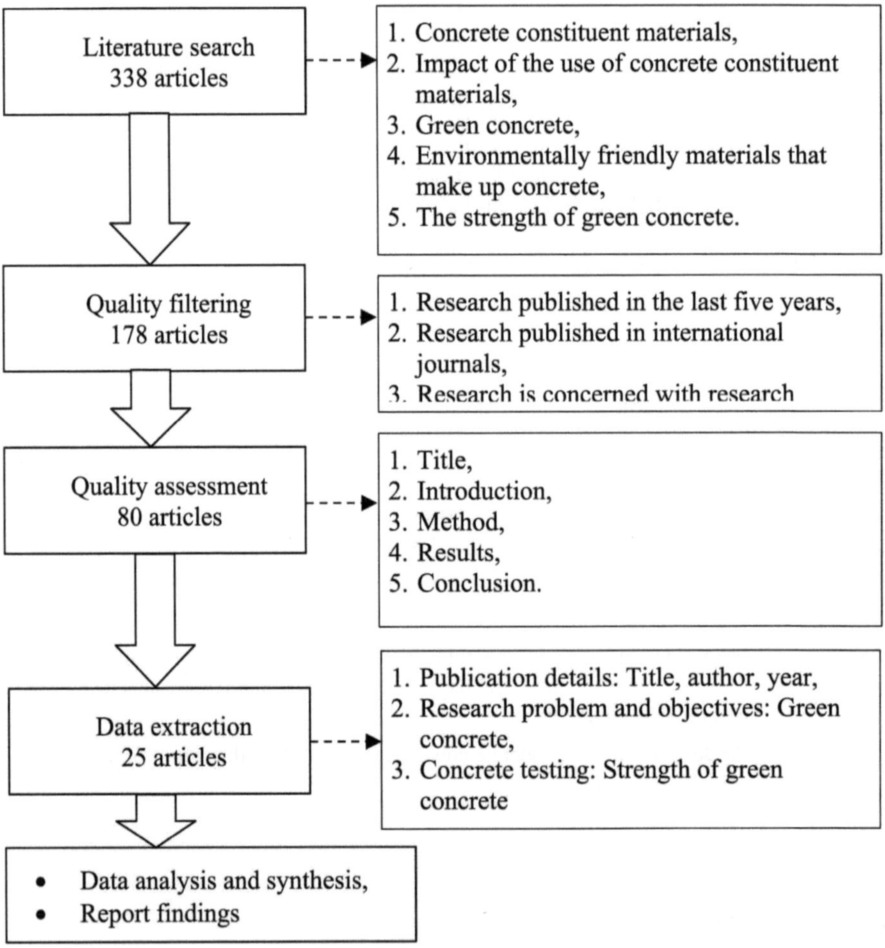

Fig. 2. Literature Review Diagram (Yu Xiao & Maria Waston 2018)

3 Results and Discussion

Portland cement is a source of CO2 emissions. Cement is most widely used as a concrete constituent for building construction. Therefore, much research has been carried out to develop and perfect green concrete. Green concrete focuses on the use of alternative materials to replace part or all of Portland cement [16]. Based on articles used as sources for literature reviews, several countries such as America, China, Thailand, India, Germany, and several countries on the European continent have been active in implementing cement substitute materials for making green concrete. These countries often have strict environmental policies and government support for innovation in the construction sector.

3.1 Rice Husk Ash

China is one of the countries that uses rice husk ash as a substitute material in green concrete mixtures. Several construction projects in China have successfully used concrete mixed with rice husk ash, showing good performance and cost efficiency [17]. China is one of the largest rice producers in the world. Therefore, China has an abundant supply of rice husks [18]. Apart from that, the use of rice husks as a substitute material in green concrete mixtures is also supported by many universities or research institutions in China which have conducted studies on the effect of rice husk ash on the mechanical properties and durability of concrete.

Apart from China, several countries that use rice husk ash as a substitute material in green concrete mixtures are Thailand, India, Indonesia, and Malaysia. The use of rice husk ash in concrete shows great potential in improving the sustainability of the construction industry [19]. Countries with significant rice production have taken the lead in research and implementation of this technology. These efforts not only help in managing agricultural waste but also contribute to reducing the carbon footprint in construction.

Rice husk ash contains high amounts of amorphous silica, which contributes to pozzolanic reactions [20]. This reaction helps fill the pores of the concrete and increases the strength of the concrete. The fine particle size of rice husk ash allows for more even distribution in the concrete mix, which can increase compressive and tensile strength thereby helping to reduce the risk of cracking. Research shows that the use of rice husk ash can increase the compressive strength of concrete by up to 20–30% compared to conventional concrete, especially in mixtures with rice husk ash substitution of up to 10–20% of the cement weight [21]. Rice husk ash can also increase the tensile and flexural strength of concrete, although the increase may not be as great as the compressive strength. Additionally, concrete with rice husk ash tends to have a good modulus of elasticity, which helps in reducing deformation and cracking.

3.2 Fly Ash

The United States has been a leader in the use of fly ash for green concrete. The success of large construction projects in America such as the Interstate Highway System uses fly ash. Fly ash not only helps reduce the environmental impact of the construction industry but also improves the quality and durability of concrete. The use of fly ash in the United States is supported by strict environmental policies and extensive research on the benefits of fly ash in concrete. The American Concrete Institute (ACI) and the American Society for Testing and Materials (ASTM) have established standards for the use of fly ash in concrete, such as ASTM C618 for fly ash classification [22]. Additionally, the Environmental Protection Agency (EPA) supports the use of fly ash as a way to reduce waste from coal-fired power plants and reduce CO2 emissions from cement production [23].

Fly ash or fly ash is a by-product of burning coal in steam-powered power plants [24]. Fly ash has been widely used as a partial replacement for cement in concrete because it has various benefits that can improve the quality and sustainability of concrete. America is one of the countries that uses fly ash in making green concrete. The use of fly ash in toll road projects shows increased durability and reduced maintenance costs.

Apart from America, several countries that use fly ash as a substitute material in green concrete mixtures are Germany, China, and India. These countries have also conducted research on the use of fly ash in construction projects which shows increased durability and reduced maintenance costs [25, 26]. Bridge projects in Germany using fly ash report increased compressive strength and resistance to aggressive environments. Research in India shows that replacing some cement with fly ash up to 30% can increase the compressive strength and durability of concrete [27].

Fly ash contains large amounts of silica (SiO2) and alumina (Al2O3), as well as other compounds such as calcium oxide (CaO) and iron oxide (Fe2O3) [28]. Fly ash has fine grains and a small ball shape which helps in filling the pores of concrete, increasing strength and durability [29]. Concrete with fly ash shows an increase in long-term compressive strength due to a longer lasting pozzolanic reaction. Fly ash can increase the tensile and flexural strength of concrete, although the increase may not be as great as the compressive strength [24]. Fly ash helps in reducing deformation and increasing the elastic modulus of concrete [30].

3.3 Fiber

Research into making concrete with substitute aggregates or additional materials is starting to appear in various countries, with the aim of concrete being designed to have excellent resistance to environmental conditions, one of which is fiber concrete [31]. Several countries have started using fiber as a substitute material in green concrete mixtures, including the United States, Canada, Germany, Japan, India, and Australia. These countries undertake research and development to improve efficiency and sustainability in the construction industry. Some types of fibers that are commonly used include: (a) Fibers used in the United States include steel fibers, recycled plastic fibers, and natural fibers such as coconut and bamboo fibers [32]; (b) Canada is also active in research into green concrete with fiber, especially polypropylene fiber and natural fibers such as hemp fiber [33]; (c) India has started using natural fibers such as coconut fiber, jute fiber, pine fiber, and jute sacks in green concrete mixes to utilize local resources and reduce carbon footprint [34, 35]; (d) Australia uses steel fibers and polymer fibers in green concrete to improve the mechanical performance and durability of concrete [36].

3.4 Glass Fiber

The United States is one of the countries that uses recycled glass as a substitute material in green concrete mixes. Many green construction projects in the United States use recycled glass in the concrete mix. For example, cities such as Seattle and Portland have implemented green building policies that encourage the use of recycled materials, including glass [37]. Apart from that, the use of recycled glass as a substitute material in green concrete mixtures is also supported by many universities or institutions in the United States conducting research on the use of glass in concrete to increase the strength and durability of green concrete.

Apart from the United States, several countries that use recycled glass as a substitute material in green concrete mixes are Canada, Australia, Germany, the Netherlands, England, Japan, and India [38]. These countries have also been actively conducting research on the benefits and challenges of using glass in green concrete.

Glass waste used as a substitute material in green concrete mixtures comes from various sources such as bottles, windows, car glass, and glass from industrial waste [39]. Glass waste is cleaned of dirt and unwanted materials, this process usually involves washing with water and cleaning chemicals. Next, the glass is crushed into a finer size according to the desired aggregate size. The use of 2.5% glass waste has a compressive strength of 25 MPa and a tensile strength of 8.4 MPa [40]. The addition of additives in the form of pozzolanic materials such as fly ash or silica fume is possible to help reduce the negative reaction between glass and alkali in cement.

Concrete with recycled glass as a partial replacement of fine aggregate (around 10–20%) can generally still achieve good compressive strength. However, if the glass proportion is too high, the compressive strength may decrease. The type of glass used (e.g., soda lime glass or borosilicate glass) also affects the strength of the concrete. Glass that is more homogeneous and cleaner tends to provide better results. Concrete with recycled glass usually has a slightly lower flexural strength than conventional concrete, but this difference can be minimized with proper mix design [41]. Meanwhile, the tensile strength of concrete with recycled glass also tends to be lower, but can be improved by using fiber or other reinforcing additives.

Concrete with recycled glass can achieve sufficient strength and durability for many construction applications, provided proper mix design and processing are carried out. In construction projects, the use of recycled glass as a building block for green concrete is carried out to reduce waste and demonstrate a global commitment to more environmentally friendly and sustainable construction practices.

3.5 Plastic Fiber

Several countries have begun to explore and implement the use of plastic as a substitute material in green concrete mixes. Several countries that have carried out research and application of plastic as a substitute material in green concrete mixtures are India, Australia, Canada, the United States, and Japan. Research produced by these countries explains that the advantages of using plastic in green concrete are reducing the amount of plastic waste that ends up in landfills, reducing raw material costs by replacing some of the aggregate with cheaper plastic waste, concrete with plastic can have insulating properties. Better thermal and acoustic and better crack resistance.

In India a pilot project using PET bottle chips as a partial replacement of fine aggregate showed that concrete with 10% plastic can achieve adequate strength for non-structural applications [42]. In Australia, research shows that the use of recycled plastic in concrete can increase concrete's resistance to cracking and improve thermal insulation properties [43].

Plastic can be obtained from various sources, such as used plastic bottles, plastic bags, and industrial waste. The types of plastic commonly used are polyethylene terephthalate (PET), polyethylene (PE), and polypropylene (PP) [33]. Plastic waste must be cleaned of dirt and other contaminants to avoid reducing the quality of the concrete. Plastic waste is cut or chopped into suitable sizes (usually granules or flakes) then mixed and stirred with dry aggregate to ensure even distribution as a concrete mixture [33].

The quality of the plastic used must be consistent to ensure the quality of the concrete produced. Plastic can affect the workability of concrete, so additives need to be added to increase workability if necessary. Excessive use of plastic can reduce the compressive strength of concrete, so the replacement proportion must be optimized based on application needs. By using plastic as a substitute material in green concrete mixes, it helps reduce plastic waste and creates a more environmentally friendly construction material.

3.6 Rubber Fiber

Several countries have used recycled rubber, especially from used tires, as a replacement material in green concrete mixes. These countries include: (a) United States, the United States has shown the potential for using recycled rubber in concrete. Universities and research institutes often conduct studies on how rubber can be used to improve concrete properties and reduce waste; (b) China, China is active in research and application of green concrete with recycled rubber. Several projects have used rubber from used tires as aggregate to reduce environmental impact and improve the mechanical properties of concrete [44]; (c) Australia, in Australia, recycled rubber has been used in various construction projects to improve sustainability and reduce tire waste; (d) Germany, Germany is one of the leaders in green technology and sustainability. The use of recycled rubber in concrete is part of their initiative to reduce waste and increase the efficiency of building materials.

The use of recycled rubber in concrete usually reduces compressive strength compared to conventional concrete. This is because rubber has different elastic properties compared to traditional mineral aggregates. Green concrete with rubber shows improvements in flexural strength and crack resistance [45]. The elastic properties of rubber help concrete be more resistant to dynamic loads and microcracks. Concrete with rubber has good resistance to abrasion and freeze-thaw cycles, because the elastic properties of rubber help reduce damage caused by expansion and contraction [46]. This makes rubber concrete suitable for applications where flexibility and crack resistance are more important than pure compressive strength, such as sidewalks, roads, and industrial floors. However, the use of rubber tends to reduce the elastic modulus of concrete, this means the concrete becomes more flexible and less stiff, making it less suitable for structures that require high stiffness.

4 Utilization of Recycled Materials

Green concrete is increasingly being applied in developed countries because it has various environmental benefits. The main reasons for implementing green concrete in these countries include, green concrete can reduce carbon dioxide (CO2) emissions by up to 8% globally, because it uses more environmentally friendly materials and a more efficient production process [47, 48]. Therefore, many developed countries have regulations and policies that encourage the use of environmentally friendly building materials as part of their commitment to reducing greenhouse gas emissions and protecting the environment.

Fig. 3. Recycled Materials.

The use of green concrete is an important step in global efforts to tackle climate change and build a more sustainable future. Awareness of the importance of maintaining a healthy environment is higher in developed countries, so that both government, industry, and society are more supportive of sustainable innovation and technology such as green concrete. The literature review used in this research produces replacement materials as in Fig. 2. However, the use of each concrete replacement material is in accordance with the desired needs (Fig. 3).

5 Conclusion

Portland cement is essential to the strength, durability, and efficiency of construction projects. However, due to the chemical nature of energy-intensive Portland cement production, especially during the heating stage of raw materials in the cement kiln, it contributes 8% of total CO2 emissions. Therefore, research is needed that examines the ingredients that make up concrete as a substitute for the proportion of Portland cement used to obtain lower carbon emissions as an important step in reducing the environmental impact of the construction industry.

The application of environmentally friendly concrete during infrastructure construction helps reduce the amount of energy consumed due to conventional concrete production. Additionally, the application of environmentally friendly concrete during the building process helps reduce the amount of energy consumed due to conventional concrete production using Portland cement. The application of green concrete also helps reduce waste products produced such as fly ash, rice husk ash, plastic fiber, glass fiber and rubber fiber.

References

1. Surahyo, A., Surahyo, L., Luby: Concrete Construction. Springer (2019)
2. Ingalkar, R.S, Harle, S.M.: Replacement of natural sand by crushed sand in the concrete. Landsc. Archit. Reg. Plan. **2**(1), 13–22 (2017)
3. Makul, N.: Advanced smart concrete-A review of current progress, benefits and challenges. J. Clean. Prod. **274**, 122899 (2020)
4. Yuhazri, M.Y., Zulfikar, A.J., Ginting, A.: Fiber reinforced polymer composite as a strengthening of concrete structures: a review. IOP Conf. Ser.: Mater. Sci. Eng. **1003**(1), 12135 (2020)
5. Almusaed, A., Almssad, A.: Introductory chapter: bricks between the historical usage and sustainable building concept. In: Masonry for Sustainable Construction, IntechOpen (2023)
6. Medeghini, L., et al.: The secret of ancient Roman hydraulic mortar: the lesson learnt from the past for future cements. Cem. Concr. Compos. **148**, 105484 (2024)
7. Örmecioğlu, H.T.: The historic concrete: construction and architecture. Eng. Technol. Manag. 210 (2021)
8. Dushimimana, A., Niyonsenga, A.A., Nzamurambaho, F.: A review on strength development of high performance concrete. Constr. Build. Mater. **307**, 124865 (2021)
9. Astutiningsih, R.A., Noermalasari, D.: Corrosion of Steel reinforcements in fly ash- and kaolin-based geopolymer concrete immersed in distilled water and ASTM seawater. Civ. Eng. Dimens. **15**(2 SE), 89–95 (2013). https://doi.org/10.9744/ced.15.2.89-95
10. Naik, T.R.: Sustainability of concrete construction. Pract. Period. Struct. Des. Constr. **13**(2), 98–103 (2008)
11. Varshney, H., Khan, R.A., Khan, I.K.: Sustainable use of different wastewater in concrete construction: A review. J. Build. Eng. **41**, 102411 (2021)
12. Garg, V., Roy, B.C.: Assessment of Construction Techniques Used to Enhance Sustainability in Affordable Housing in India BT - Sustainable Materials, pp. 239–249 (2024)

13. Emrén, A.T.: Global temperatures, CO2 concentrations and oceans. Int. J. Glob. Warm. **30**(4), 401–412 (2023)
14. Makul, N., et al.: Design strategy for recycled aggregate concrete: a review of status and future perspectives. Crystals **11**(6), 695 (2021)
15. Xiao, Y., Watson, M.: Guidance on conducting a systematic literature review. J. Plan. Educ. Res. **39**(1), 93–112 (2019)
16. Makul, N.: Cost-benefit analysis of the production of ready-mixed high-performance concrete made with recycled concrete aggregate: a case study in Thailand. Heliyon **6**(6) (2020)
17. Ozturk, E., Ince, C., Derogar, S., Ball, R.: Factors affecting the CO2 emissions, cost efficiency and eco-strength efficiency of concrete containing rice husk ash: a database study. Constr. Build. Mater. **326**, 126905 (2022)
18. Hu, L., He, Z., Zhang, S.: Sustainable use of rice husk ash in cement-based materials: environmental evaluation and performance improvement. J. Clean. Prod. **264**, 121744 (2020)
19. Gudainiyan, J., Kishore, K.: A review on cement concrete strength incorporated with agricultural waste. Mater. Today Proc. **78**, 396–402 (2023)
20. Ismail, M.H., Adnan, M.A., Deraman, R.: Short-term effect on mechanical strength and water absorption of concrete containing palm oil fuel ash exposed to chloride solution. Int. J. Sustain. Constr. Eng. Technol. **12**(3), 269–279 (2021)
21. Ahmed, A., Hyndman, F., Kamau, J., Fitriani, H.: Rice husk ash as a cement replacement in high strength sustainable concrete. Mater. Sci. Forum **1007**, 90–98 (2020)
22. Suraneni, P., Burris, L., Shearer, C.R., Hooton, R.D.: ASTM C618 fly ash specification: comparison with other specifications, shortcomings, and solutions. ACI Mater. J. **118**(1), 157167 (2021)
23. Zhao, Y., Miao, Y., Li, L., Lu, X.: Emergy evaluation of green concrete production efficiency. IOP Conf. Ser.: Mater. Sci. Eng. **677**(2), 22017 (2019)
24. Zeng, H., Li, Y., Zhang, J., Chong, P., Zhang, K.: Effect of limestone powder and fly ash on the pH evolution coefficient of concrete in a sulfate-freeze–thaw environment. J. Mater. Res. Technol. **16**, 1889–1903 (2022)
25. Gawad, T.H.: To design and study the physical properties of green concrete by using demolished concrete. Int. J. Res. Eng. Sci. Manag. **5**(1), 74–78 (2022)
26. Manjunatha, M., Seth, D., Balaji, K., Bharath, A.: Engineering properties and environmental impact assessment of green concrete prepared with PVC waste powder: a step towards sustainable approach. Case Stud. Constr. Mater. **17**, e01404 (2022)
27. Olii, M.R., Wahab, A.A., Ichsan, I., Djau, R.A., Nento, S.: Beton Hijau Menggunakan Fly ash sebagai Subtitusi Parsial Semen. Siklus J. Tek. Sipil **9**(1), 11–20 (2023)
28. Panda, L., Dash, S.: Characterization and utilization of coal fly ash: a review. Emerg. Mater. Res. **9**(3), 921–934 (2020)
29. Zhang, T., Fang, G., Chen, H., Cui, W.: Experimental study on mix proportion of green environmental protection type three-doped fair-faced concrete. IOP Conf. Ser.: Earth Environ. Sci. **787**(1), 12045 (2021)
30. Xiong, Y., Chen, Y., Wu, B., Zhao, G.: Fire resistance of composite shear walls filled with demolished concrete lumps and self-compacting concrete. Front. Mater. **9**, 1007868 (2022)
31. Lumingkewas, R.H.: BETON SERAT: Inovasi dalam Konstruksi Modern. Penerbit NEM (2023)
32. Chauhan, V., Kärki, T., Varis, J.: Review of natural fiber-reinforced engineering plastic composites, their applications in the transportation sector and processing techniques. J. Thermoplast. Compos. Mater. **35**(8), 1169–1209 (2022)
33. Safayenikoo, H.: Metalized plastic waste fiber effects on green concrete beams mechanical performance. Shock. Vib. **2022**(1), 3113841 (2022)
34. Vani, H., Arora, S.: Experimental study of concrete prepared by kota stone dust, bagasse ash, and recycled concrete. IOP Conf. Ser.: Earth Environ. Sci. **889**(1), 12040 (2021)

35. Majeed, N., Singh, B.: Experimental investigation mechanical properties of concrete mix with chopped jute fiber and fly ash. Int. J. Innov. Res. Comput. Sci. Technol. **11**(2), 9–15 (2023)
36. Natarajan, K.S., Yacinth, S.I.B., Veerasamy, K.: Strength and durability characteristics of steel fiber-reinforced geopolymer concrete with addition of waste materials. Environ. Sci. Pollut. Res. **30**(44), 99026–99035 (2023). https://doi.org/10.1007/s11356-022-22360-x
37. Kovarik, J.-B., Villarreal, A., Ramos, E.A., Alferaiheedi, Y., Modiselle, K.G.: Carbon neutral buildings and recycled materials: how cities want to solve the challenge. In: Urban 20 Mayors Summit-U20, p. 42 (2020)
38. Ahmad, W., Ahmad, A., Ostrowski, K.A., Aslam, F., Joyklad, P.: A scientometric review of waste material utilization in concrete for sustainable construction. Case Stud. Constr. Mater. **15**, e00683 (2021)
39. Alhamidi, M.: Increasing the usage of green concrete in conflict areas. Adv. Eng. Days **5**, 97–99 (2022)
40. Supardi, I., et al.: Characteristics of 'Eco Green Concrete' concrete with locagramic additional materials. Int. Res. J. Eng. IT Sci. Res. (2023). https://api.semanticscholar.org/CorpusID:266082371
41. Shayanfar, M.A., Shahrabadi, H.: Investigating the effects of recycled plastic as fibers on bending behavior of green concrete beams exposed to marine environment. Materials (Basel) **16**(17), 5912 (2023)
42. Saju, A.A., Smitha, K.K.: Structural applicability of plastic incorporated concrete. Sustain. Agri Food Environ. Res. **10**, 1–8 (2022)
43. Ahdal, A.Q., et al.: Mechanical performance and feasibility analysis of green concrete prepared with local natural zeolite and waste PET plastic fibers as cement replacements. Case Stud. Constr. Mater. **17**, e01256 (2022)
44. Liu, L., Zong, J., Hou, X., Liu, X.: Effect of the inorganic modification mode on the mechanical properties of rubber recycled concrete. Materials (Basel) **17**(10), 2217 (2024)
45. Xu, X., Zhang, Z., Hu, Y., Wang, X.: Bearing strength of crumb rubber concrete under partial area loading. Materials (Basel) **13**(11), 2446 (2020)
46. Akomah, B.B., Lawson, R.W., Jackson, E.N.: Improving labour productivity in Ghanaian building construction projects. Balt. J. Real Estate Econ. Constr. Manag. **8**(1), 152–169 (2020)
47. Fakhri, R.S., Dawood, E.T.: Limestone powder, calcined clay and slag as quaternary blended cement used for green concrete production. J. Build. Eng. **79**, 107644 (2023)
48. Siddique, U., Jawad, M., Ali, A., Cheema, S.M., Sultan, M.A., Akhtar, M.J.: Green cement valuation: an optimistic approach to carbon dioxide reduction. J. Appl. Eng. Sci. **13**(2), 259–268 (2023)

Peetest: Development of an IoT- Based Preliminary Diagnostic Kit for Urinary Tract Infection

Klyde Andrei R. Acosta(✉), Stephanie Nicole B. Arroyo, Lance Ivan Z. Avanceña, Psalm Leigh V. Llagas, Arnel M. Avelino, and Leah Q. Santos

Lyceum of the Philippines University (Cavite Campus), Governor's Drive, General Trias, 4107 Cavite, Philippines
kyldeandrei.acosta@lpunetwork.edu.ph

Abstract. Urinary tract infections (UTIs) pose a significant public health concern globally, with a notable impact on the Filipino population, ranking as the 3rd most common cause of morbidity in the Philippines. Despite advancements in healthcare, Philippine's healthcare system is still institution-centered, limiting patients' access to healthcare. Additionally, lack of awareness regarding the consequences of untreated UTIs, especially among lower to middle socioeconomic classes, has led to self- administered remedies and drugs, including herbal alternatives, without proper prescription. This study aimed to develop an IoT-based preliminary diagnostic kit that offers efficient urine reagent strip analysis to identify the levels of leukocytes, nitrite, pH level, sugar level, protein, and specific gravity. Proponents used an experimental approach to determine the model's accuracy and it was then supplemented by the authorization of M.V. Santiago Medical Center's test results. The results indicated a 94% accuracy in identifying color changes in urine test strips and interpreting data. Moreover, the prototype received positive feedback from 14 BSMT students and 4 RMTs, with agreeable ratings for the prototype and mobile application's appearance and display (3.23) and strong agreement regarding its operation (3.28), affirming its usability and potential to streamline diagnostic processes. Thes findings not only show the successful development of a preliminary diagnostic kit for UTI but also highlight the transformative potential of technology-driven healthcare innovation.

Keywords: Urinary tract infections (UTIs) · Urinalysis · Urine reagent strip · IoT (Internet of Things) · Patient-centered healthcare · Mobile Application · Medical device · Sustainable Technology

1 Introduction

Urinary tract infections (UTIs) pose a significant public health concern globally, annually prompting over 8.1 million healthcare visits and affecting approximately 88% of women and 12% of men during their lifetimes. The susceptibility to UTIs is notably higher in women, attributed to their shorter urethras, which provide a shorter path for bacteria

M. Awang (Ed.): ICCAEET 2024, PTT, pp. 87–96, 2025.
https://doi.org/10.1007/978-3-031-94347-8_8

to reach the bladder [1]. Uropathogenic Escherichia coli (UPEC) and other bacterial strains primarily induce these infections, posing substantial health risks if not promptly diagnosed and managed [2].

Globally, UTIs have emerged as the most prevalent cause of disease and mortality, with 404.61 million reported cases between 1990 and 2019. The economic burden is substantial, with over 4 billion pounds in direct medical expenses incurred each year due to the estimated 150 million cases of UTIs worldwide (National Institute of Diabetes and Digestive and Kidney Diseases).

The proposed PeeTest device represents a novel solution in the realm of diagnostic technology, particularly for addressing the significant prevalence of urinary tract infections (UTIs) in the Philippines. According to the Philippine Health Research Registry (2021) and the Philippine Health Statistics (2020), UTIs are the third most common cause of morbidity in the country, highlighting the widespread impact of this condition. Despite advancements in healthcare, the diagnostic system remains largely institution-centered, making it difficult for many patients to access timely and affordable medical care. Overcrowded medical facilities, limited accessibility to personal medical records, and the high costs associated with medical consultations and laboratory tests contribute to this challenge. As a result, many Filipinos, especially those from lower to middle socioeconomic backgrounds, often avoid hospital visits and medical examinations due to constraints related to time, patience, and financial resources. This situation is exacerbated by a lack of awareness regarding the potential complications of untreated UTIs, leading to the use of self- administered remedies or over-the-counter drugs, including herbal alternatives, without proper medical guidance.

The PeeTest device aims to address these critical issues by providing an accessible and efficient means for individuals to monitor their health through urine reagent strip analysis. Unlike existing diagnostic tools, which often require a visit to a healthcare facility for laboratory testing, the PeeTest device allows users to perform preliminary screenings in the comfort of their own homes. The device is designed to detect key urine parameters, including leukocytes, nitrite, pH level, sugar level, protein, and specific gravity, which are essential for identifying not only UTIs but also other conditions such as diabetes and kidney stones. This functionality offers a significant improvement over current diagnostic practices, which typically necessitate multiple tests and appointments to obtain similar information.

Moreover, the PeeTest device's integration with a user-friendly mobile application enhances its accessibility and convenience. Users can easily conduct the test, receive immediate results, and access detailed explanations and recommendations through the app. This approach not only empowers individuals to take proactive steps in monitoring their health but also reduces the burden on healthcare facilities by minimizing the need for routine diagnostic visits. The device's capability to store and track test results over time further contributes to its novelty, allowing users to maintain a comprehensive health record that can be shared with healthcare providers if necessary.

In comparison to existing diagnostic technologies, the PeeTest device offers a more practical and cost-effective solution, particularly for communities with limited access to healthcare facilities. By bringing reliable diagnostic capabilities closer to the patient, the PeeTest device bridges the gap between institutional healthcare and individual needs.

This innovation not only addresses the barriers of cost, accessibility, and convenience but also promotes greater awareness and understanding of health conditions among Filipinos, ultimately contributing to improved health outcomes and quality of life.

2 Methodology

The study took place at MV Santiago Medical Center and Orthopedic Hospital Inc. in Cavite, where the hospital collaborated with the researchers to collect and test urine samples, validate urinalysis results, and compare the findings with those obtained from the hospital's laboratory tests and the study's prototype device. Additionally, a total of 18 randomly selected participants were involved in analyzing color changes on simulated reagent test strips to assess the prototype's effectiveness in reducing errors in analyzing color changes compared to the participants' answers.

2.1 Project and System Flow

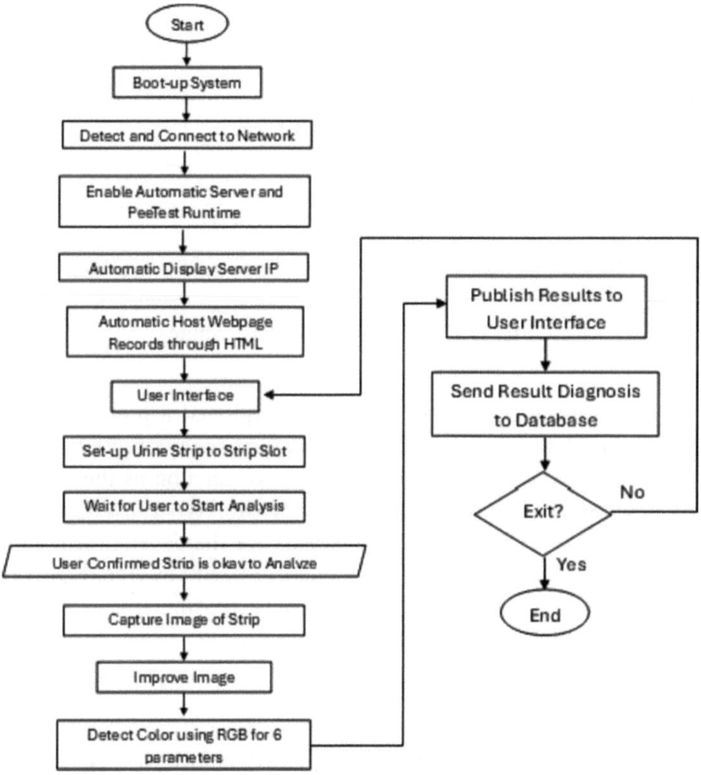

Fig. 1. PeeTest Flow

The researcher came up to the graphical representation of the whole project operation from the urinalysis, testing, evaluation, transmission, and release of results using mobile application.

Figure 1 outlines the sequence of steps for the correct operation of the PeeTest device. The process begins with inserting the sample specimen into the device for testing and concludes with using a mobile application to display results accessible to each PeeTest user. The initial five steps describe the prototype's boot-up phase before use. During the User Interface process, users must assign a Reference Name to the specimen before testing begins to facilitate easy identification of the results. This step is case-sensitive, necessitating precision. The following five steps cover the specimen's testing and analysis phase. The webpage interface is designed for the admin user, allowing them to view the entire list of current and past results. The proponents recommend that only one user/admin should access the website. After testing, all other users can access their test results using their unique ID through the PeeTest mobile application.

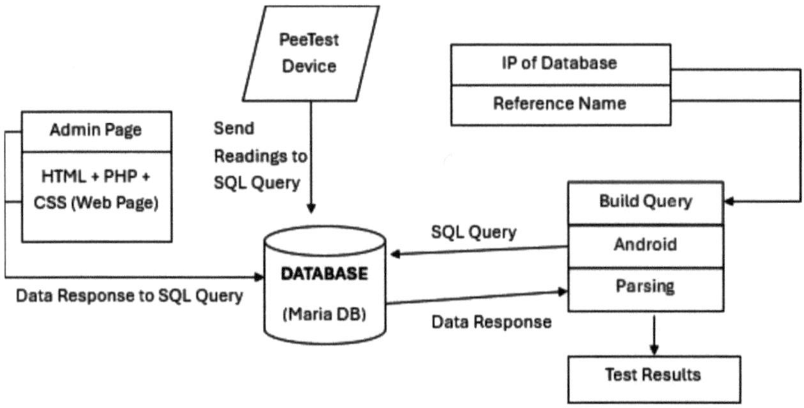

Fig. 2. PeeTest Network Infrastructure

In Fig. 2 of the thesis paper, the researchers deployed a prototype device as the client to operate the MariaDB database, an open-source solution. PHP (Hypertext Pre-processor) was selected as the scripting language to function as the indexing file for the prototype device and database. Furthermore, HTML (HyperText Markup Language) and CSS (Cascading Style Sheets) were leveraged for the development and design of the webpage user interface. Lastly, the researchers utilized Structured Query Language (SQL) for various data management tasks, including accessing, transferring, modifying, and extracting data within the prototype setup.

2.2 Block Diagram

Figure 3 shows the block diagram of the preliminary diagnostic test kit where the micro-controller used is the Raspberry Pi 4B 8G. The latter needs a power supply 5.1V/3A and a 4- channel Logic Level Controller. From the Raspberry Pi 4B, proponents will use DC motors with the aid of L298N Board to automate the movements of the sample specimen

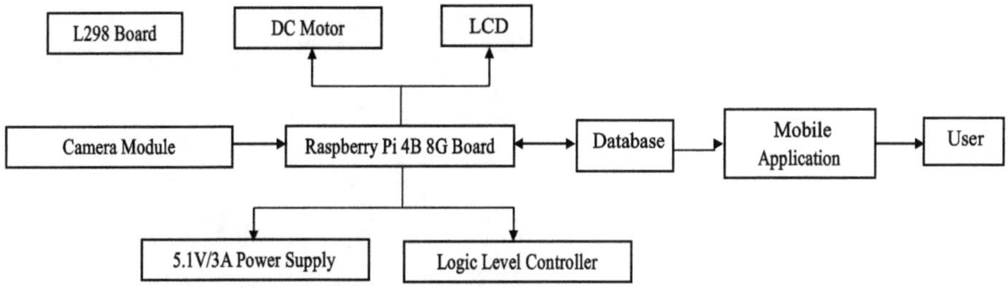

Fig. 3. Block Diagram

that were utilized in this study. Camera Module was used for detection, evaluation, and validation of analytes to be measured. The IP Address of the PeeTest operation will reflect on the LCD screen in real-time. When the numerical values of analyze levels are recorded, the data will be transmitted to the mobile application downloaded by the user in his phone.

2.3 UI Design of Mobile Application

Figure 4 illustrates the user-friendly interface of the mobile application, which has been meticulously designed to ensure intuitiveness and accessibility for all users, regardless of their technical expertise. Upon launching the application, users are prompted to enter the IP address displayed on the prototype device's LCD screen. This step is crucial, as it establishes a direct connection between the mobile application and the device, enabling effective communication and data transfer.

Furthermore, to access the most recent test results, users are required to input the unique ID associated with their specific test. This mechanism not only facilitates a straightforward retrieval process but also reinforces the application's emphasis on user privacy and security. By requiring this unique identifier, the application ensures that sensitive information is only accessible to authorized users, thereby safeguarding personal data against unauthorized access. Overall, this streamlined process is designed to provide a seamless user experience, combining ease of use with robust security measures to foster user trust and confidence in the application.

2.4 UI Design of PeeTest Administrative Portal

The image on the previous page showcases the User Interface (UI) design of the Administrative Portal for the prototype device. This portal serves as the central database, functioning as the repository where all test results (logs) are automatically stored and organized. Within the interface, there are eight (8) distinct columns, each serving a specific purpose to facilitate efficient data management.

The first column is dedicated to the Test Number, which uniquely identifies the owner of each specimen. It is important to note that the naming convention for each Test Number is case- sensitive, meaning that variations in letter casing will result in different identifiers. This attention to detail is crucial for maintaining the accuracy and integrity of the data (Fig. 5).

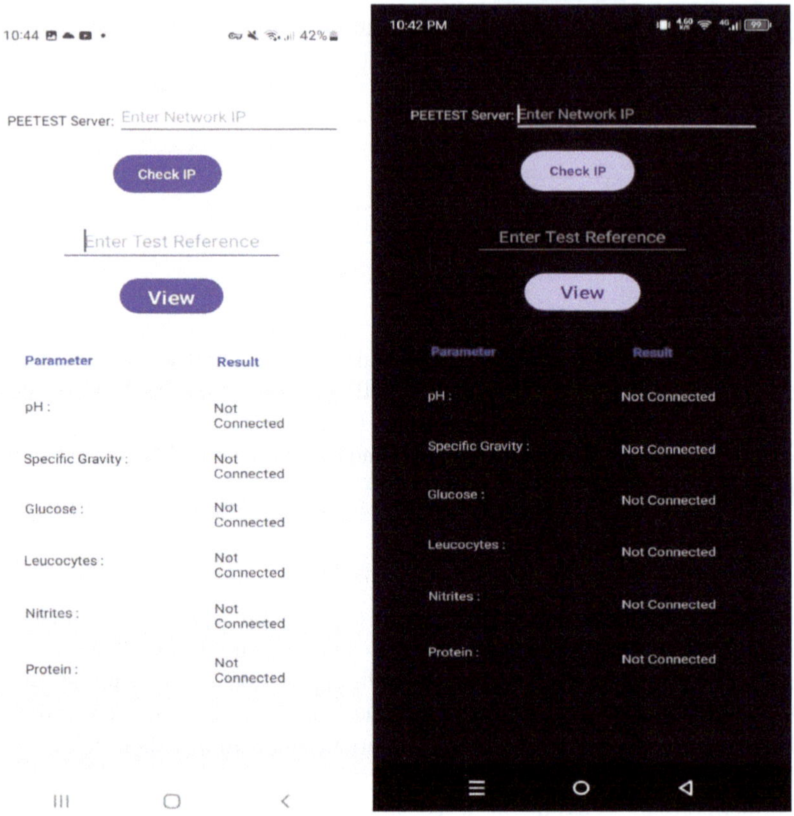

Fig. 4. GUI Design of Mobile Application

Following the Test Number, the subsequent columns present the PeeTest results for various analytes, providing a comprehensive overview of each specimen's analysis. These columns include results for key parameters such as pH, Specific Gravity, Glucose, Leukocytes, Nitrite, and Protein, respectively. Each analyte's result is clearly displayed, allowing for quick assessment and comparison of the testing outcomes. This structured layout not only enhances usability but also ensures that all pertinent information is readily accessible for administrative review and further analysis.

3 Results and Discussion

3.1 Accuracy Test, Laboratory Test V.S. Prototype Device

The average accuracy for every parameter is computed, culminating in an overall accuracy rating of 94%, which outperforms the 89.09% accuracy rate reported in a prior study [3]. This underscores the meticulous methodology adopted in the study and underscores the advancements achieved in accuracy.

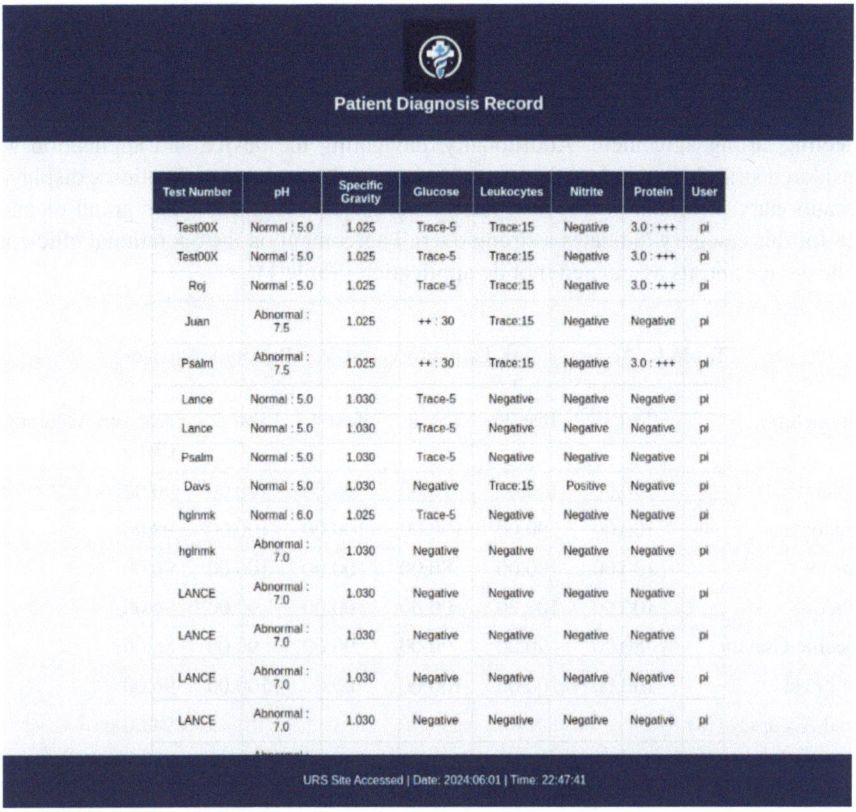

Patient Diagnosis Record

Test Number	pH	Specific Gravity	Glucose	Leukocytes	Nitrite	Protein	User
Test00X	Normal : 5.0	1.025	Trace-5	Trace:15	Negative	3.0 : +++	pi
Test00X	Normal : 5.0	1.025	Trace-5	Trace:15	Negative	3.0 : +++	pi
Roj	Normal : 5.0	1.025	Trace-5	Trace:15	Negative	3.0 : +++	pi
Juan	Abnormal : 7.5	1.025	++ : 30	Trace:15	Negative	Negative	pi
Psalm	Abnormal : 7.5	1.025	++ : 30	Trace:15	Negative	3.0 : +++	pi
Lance	Normal : 5.0	1.030	Trace-5	Negative	Negative	Negative	pi
Lance	Normal : 5.0	1.030	Trace-5	Negative	Negative	Negative	pi
Psalm	Normal : 5.0	1.030	Trace-5	Negative	Negative	Negative	pi
Davs	Normal : 5.0	1.030	Negative	Trace:15	Positive	Negative	pi
hglnmk	Normal : 6.0	1.025	Trace-5	Negative	Negative	Negative	pi
hglnmk	Abnormal : 7.0	1.030	Negative	Negative	Negative	Negative	pi
LANCE	Abnormal : 7.0	1.030	Negative	Negative	Negative	Negative	pi
LANCE	Abnormal : 7.0	1.030	Negative	Negative	Negative	Negative	pi
LANCE	Abnormal : 7.0	1.030	Negative	Negative	Negative	Negative	pi
LANCE	Abnormal : 7.0	1.030	Negative	Negative	Negative	Negative	pi

URS Site Accessed | Date: 2024:06:01 | Time: 22:47:41

Fig. 5. PeeTest Administrative Portal

3.2 Mean Response in Terms of Appearance or Display, and Operation

The evaluation results presented in Tables 2 and 3 demonstrate respondents' overall satisfaction with the device's appearance, display, and operational functionality. In terms of appearance and display, the device was generally perceived as visually pleasing, with a mean score of 3.00, indicating agreement among respondents. The device convincingly resembles a testing device, as evidenced by the highest mean score of 3.50 in this category, which is interpreted as 'Strongly Agree.' The practical and convenient design of the device also received strong approval, with a mean score of 3.28. Furthermore, the mobile application was rated positively for its user- friendly interface, scoring 3.17, and for providing clear and comprehensive explanations and recommendations, with a mean score of 3.22. The grand mean for this category is 3.23, reflecting overall agreement that the device's appearance and display meet the users' expectations.

Regarding the operational aspects, respondents evaluated various factors related to the device's usability. The secure locking mechanism, which ensures the input sample is sealed within the container, received a mean score of 3.22, indicating agreement on its effectiveness. The testing process was considered hygienic, with a mean score of 3.11, and respondents strongly agreed on the ease of attaching and removing specimen containers, as shown by the score of 3.28. Data transmission from the device to the mobile application was also rated positively, with a mean score of 3.11, suggesting that the process does not take too much time. The seamless connectivity between the device and the mobile application received the highest mean score in this section, 3.56, reflecting strong agreement. Additionally, navigating the device and application was considered straightforward, with a mean score of 3.39, and the application's display of precautionary information was also rated favorably, scoring 3.28. The grand mean of 3.28 for this category indicates a strong overall agreement on the operational efficiency of the device and its associated mobile application (Table 1).

Table 1. Accuracy Test, Laboratory Test V.S. Prototype Device

Parameters	Test 1	Test 2	Test 3	Test 4	Test 5	Average Accuracy (%)
Protein	90.00	100.00	70.00	90.00	100.00	90.00
Leukocytes	100.00	90.00	100.00	100.00	100.00	98.00
Nitrite	100.00	100.00	80.00	100.00	100.00	96.00
Glucose	100.00	100.00	90.00	100.00	90.00	96.00
Specific Gravity	80.00	90.00	90.00	90.00	90.00	88.00
pH Level	100.00	100.00	100.00	80.00	100.00	96.00
Total Accuracy (%)						94.00

Table 2. Mean Response in terms of Appearance or Display, and Operation

Factors/Indicators	Mean	Interpretation
The appearance of the device is visually pleasing	3.00	Agree
The device does look like a testing device	3.50	Strongly Agree
The device's design is convenient and practical	3.28	Strongly Agree
The mobile application has a user-friendly interface	3.17	Agree
The mobile application provides clear and comprehensive result explanations and recommendations	3.22	Agree
Grand Mean	3.23	Agree

Overall, the findings suggest a positive assessment of both the device's design and functionality. Respondents expressed satisfaction with the appearance and usability, as indicated by the grand mean scores of 3.23 and 3.28 for appearance and operation, respectively. This reflects that the device and its mobile application are well-received by users, meeting the expected standards for both visual appeal and operational effectiveness.

Table 3. Mean Response in terms of Appearance or Display, and Operation

Factors/Indicators	Mean	Interpretation
The device offers a simple locking mechanism, securely sealing the input sample within the container	3.22	Agree
The testing process is hygienic	3.11	Agree
Specimen containers can be attached/removed with ease	3.28	Strongly Agree
Data Transmission from the device to mobile application does not take too much time	3.11	Agree
The device and mobile application can be seamlessly connected or disconnected	3.56	Strongly Agree
Navigatingthedeviceandmobileapplicationis straightforward	3.39	Strongly Agree
The device and mobile application showcase precautionary information	3.28	Strongly Agree
Grand Mean	3.28	Strongly Agree

Note:
1.00–1.74 Strongly Disagree 2.50–3.24 Agree
1.75–2.49 Disagree3.25–4.00 Strongly Agree

4 Conclusion and Recommendation

4.1 Conclusion

The study successfully developed a prototype for diagnosing Urinary Tract Infections (UTIs) with a 94.00% accuracy rate, measuring key parameters such as protein, leukocytes, nitrite, glucose, specific gravity, and pH levels. Medical professionals and students provided positive feedback on the device's usability and display. This innovation addresses both global and local healthcare challenges by offering a potential solution to improve diagnostic efficiency, particularly in areas with limited access to medical facilities. However, the study's limitations, such as the sample size and environmental conditions, highlight the need for further research to ensure the prototype's reliability across diverse settings. The success of the PeeTest device demonstrates the promise of technology-driven healthcare solutions, but continued refinement and validation will be essential for its integration into routine clinical practice.

4.2 Recommendations

To enhance the performance and reliability of the PeeTest prototype, several recommendations have been made. These are listed below:

4.2.1 Select a high-quality camera that can operate consistently under various environmental conditions is crucial for accurate image capture.

4.2.2 Integrate a touchscreen interface into the Raspberry Pi 4B setup will improve user experience and operational ease.

4.2.3 To improve the accuracy of reagent test strip result interpretation, implementing machine learning algorithms is recommended. It can train the system on a comprehensive dataset.

4.2.4 Use a servo motor instead of a DC motor for the reagent test strip container's movement will provide smoother and more precise motion.

4.2.5 Testing of different LED colors is recommended to optimize lighting and enhance reagent test strip image quality, which could lead to improved illumination and precision in analyzing color changes.

References

1. Urinary Tract Infection (UTI): Symptoms, Diagnosis & Treatment - Urology Care Foundation. (n.d.). https://www.urologyhealth.org/urology-a-z/u/urinary-tract-infections-in-adults
2. Rodrigo, L.: E. Coli Infections: Importance of Early Diagnosis and Efficient Treatment (2020)
3. Huysal, K., et al.: Diagnostic accuracy of uriSed automated urine microscopic sediment analyzer and dipstick parameters in predicting urine culture test results. Natl. Libr. Med. (2019)
4. Bacârea, A.B.: Discrepancy in results between dipstick urinalysis and urine sediment microscopy. Exp. Therapeutic Med. (2021). https://www.spandidos- publications.com/https://doi.org/10.3892/etm.2021.9971#
5. Calabretta, M.Á.-D.: Nano-lantern on paper for smartphone-based ATP detection. Biosens. Bioelectron. **150** (2020). https://doi.org/10.1016/j.bios.2019.111902
6. Cruz, A., et al.: Vitaband: IoT-driven health monitoring system for home quarantine COVID-19 patient using LoRaWAN. In: IEEE International Conference on Internet of Things and Intelligence System (IOTAIS) (2022). https://ieeexplore.ieee.org/abstract/document/9975992
7. Foundation, N.K.: Urinalysis and kidney disease **16** (2023). https://www.kidney.org/sites/default/files/11-10-1815_HBE_PatBro_Urinalysis_v6.pdf

Planning for Child-Friendly Village Based on Community Based Tourism (Case Study: Kendal Village, Sekaran-Lamongan)

Mimin Aminah Yusuf(✉), Dhiah Agustina Qahar, and Muhil Frido Heriyanto

Darul 'Ulum of Islamic University, Lamongan 62253, East Java, Indonesia
mimin@unisda.ac.id

Abstract. Child Friendly Village is villages that have service mechanisms to fulfill children's rights. This can be interesting attraction to be thematic tourism village. One of the Villages that have the potential to be developed as a thematic village is Kendal, in the Lamongan district. This kampung has a unique tourism potential that is the development of Child Friendly Village and has excellent community participation. However, the development process is not yet optimally. The purpose of this research is to formulate the development concept so as Kendal Village should become a tourism destination. We proposed the community-based tourism approach being expected that the concept obtained could comply with the principles of sustainability. The first method used in this research is descriptive qualitative performed by observation and indepth-interview. Further, data obtained in the field were analyzed based on aspects of community-based tourism. The concept of area development for tourism was then formulated. The result of this research is a development concept proposal for the Kendal village that is based on community-based tourism.

Keywords: Child Friendly Village · Community Based Tourism · Tourism Village · Sustainable Tourism

1 Introduction

According to the Child-Friendly City initiated by UNICEF and UN-Habitat, one of agenda to fulfill children's right is Child Friendly City [1]. The establishment of the CFC is intended so that the city can make policies, programs and decisions by integrating children's opinions, needs, and rights [2] such as health, protection, care, education, stimulation, not being a victim of discrimination, getting to know the environment and culture [3]. Indonesia has been regulated in the Minister of State for Women's Empowerment and Child Protection Regulation regarding child-friendly city policies [4]. This policy is developed from national, provincial, district/city, village, neighborhood to family scale [5]. This is supported by the Lamongan government one of which flagship program is Child Friendly City. In 2019, Lamongan Regency received the Child Friendly City Award in Middle Category from the Ministry of Women's Empowerment and Child Protection, then in 2021 it succeeded in being promoted to the Nindya Category [6].

M. Awang (Ed.): ICCAEET 2024, PTT, pp. 97–111, 2025.
https://doi.org/10.1007/978-3-031-94347-8_9

In addition, the Lamongan Government has prepared regional regulations regarding child-friendly city in 2019 [7]. Then, considering that villages/sub-districts are children's closest environment, the Lamongan Government is actively providing outreach to villages in Lamongan and has established several indicators of child-friendly villages/sub-districts. These indicators are regulated in Lamongan Regent Regulation No. 45 of 2020 [8].

Besides that, related to tourism, after being slumped due to the Covid-19 pandemic, in 2023 tourism in Lamongan Regency began to rise, especially when the Enforcement of restrictions on community activities/PPKM policy was officially completed in December in 2022. In reviving tourism, the Lamongan Regency Government through the Lamongan Regency Tourism and Culture Office seeks to optimize Tourism through Tourism Villages [9, 10]. The tourism village is also included in the Eleven Priority Programs in an Effort to realize the Glory of Lamongan Regency with Justice in 2023 [11].

In line with the strategic plan of the Tourism Village in Lamongan, CFV is very interesting to study as a thematic village because it has attractions and attractiveness especially for children. With the existence of CFV which is also a tourism village, besides functioning to fulfill children's rights, it also improves the economy of local residents. Apart from that, it is to stimulate for other villages to immediately complete the CFV indicators.

Kendal Village proves to be suitable to respond to tourism and CFV issues. This village commited and has implemented most of the CFV indicator. This village also has tourism area and has excellent community participation. Kendal is also included in 36 Pioneer Tourism Villages in Lamongan [12].

The community-based tourism (CBT) approach is the right approach to develop the tourism potential and maintain sustainability of the tourism [13]. CBT is a tourism development model which assumes that tourism development must start from awareness of the values of community as an effort to build tourism that is more beneficial to the needs and opens up opportunities for local communities [14]. The CBT concept is prove to be the tourism approach in Indonesia because CBT is relevant in developing countries, especially in so far as it endorses strategies that favour greater benefits for the control by local communities [15].

2 Literature Review

2.1 Child Friendly City

According to UNICEF, child-friendly city is city that guarantees the rights of every child as a citizen of the city [16]. The Ministry of Women's Empowerment and Child Protection uses 24 indicators divided based on these five clusters to assess and evaluate the implementation of child-friendly cities in districts/cities. The distribution of indicators [8] is:

1. Institutional Strengthening: regulation regarding CFV, Percentage of budget for fulfilling children's rights, Involvement of community institutions, the business world and the mass media in fulfilling children's rights.
2. Children Right:

Civil rights and freedoms: Percentage of children registered at birth, children's information service, public space (reading corner), There is a Village Children's Forum which is appointed by the village head and is active in village development planning deliberations.

Basic health and wellbeing: birth in health facilities, All children are healthy, there is no malnutrition, undernutrition, overnutrition, stunting; health facilities with child friendly service; The house has access to drinking water & sanitation; Smoke-free areas are available in public spaces

Family environments and alternative care: There is no child marriage; There is a consultation institution for families regarding child care and upbringing; The existence of alternative care places/institutions; The existence of PAUD-HI services; There is a child-friendly playroom:

a. All children must attend school for 12 years
b. There are child-friendly schools
c. There is a place for creativity for children who accessible to all children

Special protection:

a. There is a place for services for children who are victims of violence and exploitation
b. Children who are victims of pornography, drugs, psychotropic substances and addictive substances and emergency situations are served
c. Underserved children with disabilities
d. Cases of children in conflict with the law are resolved, children are victims of terrorism networks, children are victims of stigmatization as a result of labeling related to the condition of their parents being served

2.2 Community Based Tourism

CBT is one of method to Sustainable Development [17]. Tourism aspects to support sustainable development as follows: tourist attractions: image, community objectives, and capacity; tourist destinations: location, target tourist, transportation, nearby facilities, and accommodation; environment: service, facilities, infrastructure, resource use, and architecture; economics: local ownership, promotion, and training; and social culture: participation, involvement, and monitoring/evaluation [13].

CBT as tourism that considering environmental, social and cultural sustainability aspects. Community based tourism has several main aspects in its development, there are 5 dimensions [18] as follows: (1) Economy: improving the community, funds are needed, jobs are created, and source of income for the local community; (2) Social: life values, gender equality including the current generation and the previous generation; (3) Culture: respect and appreciate different cultures, supporting cultural exchanges, and development based on local traditions. (4) Environment: exploring and understanding the carrying area capacity, organizing waste removal, advancing concern for environmental security. (5) Politic: community participation, expand the community, natural resources guaranteed and maintained.

3 Method

The method used in this research was qualitative performed by observation and in-depth-interview. Observation technique was applied by using documentation. The characteristics observed were about the physical condition provided by the village (existing area, potential land for playground, house condition, drainage, road, garbage, parking, traffic jam, and public facilities). Indepth-interviews were performed in relation to the stakeholder of the community that refer to the Lamongan Regent's Regulations [7], stakeholder selection as in Table 1:

Table 1. Selected Kendal's Stakeholder based on [7]

Number	Stakeholder [7]	Kendal's Stakeholder
1	Village official include neighborhood association	Kendal's village official Head of Neighborhood Association Chairman of Citizens' Association
2	Family welfare empowerment team	Kendal's family welfare empowerment team
3	Community leader, traditional leader, and religious leader	Kendal's Village Consultative Board Nahdlatul Ulama Organization Muhammadiyah Organization
4	Social organization and community organization	Kendal's Community Empowerment Institution
5	Privat business/industry located in the village	Kendal's Village-Owned Enterprises
6	Children's organizations	Children's Forum
7	Village health center officer	Assistant Health Center Integrated Health Service Post Kendal's Village health facilities
8	Non-commissioned officer for Village Development and Mass Guidance for the National Police	Village Development Officer
9	Community police partnership forum in the Village	Bhayangkara Community Security and Order Officer
10	Successors related to the interests of mother and child	Children's Forum Kendal's family welfare empowerment team
11	Mass media	Kendal's mass media team

The questions for the village official was different from the other stakeholder. For the village official, the questions were related to the implementation of child friendly village, government assistance and management. For the other stakeholder, the questions were related to the community participation, value of the local culture, and tourism preference of the community. The analysis used is descriptive qualitative by presenting facts, conditions of infrastructure, the potential attractions of Kendal village in the form of natural potential, the potential for child-friendly thematic tourist villages, and community participation to support tourism activities.

4 Result and Discussion

4.1 Current Situation and History of the Area Under Study

Kendal Village is located in Sekaran Subdistrict, Lamongan regency, East Java. This village is bordered by: Keting Village (to the north), Titik Village (to the south), solo river (to the west) and Ngarum Village (to the east). Kendal is a connector between 2 districts, that are Lamongan and Tuban (west of the Solo River). Kendal also has potential area for tourism, that is river border line area and Babat Motion Weir. This area has beautiful view because it directly faces the Solo River, therefore so many people visiting Kendal village. Apart from that, it is a gathering point because this area borders two districts.

Previously, the river border line area had a negative reputation, it is used to landfills and lots of Juvenile Delinquent like people dating, drinking alcohol, etc. Kendal was committed to developing its village so that the negative reputation disappears. Various efforts were made so that the mission can be achieved, both physical and non-physical efforts. In 2016, the village government cleaned the area so that it looked clean and also putting up banners containing rejection of thugs and juvenile delinquency hoped that people who wanted to do negative things would be embarrassed (Fig. 1). In 2018, the area was built with outbound facilities. River border line area was named by Mahogany park because so many mahogany trees there. The mahogany park made and maintained by community was managed by themselves under the auspices of Village-Owned Enterprises/BUMDES. In other area, near mahogany park, Village Government builds expression park (Fig. 6) to facilitate play-ground and reading corner in 2019 (Fig. 5).

The village government believes that by increasing the education and activities of the younger generation, the negative reputation will disappear. So in addition to non physical/infrastructure development, the village government also holds various programs to realize a happy and quality child, including;

a. SALAM MAHONI. Mahogany garden nature school (Fig. 3)
b. Clean on Friday with children (Fig. 2)
c. CANTING ARAB (Prevent Stunting Wasting with Varied Intake, Healthy, Diverse, Nutritious) is an innovation made to prevent stunting in children. The innovation consists of several activities ranging from:

- Integrated service posts (posyandu) for toddlers
- Toddler classes (Fig. 4)

Fig. 1. River border line area before 2016 (Kendal Village Government Documentation)

- Pregnant women classes (Fig. 4)
- Nutritional supplementation for toddler
- Integrative holistic early childhood education schools
- Visits to toddlers' homes to monitor the growth and development of toddlers.

d. PAAREDI. Child and Adolescent Parenting Programs in the Digital Era
e. SUJUD ATAS ANAK AKU. The Spirit of Creating a Quality Family by Reading Books
f. MAMA BELI SALMON. Let's Play and Learn Following the Montessori-Method System
g. JARKO ANTI POLAH. Stay Away from Drugs with Exercise

Fig. 2. Clean on Friday with Children

Regarding the implementation of child-friendly villages, Kendal has met almost all indicators the existence of child-friendly schools. For the commitment of the village government, Kendal is an outstanding village that has some of the achievements, obtained are:

a. Top 10 Women and Child-Friendly Villages at the Provincial Level
b. Women and child-friendly villages at the district level
c. 1st Winner Smart Village in the Literacy and Archive Development Category
d. 1st Winner Smart Village in the Tourism Order Category, Arts, Culture and Sports
e. 2nd Winner Village economic driver at the district level
f. 1st Winner in the Village Competition at the District Level

Fig. 3. Mahogany Garden Nature School

Fig. 4. Pregnant Women and Toddler Classes

Fig. 5. Reading Corner Located in Expression Park

g. 1st Winner in the Village Competition at the Regency Level
h. 1st Winner in the 10 PKK Main Programs Competition at the Regency level
i. Location of climate villages by the Minister of Environment
j. Social Inclusion-Based Literacy Village in 2022 by the Head of the Lamongan Regency Regional Archives and Library Service

4.2 The CBT Concept of Child Friendly Village

The Concept of the development of Kendal as Thematic Village was performed on five components, namely: tourist attractions, tourist destinations, environment, economy, and social culture.

Fig. 6. Mahogany Garden Nature School

Tourist Attraction

Tourist attractions are everything that has beauty, value, good in the form of a diversity, which has uniqueness, both in cultural wealth and man-made products that become Attraction factor and become a tourist destination to visit, which makes tourists motivated to take a tour of the object tour [19]. Based on the results of interviews with several stakeholders in Kendal village, it was found that the community wants tourism to be able to support activities that have been programmed by the community, not interfere with activities. In addition, tourism elements that already exist in the mahogany park are maintained such as outbound, stages, animal breeding, and literacy cafes.

Considering the preference of community, uniqueness of the village, village vision and high potential level, the concept of tourist attraction that can be developed is an educational based tourism with the tagline "Facilitate Children to Grow up to be Happy Children with Optimal Growth and Development". According to three requirements of tourist attraction [12], the development of tourist attraction is accommodated by:

a. what to see: beautiful view of solo river, babat motion weir, aviary, green house
b. what to do:
   Area 1: festival, camp, contest, reading book, photo spot, aviary, green house (Fig. 8).
   Area 2: motor cross racing, children play, outbound (Fig. 9)
c. what to buy: souvenir, book, community business product

All tourism activities are centered in Mahogany park because the community does not want tourism activities to interfere with the activities and privacy of the community as well as the safety of children.

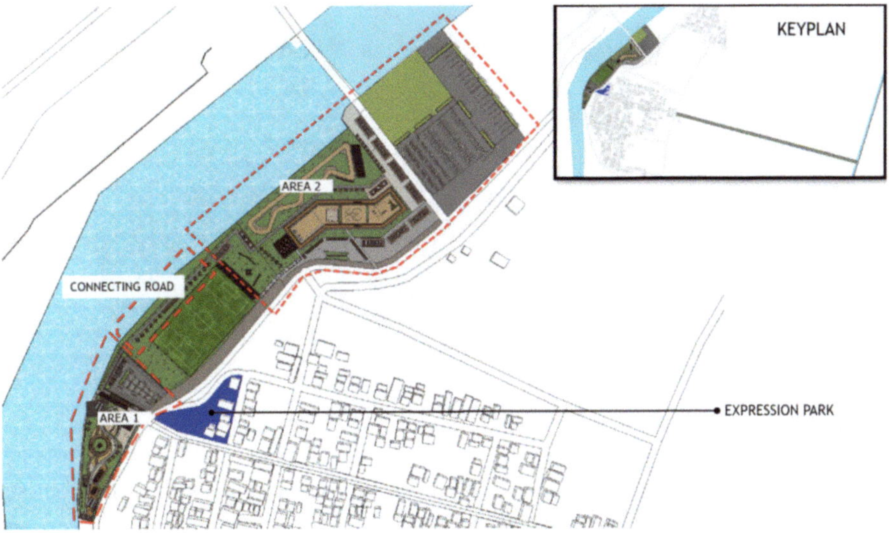

Fig. 7. The Layout Proposal of Tourist Attraction

Fig. 8. The Design of Area 1

Tourist Destination

Kendal village is strategic enough to serve as a tourism village because of its location. Kendal is a connector between 2 regencies, these are Lamongan and Tuban and also connector between 3 districts: Sekaran, Babat and Widang (Tuban Regency). It also in line with other tourism village that is Besur Agro-Education Tourism Park. In terms of transportation, the access time from Lamongan bus terminal to this Village is of 29,7 km or 43 min travel. From Lamongan rail station, the distance is of 30 km or 45 min travel. From the Babat market, which is buses stop for passengers to get on and off the bus, the distance is of 12,8 km or 28 min. But there is no mode of transportation that specifically serves the route through the Village. Kendal has bus on tour to facilitate visitor getting

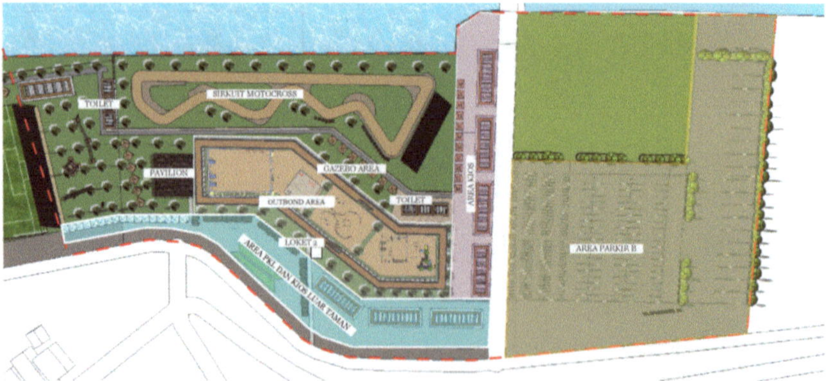

Fig. 9. The Design of Area 2

around the Kendal village. This can be improved by collaborating with Besur Tourism Village so that it can be integrated and increase the number of visitors.

The public facilities located around the village are the Babat Agrobis market within a distance of 9,7 km or 21 min. The nearest hotel is Lima Jaya Hotel within a distance of 10 km or 21 min. In this case, this is opportunity for community to facilitate homestay or lodging using their own house rooms for tourists. The advantage of homestay or lodging is in order to get to know the activities/life in Kendal village, especially to know how raise child. This may certainly improve the economy of local community.

Fig. 10. The Entrance of Kendal Village

Environment

Based on the results of the evaluation of previous research [20] on the facilities and infrastructure of Mahogany parks which is based on the function of the river boundary and also evaluation from observations in the field, to accommodate the tourist attractions, some supporting facilities are needed to improve.

a. The condition of the road to Mahony Park is damaged so it must be repaired with good and proper construction.

b. The entrance area and village access need to be rejuvenated and made better as a marker of a tourist attraction location. Design refers to the design of the gate that has been designed by the community with temple concept (Fig. 10).
c. Maximize the land for parking if there is a surge in visitors at any time and are given pavement in the form of paving blocks. The proposal using paving blocks because it accommodates function of river border line so as not to be disturbed.
d. Gazebo is multiplied to enjoy the Solo River view
e. Toilets need to be repaired immediately, separating men's and women's toilets, adding toilets so that they can be used effectively
f. The condition of the stalls is added considering that stalls are a booster of economic growth for the community and villages.
g. Providing separate trash bins for organic, inorganic and non-organic waste. This is also intended to educate visitors to be aware and care about waste.

Economy

Since Kendal become tourism village in 2020, this Village has recorded very rapid economic growth. The contribution of the Tourism Village given to Village Original Income/PADes is 35% of the total net proceeds. This is in accordance with Regulation of the Minister of Home Affairs no 66. This contribution can be used for the development of Kendal Village [21].

Currently in Mahoni Park, visitors are not charged a fee. With the addition of attractions and tourist facilities, visitors should be charged a tourist fee. The design of the tourist arrangement is divided into 2 areas, each area has a different uniqueness and function. In the design proposal (Figs. 7, 8, and 9) each area has a counter so that visitors can choose the tourist area they want to go to. In addition, the provision of connecting roads is needed to facilitate visitors' interest if they want to visit all areas.

The number of enterprises in Taman Mahogany, there are a total of 173 stalls with a variety of businesses including literacy cafes, food and beverages, children's playgrounds, and so on. In Mahogany Park, the arrangement of stalls has not been carried out, so a comprehensive arrangement needs to be carried out so that it is orderly and safe. Especially stall for children's playgrounds, it requires proper child friendly environment. Stall Design can be seen in Fig. 11.

The Village official can also get opportunities for cooperation with investors, academics and other parties in order to accelerate the tourist villages. In addition, the village government can also seek Corporate social responsibility/CSR funds from companies around the village. Investor also can be funded from community itself. People can invest a certain amount, then receive a profit share according to the funds invested.

To enhance the promotion of the tourism, through various media, both offline and online actions must be developed. An example of offline media is by distributing brochures, pamphlets, billboards, etc., while online media should be done through websites, youtube and social media (tiktok, Instagram, facebook, etc.). In addition, the endorsement method is also currently very widely used by business actors so that it can also be used for tourist villages because it is quite effective in increasing traffic.

Fig. 11. The Design of Stalls

Social and Culture

Social Culture aspects include life values, gender equality, respect and appreciate different cultures, supporting cultural exchanges, and development based on local traditions. Life value of community refers to village vision that is Realizing a Happy Community. This vision is then implemented in various existing activities that have been mentioned in the previous chapter. From the life value of community, the concept of tourism is formed.

The gender equality including man and woman, the current generation and the previous generation. In this village has organization that accommodate all gender and generation. Especially for children, there is specific organization for children, it is children forum. This organization functions as a reporter and pioneer to support the child friendly village [22]. This organization can also be involved in tourism activities, for example when there is an event/festival.

Related to culture, especially related to tourism and children's activities, the community wants that with the existence of tourist villages, children's activities will not decrease, in fact, with tourism, children's activities will be more facilitated. The community hopes that the childcare activities initiated in Kendal village can inspire other people/parties. The community also wants tourism activities to be centered in mahogany parks only. This is intended to anticipate the negative effects of tourism that originate from the crowded tourist visits. Villages that were originally quiet, peaceful and peaceful can become very crowded so that some people will feel disturbed, especially children, children will feel that their environment is not safe because there are many strangers. So the development proposal of tourism is centered in Mahogany park.

Politic

The purpose of tourism village governance is to increase the number of visitors, the length of visits, satisfaction, community business opportunities, environmental quality and stakeholder involvement. Governance that needs to be carried out by tourism villages includes tourism product governance, stakeholder governance, visitor governance, community governance, and marketing and promotion governance [12]. Kendal has a Village-Owned Enterprises/BUMDES called BUMDES Mandiri was established in 2016. The function of Bumdes mandiri is not only to get profit but also to communities are involved and receive many benefits, communities can contribute in providing an inexpensive educational place, evoking a sense of belonging and pride to the their village, raise the good reputation of village which was once known to be negative, and reduce juvenile delinquency by accomodate creativity in the management of Bumdes [12].

The Bumdes job description is overlapping with the Tourism Awareness Group/Pokdarwis. For the further development, the job description can be separated between Bumdes and Pokdarwis. Bumdes focused on the Improving the Village Economy, maximizing the benefits of village assets, maximizing community business, collaborating with other institutions, creating business opportunities and market networks, improving public services for economic growth and equity, and increase the original income of the village and the community [23].

While the function of Pokdarwis is (1) increasing the knowledge and insight of Pokdarwis members; (2) improvement of abilities and skills in members in managing the business tourism; (3) encourage or motivate the community to be a good host for tourist; (4) encourage or motivate the community to increase tourism attractiveness local through efforts to realize Sapta Pesona; (5) Collecting, processing and providing tourism information services to tourists and local communities; (6) Providing input to government officials who are authorized in the field of Tourism [24].

In terms of human resource development, some of the trainings that can be carried out include tourism training, tourism culture awareness training, tour guide training, marketing training and also training for Children's Forum to contribute in tourism. In addition, coaching and mentoring tourism villages can be collaborated with academics, universities, and the government.

5 Conclusions

The development concept of Kendal Village into a tourism Village is an educational based tourism with the tagline "Facilitate Children to Grow up to be Happy Children with Optimal Growth and Development". Some of the aspects developed include tourist attraction, tourist destination, environment, economy, social and culture, and politic. The concept of tourist attraction include: (1) what to see: beautiful view of solo river, Babat Motion Weir, Animal breeding; (2) what to do: outbound, festival, camp, contest, motoric and sensory stimulation of children, children play, reading book, photo spot. Exercise, gymnastics, motor cross racing; (3) what to buy: souvenir, book, community business product. The tourist destination accommodates Bus on Tour and homestay/lodging. Environments develop road, entrance, parking, gazebo, toilet, stall and trash cans. The social culture considers the preference of community, so tourism activity is centered in Mahogany Park. The economy aspect includes providing stall, seek investor and CSR, also promoting both online and offline media. The Politic aspect suggest separation of job description between Bumdes and Pokdarwis.

References

1. Ministry of Women's Empowerment and Child Protection. https://www.kemenpppa.go.id/index.php/page/read/31/612/kota-ramah-anak-apa-mengapa-bagaimana. Accessed 30 Mar 2023
2. Childfriendlycities.org. https://childfriendlycities.org/. Accessed 30 Mar 2023
3. Riggio, E.: Child friendly cities: good governance in the best interest of the child. Environ. Urban. **14**(2), 45–13 (2002)
4. KPPPA. Peraturan Menteri Negara Pemberdayaan Perempuan dan Perlindungan Anak Nomor 02 Tahun 2009 mengenai Kebijakan Kabupaten/Kota Layak Anak/KLA. Jakarta: KPPPA (2009)
5. Rosalin, L., et al.: Kabupaten/Kota Layak Anak. Bahan advokasi Kebijakan KLA. Jakarta: Ke-menpppa (2015)
6. Klik Jatim. https://klikjatim.com/lamongan-terima-penghargaan-kabupaten-layak-anak/. Accessed 30 Mar 2023
7. Regional Regulations of Lamongan Regency no 3 in 2019 tentang Kabupaten Layak Anak. Lamongan (2019)
8. Peraturan Bupati Lamongan No.45 tahun 2020 tentang Indikator Desa/Kelurahan Layak Anak. Lamongan (2020)
9. East Java Provincial Communication and Information Service. http://kominfo.jatimprov.go.id/berita/Pemkab-Lamongan-Optimalkan-Desa. Accessed 20 Apr 2024
10. Lamongan Megilan. http://lamongankab.go.id/beranda/portal/post/7919. Accessed 20 Apr 2024
11. Nasional Tempo.co. http://nasional.tempo.co/read/1742368/Sebelas-Program-Prioritas-Upaya-Wujudkan-Kejayaan-Lamongan-yang Berkeadilan. Accessed 20 Apr 2024
12. Alfan, A., Basith, A., Priyasmika, R.: Workshop Tata Kelola Desa Wisata Sebagai Langkah Awal Pengembangan Wisata Taman Mahoni Desa Kendal, Sekaran. Lamongan. JUKESHUM **4**(1), 205–213 (2024)
13. Yusuf, M.A., Ari, R., Kisnarini, R., Septanti, D., Santosa, H.R.: Planning for sustainable tourism. Case study: kampung of cookies, Surabaya, Indonesia. J. Settle. Spatial Plan. **10**(1), 49–11 (2019)
14. Ahsani, R.D.P.: dkk. Penerapan Konsep Community Based Tourism (Cbt) Di Desa Wisata Candirejo Borobudur Mewujudkan Kemandirian Desa. PUBLISIA: Jurnal Ilmu Administra-si Publik. **3**(2) (2018)
15. Chaudhary, M., Lama, R.: Community based tourism development in sikkim of India – a study of Darap and Pastanga villages. Transnatl. Corpor. Rev. **6**(3), 228–237 (2014)
16. Child Friendly Cities Initiative. http://www.unicef.org/childfriendlycities/. Accessed 2 Aug 2024
17. Wijaya, N.S., dan Sudarmawan, I.W.E.: Community based tourism (CBT) as a sustainable tourism development strategy in Ceking DTW, Tegallantang Village. Jurnal Ilmiah Hospi. Manag. **10**(1), 77–98 (2019)
18. Syarifah, R., Rochani, A.: Studi literatur: pengembangan desa wisata melalui community based tourism untuk kesejahteraan masyarakat. Jurnal Kajian Ruang **1**(1), 109–130 (2021)
19. Fitroh, S.K.A., Hamid, D., Hakim, L.: Pengaruh atraksi wisata dan motivasi wisatawan terhadap keputusan berkunjung (survei pada pengunjung wisata alam kawah ijen). J. Admin. Bisnis (JAB) **42**(2), 18–25 (2017)
20. Putra, S.F.P., Raidi, S.: Identifikasi Sarana Dan Prasarana Taman Mahony Kendalifornia Sekaran Lamongan. In: SEMINAR ILMIAH ARSITEKTUR IV, pp. 371–380. Universitas Muhammadiyah Surakarta. Surakarta (2023)

21. Devi, F., Astuti, S.Y.: Determination tourism village economy: does Covid-19 have an impact onpades and village community income? JurnalAkademiAkuntansi **6**(3), 359–369 (2023)
22. Women's Empowerment and Child Protection of the Republic Of Indonesia. Lampiran Kepmen Daftar Pertanyaan Kab Kota (2020)
23. BumDESA. https://bumdesa.id/. Accessed 2 Aug 2024
24. Putra, T.R.: Peran Pokdarwis dalam Pengembangan Atraksi Wisata di Desa Wisata Tembi, Kecamatan Sewon-Kabupaten Bantul. Jurnal Pengembangan Wilayah dan Kota **9**(3), 225–235 (2013)

The Influence of Availability of Infrastructure and Work Environment on Increasing Employee Performance – Employee Job Satisfaction as Moderation Test

Hadi Kurniawan[1], Moch Kalam Mollah[2](✉), Surjo Hadi[3], and Indung Sudarso[1]

[1] Department of Magister in Industrial Engineering, Faculty of Industrial Technology, Institut Teknologi Adhi Tama Surabaya, Surabaya, Indonesia

[2] Department of Civil Engineering, Faculty of Civil and Planning, Institut Teknologi Adhi Tama Surabaya, Surabaya, Indonesia
kalam@itats.ac.id

[3] Department of Industrial Engineering, Faculty of Industrial Technology, Universitas Yos Soedarso, Surabaya, Indonesia

Abstract. Employees are an essential asset for an organization or company because employees are dynamic in various things, such as innovation, adaptation, implementation, design, and so on, which can support the organization's progress and development. Institutions or organizations must be able to support employee performance in various ways or methods. This research was motivated by indications of a decline in employee performance within the Pasuruan Regency Government, such as increased indiscipline, absence from work, delays in carrying out tasks, and so on. This research explores two critical variables that theoretically influence employee performance, namely infrastructure and work environment, but there is an intervening variable as a moderator, namely job satisfaction. The respondents of this research were 136 people taken from a population of 427 or the total number of Pasuruan Regency Government employees. The research method used is path analysis with a PLS (Partial Least Square) approach. This research shows that the infrastructure variable has the most decisive influence on job satisfaction compared to the work environment. With indicators of machine and equipment availability, namely all equipment used to support work processes in the office (X1).

Meanwhile, suppose. In that case, the job satisfaction variable moderates the influence of exogenous variables on employee performance. Infrastructure still has a more decisive influence than the work environment, with the availability of transportation facilities to support work (X1) and the existence of representative physical buildings to support employee performance and job satisfaction (X5). Leaders in the Pasuruan Regency Government must pay attention to the infrastructure and work environment variables, especially indicators with high loading factor values.

Keywords: sustainable infrastructure · satisfaction · performance

M. Awang (Ed.): ICCAEET 2024, PTT, pp. 112–124, 2025.
https://doi.org/10.1007/978-3-031-94347-8_10

1 Introduction

In the Industry 4.0 era, business competition is increasingly fierce and has complex problems, which cannot be separated from the increasingly rapid growth of the manufacturing and service industries. A company with less value than its competitors will have low value in the eyes of consumers [1, 2]. A company's success can be seen from the achievement of planned goals. What is done so that this goal can be achieved by utilizing various resources owned by the company, such as office facilities and infrastructure [3, 4]. These facilities and infrastructure will produce conditions that indicate the success of management activities in achieving goals, including quantity of work, quality of work, and timeliness in completing work [5, 6].

Several problems in the Pasuruan Regency Government are related to the work environment, including a spatial layout that needs to be more comfortable. The relationship between workers looks quite good, and a pleasant and conducive atmosphere is created [7]. However, during working hours after 10.00 and above, the district government office will feel empty and quiet, so that employees will feel a little bored and their performance will decrease slightly [8]. Workers tend to be busy with their respective activities, such as playing with cellphones and computers [9]. Another activity carried out by Pasuruan Regency Government employees is that many employees still leave the room and look for other activities, such as meeting with several colleagues in different departments or leaving the office during working hours. This also cannot be separated from the fact that many rooms still do not use air conditioning so that the room temperature feels hot and the sound of vehicles and other noises can still be heard [10].

The phenomenon of job satisfaction in the Pasuruan Regency Government is that several employees in the Pasuruan Regency Government do not get job satisfaction because they feel that the infrastructure could be more supportive. The work environment is supposed to be not conducive, resulting in negative attitudes or behavior, such as not doing work on time. Often arrive late and need to be present with notice. Several employees in the Pasuruan Regency Government feel that the leadership needs to pay more attention to the results of their work, which causes frustration and will hurt the company. This research aims to understand the influence of the availability of infrastructure and work environment on improving employee performance through job satisfaction as a moderating variable by taking a case study in the Pasuruan Regency Government - East Java. The method used is PLS (Partial Least Square), so indicators will likely be obtained that must be prioritized to increase employee performance.

2 Methods

This research obtains data through data collection based on answers from respondents through a questionnaire containing instruments regarding statements addressed to respondents based on the variables included in this research, namely, the dimensions of infrastructure, dimensions of the work environment, dimensions of job satisfaction, and dimensions of employee performance. This research used 136 respondents at the Pasuruan Regency Government Office. The sampling technique used is a purposive sampling technique, namely where the samples used as respondents for this research will be selected based on specific criteria and taken in each department or section. Some of these criteria include that the respondent has work experience of more than 5 (five) years and the respondent is a permanent employee or ASN within the Pasuruan Regency Government. Data processing was carried out using PLS (Partial Least Square) because this method has advantages, including [11]: (1) PLS can handle many independent variables, even if multicollinearity occurs between the independent variables [12]; (2) PLS is reflexive but can also be used for formative relationships [13]; (3) PLS can be used for relatively small sample sizes [14]; (4) PLS can be used for very complex models; and (5) PLS can be used when the distribution is skewed [15] (Fig. 1).

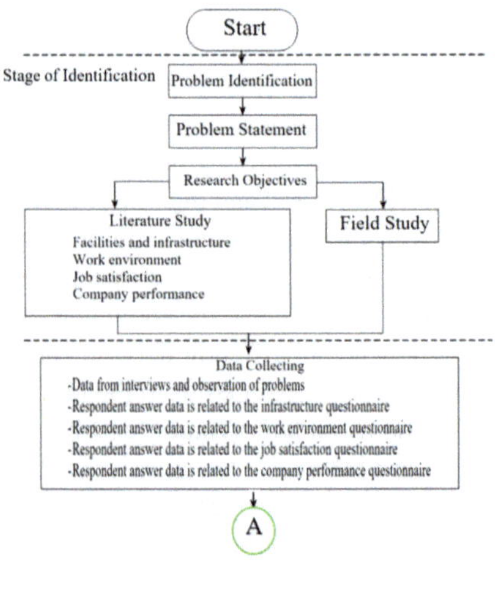

Fig. 1. Research Methodology

Meanwhile, indicators related to exogenous and endogenous variables can be described in Table 1 as follows:

3 Result and Discussion

3.1 Convergent Validity

In concurrent validity testing, the aim is to evaluate the indicators for the variables by looking at the value of the outer loading where the required quantity is > 0.7 to be said to be good or valid (Fig. 2).

Table 1. Indicators of Research Variables

No.	Variable	Indicator	Symbol
1	Facilities and Infrastructure	There are machines and equipment that support the work process in the office Aryanta et al. [7], Riyanto et al. [9]	$X_{1.1}$
		The office has equipment and inventory to support the smooth running of work Zacharias et al. [8]	$X_{1.2}$
		Work activities in the office are supported by complete equipment (Purwanto 2020; Basalamah 2021)	$X_{1.3}$
		The Pasuruan Regency Government has a representative building to carry out work for its employees (Parashakti et al. 2020; Hasan 2021)	$X_{1.4}$
		Pasuruan Regency Government has adequate means of transportation to support work activities (Kurniawan and Heryanto,2019)	$X_{1.5}$
2	Work Environment	Every Pasuruan Regency Government office employee has clear duties because the Civil Service Agency has regulated the job description (Riyadi 2019; Prasetyo, et al. 2021)	$X_{2.1}$
		Bright workplace lighting (Prasetyo, et al. 2021; Tahir 2023)	$X_{1.2}$
		Air circulation in the workplace is excellent, so the room is not stuffy (Budie, et al. 2019; Putra, et al. 2020)	$X_{2.3}$
		The room temperature is perfect, so employees do not feel too hot or cold (Putra, et al. 2020; Pusparani, et al. 2021)	$X_{2.4}$
		Workers feel comfortable at work (Riyadi 2019;)	$X_{2.5}$
		Workers receive security guarantees at work (Aruldoss, et al., 2022)	$X_{2.6}$
		Leaders respect their subordinates (Basalamah, 2021; Hasan, 2021)	$X_{2.7}$
		Coworkers support each other (Budie, et al., 2019; Purwanto 2020)	$X_{2.8}$

(*continued*)

Table 1. (*continued*)

No.	Variable	Indicator	Symbol
3	Job Satisfaction	Employees receive salaries from Pasuruan Regency Government institutions for contributions made as expected (Pusparani, et al. 2021)	Z_1
		Employees receive assignments that suit the employee's abilities so that employees carry out the work with high dedication (Parashakti, et al. 2020)	Z_2
		The work culture in Pasuruan Regency Government institutions is influenced by employee behavior and is in line with what is expected of Pasuruan Regency Government employees (Zacharias, et al. 2021)	Z_3
		The Pasuruan Regency Government has fair and reasonable employee promotion regulations and assessments (Aryanta, et al., 2019; Tahir, 2023	Z_4
		The Pasuruan Regency Government has a good coaching concept for employees who are still considered to be underperforming in their work (Kurniawan and Heryanto 2019; Aruldoss, et al. 2022)	Z_5
4	Employee Performance	Employees can carry out their duties and obligations to accelerate and facilitate the achievement of Pasuruan Regency Government goals, both long-term and short-term organizational goals (Riyanto, et al. 2021)	Y_1
		Pasuruan Regency Government employees carry out their duties and obligations by established standards (Putra, et al. 2020; Parashakti, et al. 2020)	Y_2
		Pasuruan Regency Government employees can provide good feedback if they receive an assignment from the company (Aruldoss, et al. 2022)	Y_3

(*continued*)

Table 1. (*continued*)

No.	Variable	Indicator	Symbol
		Pasuruan Regency Government employees can use equipment or facilities provided by the institution (Riyadi 2019; Pusparani, et al. 2021)	Y_4
		Pasuruan Regency Government employees have good competence in carrying out the work they are given (Basalamah 2021)	Y_5
		Pasuruan Regency Government employees have good motivation in carrying out the work they are given (Purwanto 2020; Zacharias, et al. 2021)	Y_6
		Pasuruan Regency Government employees can create opportunities related to achieving organizational goals (Prasetyo, et al. 2021; Tahir 2023)	Y_7
		Employees can carry out their duties and obligations to accelerate and facilitate the achievement of Pasuruan Regency Government goals, both long-term and short-term organizational goals (Aryanta, et al. 2019)	Y_8

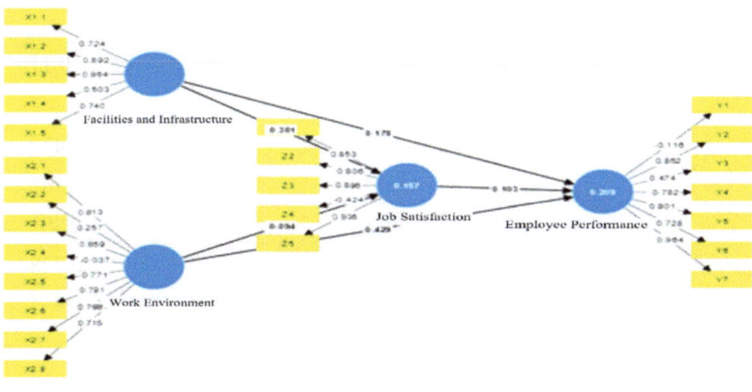

Fig. 2. Initial Stage Model Test Results

From the test results of this model, it can be seen that in this model, there are still loading factor indicator values that are low, below 0.7. So, this indicator can be said to be unable to reflect the latent variable. Based on Table 1 above, the results of concurrent validity testing can be described as follows:

Table 2. Outer Loading Test Results at the Initial Stage

Indicator	Variable			
	Facilities and Infrastructure	Work Environment	Job Satisfaction	Employee Performance
$X_{1.1}$	0,724			
$X_{1.2}$	0,892			
$X_{1.3}$	0,864			
$X_{1.4}$	**0,503**			
$X_{1.5}$	0,740			
$X_{2.1}$		0,813		
$X_{2.2}$		**0,257**		
$X_{2.3}$		0,859		
$X_{2.4}$		**0,037**		
$X_{2.5}$		0,771		
$X_{2.6}$		0,791		
$X_{2.7}$		0,788		
$X_{2.8}$		0.715		
Z_1			0.853	
Z_2			0.806	
Z_3			0.896	
Z_4			**–0,424**	
Z_5			0.935	
Y_1				**–0,116**
Y_2				0,852
Y_3				**0,474**
Y_4				0,782
Y_5				0,801
Y_6				0,728
Y_7				0,954

In the facilities and infrastructure variable (X1), it is known that there is still one indicator that does not meet the prerequisites for an outer loading value of > 0.7, namely, the Pasuruan Regency Government has a representative building for carrying out work for its employees (X1.4) with a value of 0.503. In the work environment variable (X2), there are still two indicators that do not meet the prerequisites for an outer loading value, namely bright lighting in the workplace (X2.2) of 0.267 and a perfect room temperature so that employees do not feel too hot or too cold (X2.4) is fair at –0.037. In the work satisfaction variable (Z), there are still two indicators that do not meet the prerequisites for the outer loading value, where the indicators are employees receiving salaries from Pasuruan Regency Government institutions for contributions given as expected (Z1) of –0.116 and work culture in government institutions Pasuruan Regency is by employee work behavior and is in line with what is expected by Pasuruan Regency Government employees (Z3) of 0.474. Meanwhile, in the Employee Performance (Y) variable, there is still one indicator that does not meet the prerequisites for an outer loading value, namely the indicator for the availability of Pasuruan Regency Government employees who can use equipment or facilities provided by the institution (Y4) of –0.424. So, these indicators must be eliminated because the outer loading value is less than 0.7.

Next, retesting will be carried out to obtain a good loading factor value. So, the test results can be seen in Fig. 3 below.

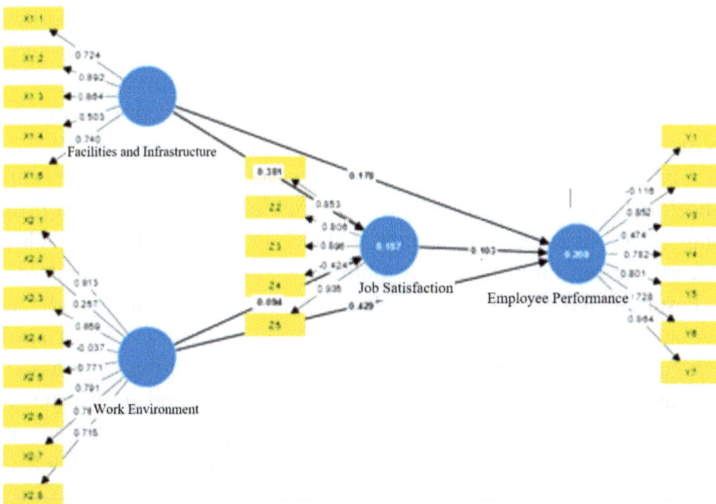

Fig. 3. Final Stage Model Test Results

The results of testing this final stage of the model show that the loading factor value on the indicator meets the requirements, namely >0.7. So, this indicator can be said to reflect the latent variable.

Based on Table 2, the results of concurrent validity testing at the final stage after modification can be described as follows:

In the facilities and infrastructure variable (X1), which has been retested, it can be seen that the indicators in the facilities and infrastructure variable have met the prerequisites for an outer loading value of >0.7. These indicators include: (1) there are machines and equipment that support work processes in the office (X1.1) of 0.872; (2) the office has equipment and inventory to support the smooth running of work (X1.2) of 0.941; (3) work activities in the office are supported by complete equipment (X1.3) amounting to 0.933 and (4) Pasuruan Regency Government has adequate means of transportation to support work activities (X1.5) amounting to 0.986.

In the work environment variable (X2), which has been retested, it can be seen that the indicators in the work environment variable have met the prerequisites for an outer loading value >0.7. These indicators include: (1) every employee in the Pasuruan district government office has clarity in their duties because the job description has been regulated by the Civil Service Agency (X2.1) of 0.831; (2) air circulation in the workplace is very good so that the room is not stuffy (X2.3) of 0.888; (3) the company meets the need to do work (X2.5) of 0.925; (4) workers receive security guarantees at work (X2.6) of 0.807; (5) leaders appreciate subordinates (X2.7) amounting to 0.817; and (6) coworkers support each other (X2.8) of 0.892.

In the work satisfaction variable (Z), which has been retested, it can be seen that the indicators in the work satisfaction variable have met the prerequisites for an outer loading value >0.7. These indicators include: (1) employees receive salaries from Pasuruan Regency Government institutions for contributions provided by expectations (Z1) of 0.925; (2) Employees receive assignments that suit the employee's abilities so that employees carry out the work with high dedication (Z2) of 0.908; (3) work culture in Pasuruan Regency Government institutions is by employee work behavior and is in line with what is expected by Pasuruan Regency Government employees (Z3) of 0.923; (4) The Pasuruan Regency Government has a good coaching concept for employees who are still considered not optimal in contributing to their work (Z5) of 0.899.

For the employee satisfaction variable (Y), which has been retested, it can be seen that the indicators for the employee satisfaction variable have met the prerequisites for an outer loading value >0.7. These indicators include (1) Pasuruan Regency Government employees carrying out their duties and obligations by established standards (Y2) of 0.920; (2) Pasuruan Regency Government employees can provide good feedback if they receive an assignment from the company (Y3) of 0.848; (3) Pasuruan Regency Government employees have good competence in carrying out the work assigned (Y5) of 0.822; (4) Pasuruan Regency Government employees have good motivation in carrying out the work assigned (Y6) of 0.812; and (5) Pasuruan Regency Government employees can create opportunities related to achieving organizational goals (Y7) of 0.976 (Table 3).

3.2 Discriminant Validity

The discriminant validity test stage aims to evaluate each construct variable's correlation with other variables. This can be known through the value of each indicator's cross-loading. This test can also be seen from the AVE value, which is compared with the root AVE value. The model is good if the AVE value is >0.5.

Based on Table 4 above, the AVE value obtained for each variable construct is >0.5, which can be said to be excellent or valid because the variable can explain the variance of the indicators. Apart from that, in testing discriminant validity, it can be determined by comparing the AVE root values for each variable with the condition that the value is >0.7. The comparison results above show that the root AVE value for each variable has a value greater than 0.7. So, this model is said to be excellent or valid and meets the requirements of discriminant validity.

Table 3. Outer Loading Test Results at the Final Stage

Indicator	Variable			
	Facilities and Infrastructure	Work Environment	Job Satisfaction	Employee Performance
$X_{1.1}$	0,872			
$X_{1.2}$	0,941			
$X_{1.3}$	0,933			
$X_{1.5}$	0,986			
$X_{2.1}$		0,832		
$X_{2.3}$		0,888		
$X_{2.5}$		0,925		
$X_{2.6}$		0,807		
$X_{2.7}$		0,811		
$X_{2.8}$		0.892		
Z_1			0.925	
Z_2			0.908	
Z_3			0.923	
Z_5			0.899	
Y_2				0.920
Y_4				0.848
Y_5				0.822
Y_6				0.812
Y_7				0.976

Table 4. Comparison of AVE Values and AVE Roots

Construct	AVE	Akar AVE
Facilities and infrastructure	0,722	0,850
Work Environment	0,671	0,819
Job Satisfaction	0,634	0,796
Employee Performance	0,752	0,851

3.3 Composite Reliability

Reliability testing can be done by looking at the output value of composite reliability, provided the value is >0.7. Based on Table 5, the composite reliability value for each variable in the study is more significant than 0.7. So, the model in this research is reliable.

Table 5. Composite Reliability

Construct	*Composite Reliability*
Facilities and Infrastructure	0,862
Work Environment	0.765
Job Satisfaction	0,812
Employee Performance	0,843

3.4 Structural Model (Inner Model)

The path coefficient in the model demonstrates the significance of the influence between these variables. The complete results of this test are also available by looking at the output of the bootstrapping report. This inner model tested several things, which will be described below.

R-Square (R2). In this case, the value obtained from R-Square can measure the magnitude of the structural model predictions. Apart from that, R-Square functions to show the ability of exogenous latent variables to explain endogenous latent variables. This measurement is divided into 3 (three) parameters, including R-Square, with a value of 0.75, the model is declared strong; with a value of 0.50, the model is declared moderate; and with a value of 0.25, the model is declared weak. The results of the R-Square in this research can be seen in Table 6 below.

Table 6. Nilai R-Square dan R-Square Adjusted

Construct	R-*Square*	R-*Square Adjusted*
Job Satisfaction	0,639	0,419
Employee Performance	0,798	0,678

Based on Table 6 above, it can be seen that the R-Square value of the job satisfaction variable is 0.639, where it can be concluded that job satisfaction can be explained by the work environment and facilities and infrastructure at 63.9% $\approx$ 64%. In comparison, the rest variables outside this research influence 36%. The Employee Performance variable is 0.798, where it can be concluded that Employee Performance can be explained by facilities and infrastructure, work environment, and job satisfaction amounting to 79.8% $\approx$ 80%. In comparison, variables outside this research influence the remaining 20%.

Inner Weight. In inner weight, this functions to determine the effect of the relationship between exogenous latent variables and endogenous latent variables or moderating variables (Table 7).

Table 7. Path Coefficient

Correlate	Original Sample	T Statistic	P *Value*
FI → JS	0,112	2.605	0,000
WE → JS	0,164	2.025	0,014
FI → EP	0,222	1,982	0,031
WE → EP	0,352	2,214	0,001
JS → EP	0,368	2,091	0,017

Where:
FI: Facilities and Infrastructure
WE: Work Environment
JS: Job Satisfaction
EP: Employee Performance (Table 8)

Table 8. Path Coefficient with Moderation Effect

Korelasi	*Original Sample*	T Statistik	P *Value*
FI → JS → EP	0,173	2,582	0,000
WS → JS → EP	0,427	2,224	0,021

The original sample's value shows that the relationship's direction is positive or negative depending on the size of the value. Meanwhile, the t statistic shows how much influence these variables have, and the p-value aims to determine the significance level of the relationship between exogenous and endogenous latent variables.

4 Conclusion

Facilities, infrastructure, and work environment positively and significantly influence job satisfaction. Likewise, facilities and infrastructure, work environment, and job satisfaction positively and significantly influence job satisfaction in the Pasuruan Regency Government. This is because the indicators of the facilities and infrastructure variable in this institution have several important things, namely that there are machines and equipment that support the work process in the office, and work activities in the office are supported by complete equipment. Apart from that, the work environment implemented is by the employees' character, where every employee in the Pasuruan Regency Government office has clarity in their duties because the civil service agency has regulated the job description. Apart from that, there are indicators for workers to obtain security guarantees at work.

Meanwhile, job satisfaction as an intervening variable can moderate well so that it can act as a variable that strengthens the relationship between the variables of facilities and infrastructure and work environment on employee satisfaction. This is because supporting facilities and infrastructure and a good work environment can minimize aspects that trigger job satisfaction, among others, employees. Obtain assignments that suit the employee's abilities so that employees carry out their work with high dedication, and the Pasuruan Regency Government has a good coaching concept for employees who still need to be more optimal in contributing to their work.

References

1. Budie, B., et al.: Employee satisfaction with the physical work environment: the importance of a need based approach. Int. J. Strateg. Property Manage. **23**(1), 36–49 (2019)
2. Tahir, M.: Employee performance and the impact of workplace facilities and discipline. J. Manage. Bisnis **10**(2), 417–425 (2023)
3. Parashakti, R.D., et al.: The influence of work environment and competence on motivation and its impact on employee performance in health sector. In: 3rd Asia Pacific International Conference of Management and Business Science (AICMBS 2019). Atlantis Press (2020)
4. Prasetyo, I., et al.: Discipline and work environment affect employee productivity: evidence from Indonesia. Int. J. Entrepreneurship **25**(5), 1–32 (2021)
5. Kurniawan, H., Heryanto, H.: Effect of work discipline and work environment on employee performance with work motivation as an intervening variable in department of tourism, youth and sport of Padang District. Archiv. Bus. Res. **7**(7), 88–101 (2019)
6. Riyadi, S.: The Influence of job satisfaction, work environment, individual characteristics and compensation toward job stress and employee performance. Int. Rev. Manage. Market. **9**(3), 93 (2019)
7. Aryanta, I.K., Ni Wayan, S., Putu Ngurah Suyatna, Y.: Influence of motivation on job stress, job satisfaction and job performance at alam puri villa art museum and resort Denpasar. Jurnal Ekonomi Dan Bisnis Jagaditha **6**(2), 113–120 (2019)
8. Zacharias, T., Mohamad Arsad, R., Yusriadi, Y.: Cultural reconstruction and organization environment for employee performance. J. Ethnic Cult. Stud. **8**(2), 296–315 (2021)
9. Riyanto, S., Endri, E., Abdul, H.: The influence of transformational leadership and the work environment on employee performance: mediating role of. Acad. Entrepreneurship J. **27**(6), 1–11 (2021)
10. Hasan, H.: Effect of work experience and work environment on job satisfaction. BJRM (Bongaya J. Res. Manag. **4**(2), 7–10 (2021)
11. Pusparani, M., Shopia, A., Hapzi, A.: The effect of work environment and job satisfaction on employee performance with organizational commitment as an intervening variable at the department of population control and family planning Sarolangun regency. Dinasti Int. J. Manag. Sci. **3**(2), 202–219 (2021)
12. Basalamah, S.A.: The role of work motivation and work environment in improving job satisfaction. Golden Ratio Hum. Resour. Manage. **1**(2), 94–103 (2021)
13. Aruldoss, A., et al.: The relationship between work-life balance and job satisfaction: the moderating role of training and development and work environment. J. Adv. Manage. Res. **19**(2), 240–271 (2022)
14. Purwanto, A.: The role of job satisfaction in the relationship between transformational leadership, knowledge management, work environment and performance, Solid State Technology (2020)
15. Putra, A.R., et al.: Role of work environment and organizational culture to job performance. J. Islamic Econ. Perspect. **1**(2), 1–13 (2020)

Chem U?: An Android Interactive Chemistry-Learning App for STEM Students

Mico Wilson S. Punzalan(✉), Niña Mari Jose, Jan Wilfred P. Malamug, Mark Justin M. Santos, Genson P. Mendoza, and Amanda Jane C. Menta

Lyceum of the Philippines University Cavite, Gen. Trias, Cavite, Philippines
eko_prihartanto@borneo.ac.id
https://cavite.lpu.edu.ph

Abstract. Recognizing that students are losing interest and facing learning difficulties in chemistry, there is a necessity to implement a new instructional approach. Using Unity, C#, and Firebase, the developers created Chem U? – an innovative Android e-learning app designed to enhance the learning experience of STEM students taking General Chemistry 1. The app introduces a gamified approach to learning, featuring leaderboards, scoring systems, badges, interactive modules, and game-based assessments for students, with learning and content management tools for teachers. It is accessible on any Android device running version 8.0 (Oreo) or higher with at least 400 MB available storage. Chem U? achieved ratings of 99.94% in functionality testing and 100% in compatibility. The app underwent evaluation using the Mobile Application Rating Scale, garnering a highly acceptable rating ($M = 3.76$, $SD = 0.46$). With its efficacy evaluated at the University of Perpetual Help Molino, independent samples t-test revealed that students using Chem U? as a supplementary resource exhibited significantly improved exam results compared to those relying solely on classroom instruction, $t(86) = 10.02$, $p < .05$. Similarly, paired samples t-test showed a significant improvement in exam scores of students using Chem U? as a self-learning tool, $t(21) = -13.30$, $p < .05$. Chem U? emerges as a valuable tool for both supplementary and self-directed learning, effectively addressing the challenges in teaching and learning General Chemistry 1, while promoting student engagement and academic success.

Keywords: General Chemistry · E-Learning · STEM Education

1 Introduction

In the Philippines, General Chemistry 1 is one of the specialized subjects of senior high school students in Science, Technology, Engineering, and Mathematics (STEM) program. It serves as a foundation for more advanced Chemistry-related courses offered at the tertiary level (Reed et al. 2017). Having competent knowledge and mastery of this course can pave the way for more research and career opportunities (Soken-Huberty 2023). However, Swoope (2020) discovered that many students are demotivated and disillusioned to continue learning chemistry due to poor academic performance despite their initial interest and enthusiasm to learn the field.

M. Awang (Ed.): ICCAEET 2024, PTT, pp. 125–133, 2025.
https://doi.org/10.1007/978-3-031-94347-8_11

According to a study conducted by Sadera et al. (2020), students face difficulty in learning Chemistry due to a lack of motivation, a poor learning environment, and the unavailability of appropriate studying resources, which results in students' poor academic performance on paper-based assessments. Although this problem still exists and is a recurring dilemma, the latest work of Bedada and Fita (2023) revealed that schools had yet to make significant progress in alleviating it in an attempt to help students improve their academic performances.

Moreover, despite the existence of interactive learning applications for chemistry, the majority still have not become effective, as they only focus on reinforcing concepts and are not friendly to students without prior knowledge of these topics (Jones et al. 2018). In line with this, the majority of e-learning apps deviate from the learning competencies outlined in the curriculum being followed by educational institutions. One evident example is the Android learning application developed by Barraquias et al. (2019) called "Turo Agham." Although the application shows a potential to serve as an alternative learning resource for kindergarten students, its learning contents are not aligned with any DepEd curriculum for the said grade level.

Aside from that, many of these chemistry-learning apps lack experimental validations. Although they underwent app quality evaluations, such as the Mobile Application Rating Scale (MARS), the app's effectiveness in enhancing students' academic performances has not been rigorously tested in a controlled experimental setting. Considering all these, there is indeed a heightened need to introduce an effective, innovative, and interactive way for students to learn Chemistry.

The developed project aims to immerse the students in a modern, interactive way of learning chemistry topics. It serves as a tool for reinforcement or supplementary learning and a platform for self-learning. Moreover, the app serves as a student progress monitoring and evaluation tool for teachers in chemistry topics aligned with the current K to 12 Basic Education Curriculum.

2 Methodology

2.1 App's Core Features

The application was designed to combine learning and entertainment. For students, the app offers interactive learning modules aligned with the Department of Education of the Philippines and gamified assessments which can be utilized in studying the subject. All pre-defined assessments within the application were consulted and validated by General Chemistry 1 teachers at the University of Perpetual Help Molino to ensure that they met the stated learning competencies. It is aided with other gamification techniques such as star scoring system, accomplishment badges and leaderboards to promote engagement and motivation.

Adopting Hiter's (2023) suggestion, learning management system functionalities are integrated into the application. This includes the ability of teachers to create classes and invite students via code, set quizzes, create personalized assessments, and monitor student quiz scores and progress in completing the learning modules and gamified assessments in the application.

2.2 Test Procedures

The app was tested to determine bugs or errors and address them immediately to improve its overall quality. Test instruments were devised to test its functionality and compatibility. Functionality test subjected the app to different conditions according to its intended purpose or use. Each module was thoroughly verified to ensure that it work accordingly. On the other hand, compatibility test examined the app if it works on various platform conditions. It ensures that the app is compatible with different Android versions and smartphone resolutions. Both tests involved one (1) technical adviser, one (1) consultant, and three (3) IT experts who deliberately tested all the criteria or test cases. Test results serves as the basis to improve the app before subjecting it to further evaluation.

2.3 Evaluation Procedures

The Mobile Application Rating Scale (MARS) was adapted to evaluate the user acceptance of the application. It comprises four primary criteria. These are Engagement, Functionality, Aesthetics, and Information. Using a 4-Point Likert Scale, scores are interpreted with the following numerical ranges: 1.00–1.75 as "unacceptable," 1.76–2.50 as "fairly acceptable," 2.51–3.25 as "acceptable," and 3.26–4.00 as "highly acceptable." Evaluators consist of ten (10) IT experts and forty (40) end-users, which are composed of thirty-five (35) students and five (5) chemistry teachers. The tool was used to determine the overall quality of the app as well as if it met the requirements and objectives set on the project.

In addition, the developers evaluate the app's effectiveness in two ways: (1) as a supplementary learning platform and (2) as a self-learning platform. The first evaluation involves 88 student participants from the University of Perpetual Help Molino, who were conveniently selected and completed pretest and posttest paper-based assessments. The app was utilized after the pretest and throughout the first academic quarter. The developers collected and analyzed student exam scores using an independent samples t-test. On the other hand, the second evaluation involves 22 student participants who took the pretest and posttest assessments directly through the app, with the data analyzed using a paired samples t-test. The two statistical tests were also conducted to determine if the application has a significant positive effect to the academic performance of the students (Figs. 1 and 2).

Fig. 1. Chem U?'s User Interface for Students.

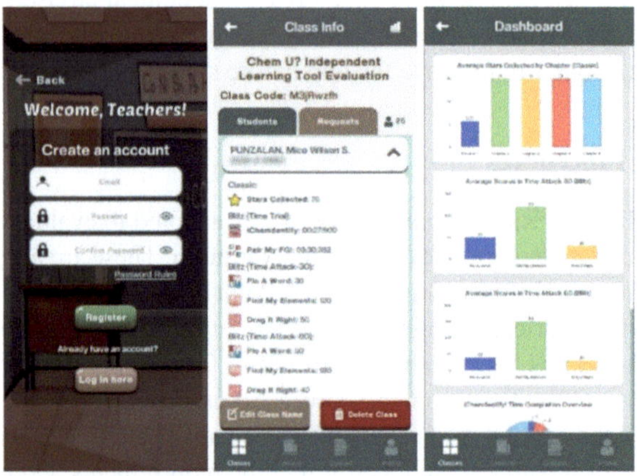

Fig. 2. Chem U?'s User Interface for Teachers.

Table 1. Functionality Test Results.

Test Respondent	Pass	Fail	Test Criteria	Percentage
Technical Adviser	307	1	308	99.68%
Consultant	308	0	308	100%
IT Expert 1	308	0	308	100%
IT Expert 2	308	0	308	100%
IT Expert 3	308	0	308	100%

2.4 User Interface

3 Results and Discussions

3.1 Test Results

The technical adviser, consultant, and three IT experts tested the app using 308 criteria (Tables 1 and 2). Overall, the application achieved a 99.94% passing rate. This shows that the user interface elements and functionalities of the app are working as intended.

Table 2. Compatibility Test Results.

Test Respondent	Pass	Fail	Test Criteria	Percentage
Technical Adviser	8	0	8	100%
Consultant	8	0	8	100%
IT Expert 1	8	0	8	100%
IT Expert 2	8	0	8	100%
IT Expert 3	8	0	8	100%

All test respondents tested and marked a total of eight (8) criteria as passed, achieving a 100% passing rate. This shows that the application can run smoothly in Android versions 10, 11, 12, and 13, particularly those with screen resolutions of 1520 × 720, 1920 × 1080, 2340 × 1080, and 2400 × 1080 (Table 3).

Table 3. Overall Evaluation Results from forty (40) End-users and ten (10) IT experts.

Criteria	Mean	SD	Interpretation	Rank
Engagement	3.76	0.47	Highly Acceptable	2
Functionality	3.70	0.50	Highly Acceptable	4
Aesthetics	3.81	0.41	Highly Acceptable	1
Information	3.76	0.44	Highly Acceptable	2
Grand Mean and SD	**3.76**	**0.46**	**Highly Acceptable**	

Evaluators rated Aesthetics as the highest among the four criteria, with a mean of 3.81 and a standard deviation of 0.41. Engagement and Information were rated equally second, each with a mean of 3.76. The former has a 0.47 standard deviation, while the latter results in 0.44. Finally, Functionality ranked fourth among the criteria, with a mean of 3.70 and 0.50 standard deviation. All criteria can be interpreted as "Highly Acceptable." Overall, the evaluation result has a grand mean of 3.76 and 0.46 grand standard deviation. Hence, the app is "Highly Acceptable" in terms of overall performance (Table 4).

Table 4. Independent Samples T-Test Results Comparing the Pretest Scores of Groups A and B

	Mean	SD	df	T-Statistic	Critical Value	P-Value
Group A	15.66	4.08	86	0.58*	1.99	0.56
Group B	15.14	4.37				
Difference	0.52	0.29				
Decision			Accept H_0			
Verbal Interpretation			No Significant			

Note. ** - The test statistic (t-test) is significant at 0.05 level. * - The test statistic (t-test) is not significant at 0.05 level.

Independent samples t-test was performed to statistically determine if groups A and B had similar levels of knowledge at the initial phase of the experiment. The results revealed that participants in Group A ($M = 15.66$, $SD = 4.08$) and Group B ($M = 15.14$, $SD = 4.37$) demonstrated similar levels of knowledge upon taking the pre-assessment. Hence, accepting the null hypothesis wherein the difference between their pretest scores was not statistically significant ($t(86) = 0.58$, $p = .56$). This finding suggests that, initially, there is no significant difference in knowledge levels between the two groups, providing a balanced baseline for subsequent analyses. With that, the developers proceeded to analyze the data related to the score improvements of the two groups (Table 5).

Table 5. Independent Samples T-Test Results Comparing the Mean Score Improvements of Groups A and B in Pretest-Posttest Exam.

	Mean	SD	df	T-Statistic	Critical Value	P-Value
Group A	19.34	5.66	86	10.02**	1.99	4.22E-16
Group B	9.30	3.49				
Difference	10.04	2.17				
Decision			Reject H_0			
Verbal Interpretation			No Significant			

Note. ** - The test statistic (t-test) is significant at 0.05 level. * - The test statistic (t-test) is not significant at 0.05 level.

By observing the means of the group's improvement scores, it is clearly evident that group A, the students who utilized Chem U? as a supplementary resource, has more improved results than the other group. Students who utilized the app as a supplementary learning platform gained a difference of 10 points when rounded off.

Moreover, the independent samples t-test was performed to statistically determine if integrating Chem U? as a supplementary learning platform for students has a significant improvement in the student's scores. The results showed that the student participants who utilized the app supplementary to the classroom instruction ($M = 19.34$, $SD = 5.66$)

reported greater score improvements than participants who only relied on classroom learning ($M = 9.30$, $SD = 3.49$). Hence, the difference was significant ($t(86) = 10.02$, $p < .05$).

Based on the results, the developers infer that the null hypothesis has been rejected. Therefore, it can be inferred that there is a significant difference in score improvements between students who utilized Chem U? alongside classroom learning compared to those who relied only on classroom instruction (Table 6).

Table 6. Paired Samples T-Test Results Comparing the Pretest and Posttest Mean Scores of Students for Chem U's Evaluation as a Self-Learning Platform.

	Mean	SD	df	T-Statistic	Critical Value	P-Value
Pretest	16.91	4.57	21	−13.30**	2.080	1.07E-11
Posttest	29.05	5.31				
Difference	12.14	4.28				
Decision			Reject H_0			
Verbal Interpretation			No Significant			

Note. ** - The test statistic (t-test) is significant at 0.05 level. * - The test statistic (t-test) is not significant at 0.05 level.

By observing the mean of the pretest scores and mean of the posttest scores, it is clearly evident that the latter is higher, which indicates that the students have improved their scores after using the app as a self-learning tool. The app helped students gain a difference of 12 points when rounded off.

Moreover, the paired samples t-test was performed to statistically determine if utilizing Chem U? as a self-learning platform for students has a significant improvement in their scores. The results showed that the students' posttest scores ($M = 29.05$, $SD = 5.31$) were greater than the students' pretest scores ($M = 16.91$, $SD = 4.57$). This difference was significant ($t(21) = -3.30$, $p < .05$).

Based on the result, it is inferred that the null hypothesis has been rejected. Therefore, there is a significant difference between student scores before and after using Chem U? as a self-learning tool.

4 Conclusion

The project has successfully developed Chem U? with the intention of providing students with a platform that combines chemistry education and entertainment. It aims to alleviate or eliminate the learning difficulties faced by chemistry students (Capanzana 2020; Rafanan et al. 2020; Sibomana et al. 2021). It also seeks to fill the gap identified by Partovi and Razavi (2019) regarding the rising need to integrate electronic learning applications in the classroom setting. Moreover, the project is also an answer to the heightened call for a new instructional and teaching approach for General Chemistry courses (Capanzana 2020).

Through functionality and compatibility testing, the app is proven to be functional and runs smoothly on Android versions 10 or higher on various smartphone screen resolutions such as 1520 × 720, 1920 × 1080, 2340 × 1080, and 2400 × 1080. Moreover, based on the evaluation results, the app is found to be highly engaging, functional, user-friendly, visually appealing, entertaining, and informative in learning General Chemistry.

Ultimately, this project has evaluated the effectiveness of the developed application, which many studies failed to perform (Abuhammad et al. 2021; Yusuf et al. 2022). Unlike other e-learning apps, Chem U? is statistically tested and proven to be effective both as a supplementary and self-learning resource in improving the student's academic performances. Significantly, this project not only establishes Chem U?'s own effectiveness but also contributes significantly to the research community by providing robust evidence supporting that e-learning applications have positive impacts on the academic performance of students in chemistry.

5 Recommendations

The current study can be interpreted as an initial step in researching the use of e-learning applications to enhance students' academic performance in chemistry. However, the results should be interpreted with caution due to the relatively small sample size used for app testing and evaluation. For other developers and researchers working on similar projects in the future, the researchers recommend expanding the sample size in subsequent evaluations. This will help improve the generalizability of the results and provide a more comprehensive understanding of the application's effectiveness. To further validate the effectiveness of the developed app, a comparative analysis with the existing chemistry app must also be made by the future researcher as the basis of app improvement. Finally, for the long-term impact or retention of data of this study, proper coordination must be done with the Department of Education to adopt and integrate the developed learning tool to the different high school students studying chemistry in the Philippines.

References

Abuhammad, A., et al.: "MedChemVR": a virtual reality game to enhance medicinal chemistry education. Multimodal Technol. Interact. **5**(3), 10 (2021). https://doi.org/10.3390/mti5030010

Barraquias, A., et al.: Turo Agham: Science Android Application for Kindergarten with Augmented Reality. Lyceum of the Philippines University Cavite (2019)

Bedada, T., Fita, K.K.: Causes of poor academic performance in chemistry at primary school students in Chiro Town, Ethiopia. Eur. J. Educ. Dev. Psychol. **11**, 1, (2023). https://doi.org/10.37745/ejedp.2013/vol11n15372

Capanzana, C.: Exploring STEM students' perceived Self-Efficacy and attitudes in general chemistry (2020). https://www.semanticscholar.org/paper/Exploring-STEM-Students%E2%80%99-Perceived-Self-Efficacy-in-Capanzana/cad22e9558b4a0f0c3c0bbd82a252e7abd0ff057

Hiter, S.: LMS features you need for a great E-Learning program in 2024 (2023). https://technologyadvice.com/blog/human-resources/important-learning-management-system-lms-features/

Jones, O.A.H., et al.: Chirality-2: development of a multilevel mobile gaming app to support the teaching of introductory undergraduate-level organic chemistry. J. Chem. Educ. **95**(7), 1216–1220 (2018). https://doi.org/10.1021/acs.jchemed.7b00856

Partovi, T., Razavi, M.R.: The effect of game-based learning on academic achievement motivation of elementary school students. Learn. Motiv. **68**, 101592 (2019). https://doi.org/10.1016/j.lmot.2019.101592

Rafanan, R.J.L., et al.: Pursuing STEM careers: perspectives of senior high school students. Particip. Educ. Res. **7**(3), 38–58 (2020). https://doi.org/10.17275/per.20.34.7.3

Reed, J.J., et al.: What we don't test: what an analysis of unreleased ACS exam items reveals about content coverage in general chemistry assessments. J. Chem. Educ. **94**(4), 418–428 (2017). https://doi.org/10.1021/acs.jchemed.6b00863

Sadera, J.R.N., et al.: Challenges encountered by junior high school students in learning science: basis for action plan. Univ. J. Educ. Res. **8**(12A), 7405–7414 (2020). https://doi.org/10.13189/ujer.2020.082524

Sibomana, A., et al.: Factors affecting secondary school students' academic achievements in chemistry. Int. J. Learn. Teach. Educ. Res. **20**(12), 114–126 (2021). https://doi.org/10.26803/ijlter.20.12.7

Soken-Huberty, E.: 10 reasons why chemistry is important (2023). https://openeducationonline.com/magazine/10-reasons-why-chemistry-is-important/

Swoope, M.W.: What impacts the student outcome in general chemistry. Modern Chem. Appl.. **8**(1), 1–3 (2020). https://doi.org/10.35248/2329-6798.20.8.270

Yusuf, A., et al.: VC.LAB (virtual chemistry laboratory): design and build an android-based acid-base titration game application as a chemistry interactive learning medium in high school. Int. J. Ethnosci. Bio-inform. Innov. Invent. Techno-sci. **2**(01), 22–32 (2022). https://doi.org/10.54482/ijebiiits.v2i01.187

Identifying Infrastructure Needs for Sustainable Cultural Heritage Tourism Development in Malinau Regency, North Kalimantan, Indonesia

Efraim Bavo Priyana[1] and Eko Prihartanto[2](✉)

[1] Department of Management, Open University, 77111 Tarakan, Indonesia
[2] Department of Civil Engineering, Universitas Borneo Tarakan, 77111 Tarakan, Indonesia
eko_prihartanto@borneo.ac.id

Abstract. Malinau Regency, with its rich cultural heritage, has great potential to develop sustainable cultural tourism. This research aims to identify the needs of facilities and infrastructure needed for cultural heritage conservation through tourism development in Malinau Regency. This study uses a qualitative method with purposive sampling of 20 respondents from various circles. The results of the study show that the condition of tourism facilities and infrastructure in Malinau is quite good, but it needs to be improved in several aspects, such as public transportation accessibility, variety and quality of accommodation, as well as information and promotion regarding recreational and educational facilities. In addition, conservation support facilities need to be more socialized and improved in quality. This study recommends the improvement and improvement of tourism facilities and infrastructure in Malinau to support the conservation of cultural heritage and sustainable tourism development.

Keywords: Conservation · Cultural heritage · Infrastructure · Sustainable cultural tourism · Tourism facilities

1 Introduction

1.1 Background

Malinau Regency, located in the northern border region of Indonesia-Malaysia, is an area rich in cultural heritage. The cultural diversity that Malinau has, ranging from oral traditions, traditional rituals, to archaeological sites, is an invaluable asset that has great potential to be developed as a sustainable cultural tourism attraction. Sustainable cultural tourism can not only provide economic benefits to local communities, but also play an important role in preserving and conserving existing cultural heritage [1].

Malinau has great potential for developing sustainable cultural tourism due to its rich cultural heritage, including various Dayak tribes such as Kenyah, Kayan, and Punan, each with their own unique traditions. Additionally, historical sites like Liang Kabut and the Mettulang Ancient Stone are evidence of past civilizations in the region. However,

M. Awang (Ed.): ICCAEET 2024, PTT, pp. 134–144, 2025.
https://doi.org/10.1007/978-3-031-94347-8_12

challenges in preserving cultural heritage include a lack of public awareness, limited human resources and funding, and social and cultural changes that may threaten its sustainability. Therefore, comprehensive and sustainable efforts are needed to address these challenges.

To develop sustainable cultural tourism in Malinau, the role of facilities and infrastructure is very crucial. Adequate facilities and infrastructure will facilitate tourist access to cultural tourism sites, increase tourist comfort during visitation, and support cultural heritage preservation and conservation activities. The facilities and infrastructure needed include transportation infrastructure, accommodation, interpretation facilities, information centers, and other supporting facilities [2]. The development of facilities and infrastructure also needs to pay attention to sustainability aspects, such as the use of renewable energy, good waste management, and conservation of natural resources [3].

This research aims to identify the needs of facilities and infrastructure needed for cultural heritage conservation through tourism development in Malinau Regency. By identifying the need for the right facilities and infrastructure, it is hoped that the development of sustainable cultural tourism in Malinau can run optimally. The results of this study are expected to provide useful recommendations for local governments, tourism actors, and the community to preserve Malinau's cultural heritage while improving community welfare through tourism.

2 Literature Review

2.1 Tourism

Tourism is the activity of traveling for fun or work. It involves tourists visiting destinations and using services from the tourism industry [4]. Tourism has both positive and negative impacts on the environment, economy, and local communities [5].

Types of Tourism

- **Leisure tourism:** For fun and leisure [6].
- **Cultural tourism:** To learn about other cultures through visits to historical sites, museums, or festivals [7].
- **Nature tourism:** To enjoy the beauty of nature [8].
- **Adventure tourism:** To try challenging activities [9].

Principles of Sustainable Tourism
Sustainable tourism aims to balance economic benefits with environmental and cultural preservation [10]. This means:

- **Economic Sustainability:** Tourism must provide economic benefits to local communities [11].
- **Social Sustainability:** Tourism must respect the local culture and involve the community in decision-making [12].
- **Environmental Sustainability:** Tourism must minimize negative impacts on the environment and conserve natural resources [13].

2.2 Tourism Facilities and Infrastructure

Tourism facilities and infrastructure refer to all facilities that support tourism activities, both physical and non-physical [14]. Tourism facilities include facilities that are directly used by tourists, such as accommodations, restaurants, and recreational places [15]. Meanwhile, tourism infrastructure includes supporting facilities for tourism facilities, such as road networks, electricity supplies, and sanitation systems [16].

Classification of Tourism Facilities and Infrastructure:

1. **Infrastructure Accessibility:** Availability and quality of roads, bridges, public transportation, and airports/ports that are adequate to reach tourist destinations [17].
2. **Visitor Service Facilities:** The availability of information centers, public toilets, adequate parking lots, and comfortable rest areas for visitors [18].
3. **Activity Support Facilities:** The existence of recreational, sports, and entertainment venues that are in accordance with the interests and preferences of tourists [19].
4. **Environmental Physical Condition:** Includes the cleanliness, aesthetics, and environmental safety of tourist destinations [20].
5. **Conservation Support Facilities:** The existence of museums, interpretation centers, and other facilities that provide comprehensive information regarding cultural heritage and the importance of its preservation [21].
6. **Accommodation Facilities:** There are a variety of accommodation options, such as hotels, inns, homestays, and other types of lodging that suit the needs and budgets of tourists [22].
7. **Tourist Activity Support Facilities:** Including equipment rental, professional tour guide services, and other services that support tourist activities [19].
8. **Educational and Interpretation Facilities:** The existence of educational programs, workshops, and other activities that aim to increase the understanding of tourists and local communities about cultural heritage [23].

Significance of Tourism Facilities and Infrastructure in Cultural Heritage Conservation

The availability of adequate tourism facilities and infrastructure contributes significantly to cultural heritage conservation efforts [14]. Good accessibility, convenient facilities, and comprehensive information will increase tourists' interest in visiting cultural heritage sites [24]. This can increase public awareness of the importance of preserving cultural heritage and generate income for local communities to support conservation efforts.

2.3 Conservation of Cultural Heritage

Cultural heritage conservation is an effort to preserve and protect cultural heritage, both tangible such as buildings, archaeological sites, and artifacts, as well as intangible such as oral traditions, rituals, and traditional knowledge [4]. The main goal of conservation is to maintain the sustainability of cultural heritage so that it can be passed on to future generations, while providing benefits to the current community [25].

The concept of cultural heritage conservation includes several key principles, including:

- **Authenticity:** Maintaining the authenticity of cultural heritage by avoiding changes that are not in accordance with its original values and character [26].
- **Integrity:** Maintaining the integrity of cultural heritage as a whole and inseparable whole [27].
- **Representativeness:** Ensuring that the preserved cultural heritage represents the diversity and cultural richness of a society [28].
- **Participation:** Involving local communities in the conservation process, as they are the main owners and heirs of cultural heritage [29].

The Importance of Cultural Heritage Conservation in Tourism Development
Cultural heritage conservation has an important role in the development of tourism, especially cultural tourism [30]. A well-preserved cultural heritage can be a major attraction for tourists, which in turn can provide economic benefits to the local community. In addition, cultural tourism can also increase public awareness of the importance of cultural heritage and encourage them to appreciate and preserve it more [31].

Some of the benefits of cultural heritage conservation in tourism development include:

- **Increasing the attractiveness of tourist destinations:** Unique and authentic cultural heritage can be a major attraction for tourists, especially those who are interested in history, culture, and traditions [32].
- **Creating jobs and economic opportunities:** Cultural tourism can create jobs in various sectors, such as accommodation, transportation, culinary, and handicrafts [33].
- **Increase local community income:** Cultural tourism can increase local community income through the sale of products and services, as well as tourism taxes and levies [34].
- **Encouraging cultural preservation:** Cultural tourism can increase public awareness of the importance of cultural heritage and encourage them to be more active in its preservation efforts [31].
- **Participatory approach:** Involving local communities in all stages of the conservation process, from planning, implementation, to monitoring and evaluation [35].

The approach taken is an effort to improve cultural development optimally.

3 Research Methods

This chapter explains in detail the research methodology used to achieve the research objectives, namely identifying the need for facilities and infrastructure for cultural heritage conservation through tourism development in Malinau Regency.

3.1 Research Location

This research will focus on two tourist villages in Malinau Regency, namely Pulau Sapi Tourism Village and Settulang Tourism Village. The selection of these two locations is based on the consideration that the two villages have significant cultural heritage potential and have developed culture-based tourism.

3.2 Research Object

The research object in this study includes the aspects evaluated including:

- **Infrastructure Accessibility:** Availability and quality of roads, transportation, and other supporting facilities that allow tourists to reach cultural heritage sites easily.
- **Visitor Service Facilities:** The availability of information centers, public toilets, parking lots, and comfortable rest areas for visitors.
- **Activity Support Facilities:** The existence of museums, interpretation centers, and other facilities that provide information about cultural heritage and support cultural tourism activities.
- **Physical Condition of the Environment:** Cleanliness, aesthetics, and safety of the environment around the cultural heritage site.
- **Conservation Support Facilities:** Availability of equipment and technology used to protect and care for cultural heritage.
- **Accommodation Facilities:** The availability and quality of accommodation options around cultural heritage sites, such as hotels, inns, or homestays.
- **Tourism Activity Support Facilities:** The availability of tour guide services, equipment rental, and other facilities that support tourism activities.
- **Educational and Interpretation Facilities:** The existence of educational programs, workshops, or interpretation materials that help visitors understand the value and importance of cultural heritage.

3.3 Sampling Techniques

While purposive sampling was employed, resulting in a focused sample size of 20 respondents, this approach allowed for the collection of rich, in-depth qualitative data from key stakeholders possessing extensive knowledge and experience related to the research objectives. Respondents were selected as many as 20 people consisting of:

- **Local communities:** Traditional leaders, community leaders, and residents involved in cultural tourism activities.
- **Tourism actors:** Tourism village managers, tour guides, homestay owners, and other tourism entrepreneurs.
- **Local government:** Tourism office officials and other related agencies responsible for tourism development and cultural heritage preservation.

3.4 Data Collection Methods

Data will be collected through several methods, namely:

- **Questionnaire:** Used to collect quantitative data from respondents about their perception of the condition of tourism facilities and infrastructure, as well as their needs and expectations related to the development of cultural tourism.
- **Interviews:** Used to collect more in-depth qualitative data from respondents about their experiences in managing and utilizing cultural heritage, as well as the challenges and opportunities faced in the development of cultural tourism.
- **Observation:** Used to directly observe the condition of tourism facilities and infrastructure in the field, as well as the behavior of tourists and local communities in interacting with cultural heritage.

3.5 Data Analysis Methods

The collected data will be analyzed using a combination of qualitative and quantitative descriptive analysis.

- **Qualitative descriptive analysis:** Used to analyze qualitative data from interviews and observations. This analysis will be carried out by identifying the main themes, patterns, and meanings that emerge from the data.
- **Quantitative descriptive analysis:** Used to analyze quantitative data from questionnaires. This analysis will be carried out using descriptive statistics

4 Results and Discussion

This study identified the needs of facilities and infrastructure to support sustainable cultural heritage tourism development in Malinau Regency. The results are presented and discussed in this section.

4.1 Current Conditions of Tourism Facilities and Infrastructure in Malinau

The assessment of the current conditions of tourism facilities and infrastructure in Malinau revealed a mixed picture. While some facilities, such as hotels and restaurants, are relatively well-developed, others, such as cultural heritage sites and public transportation, require significant improvement (Fig. 1).

4.1.1 Current Conditions of Tourism Facilities and Infrastructure in Malinau

The existing accommodation facilities in Malinau are generally adequate in terms of quantity, but there is a need for improvement in quality and diversity. While several hotels and guesthouses are available, they often lack proper maintenance and amenities, leading to visitor dissatisfaction. Moreover, there is a scarcity of accommodation options that cater to different budget levels and preferences, limiting the appeal of Malinau to a wider range of tourists.

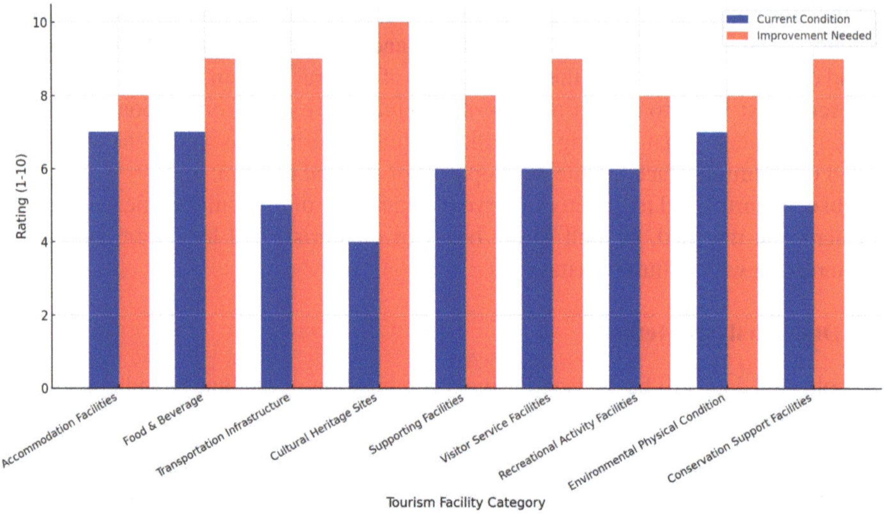

Fig. 1. Current Condition and Improvement Needs for Tourism

4.1.2 Food and Beverage

The food and beverage sector in Malinau is relatively well-developed, with a variety of restaurants and cafes offering local and National cuisine. However, there is room for improvement in terms of hygiene standards and service quality. Additionally, many restaurants lack a distinct focus on showcasing the unique culinary heritage of Malinau, missing an opportunity to enhance the cultural tourism experience.

4.1.3 Transportation

The transportation infrastructure in Malinau poses a significant challenge to tourism development. The road network is limited and often in poor condition, hindering accessibility to many cultural heritage sites. Public transportation options are also limited, making it difficult for tourists to navigate the region independently. This lack of adequate transportation infrastructure not only affects visitor experience but also hampers the economic benefits that tourism can bring to local communities.

4.1.4 Cultural Heritage Sites

Malinau boasts a rich cultural heritage, but many of its cultural sites are in a state of neglect. Lack of proper maintenance and preservation efforts has led to the deterioration of these valuable assets. Moreover, many sites lack basic visitor facilities, such as information centers, restrooms, and signage, further diminishing their appeal to tourists.

4.1.5 Other Supporting Facilities

Other supporting facilities, such as souvenir shops, tour operators, and information centers, are also underdeveloped in Malinau. The lack of these facilities limits the opportunities for tourists to engage with the local culture and economy, hindering the growth of sustainable tourism.

4.2 Infrastructure Needs for Sustainable Cultural Heritage Tourism Development

Based on the assessment of the current conditions and the interviews with stakeholders, the following infrastructure needs were identified for sustainable cultural heritage tourism development in Malinau:

Improved accessibility to cultural heritage sites: This includes upgrading the road network, developing public transportation options, and providing clear signage and information.
Enhanced preservation and maintenance of cultural heritage sites: This involves implementing conservation plans, providing training to local communities on heritage preservation, and establishing a dedicated fund for maintenance.
Development of quality accommodation facilities: This includes encouraging investment in new hotels and guesthouses, improving the quality and standards of existing facilities, and promoting the development of homestays and other community-based accommodation options.
Promotion of local cuisine and culinary heritage: This can be achieved through supporting the development of restaurants and cafes that specialize in local cuisine, organizing culinary festivals and events, and providing training to local communities on food preparation and presentation.
Establishment of supporting facilities: This includes developing souvenir shops, tour operators, and information centers that cater to the needs of cultural heritage tourists.
Capacity building for local communities: This involves providing training and education programs to local communities on tourism management, hospitality, and cultural interpretation.

4.3 Challenges and Recommendations

The development of sustainable cultural heritage tourism in Malinau faces several challenges, including:

Limited financial resources: The local government has limited financial resources to invest in infrastructure development.
Lack of awareness and understanding: Many local communities lack awareness and understanding of the importance of cultural heritage preservation and the potential benefits of tourism.
Environmental concerns: The development of tourism infrastructure can have negative impacts on the environment if not managed properly.

To address these challenges, the following recommendations are proposed:

- **Seek partnerships and collaboration**: The local government should actively seek partnerships and collaboration with the private sector, NGOs, and international organizations to secure funding and expertise for infrastructure development.
- **Conduct awareness-raising campaigns**: The local government should conduct awareness-raising campaigns to educate local communities about the importance of cultural heritage preservation and the potential benefits of tourism.
- **Implement sustainable tourism practices**: The local government should ensure that tourism development is carried out in a sustainable manner, minimizing negative impacts on the environment and maximizing benefits for local communities.

5 Conclusions

The development of sustainable cultural heritage tourism in Malinau Regency requires significant investment in infrastructure development. By addressing the identified infrastructure needs and challenges, Malinau can unlock its full potential as a cultural tourism destination, generating economic benefits for local communities while preserving its rich cultural heritage for future generations.

References

1. Muhamad, Sopjan, D.: Religious, cultural, and artistic tourism in sustainable tourism. Formosa J. Multidisc. Res. **1**, 13–28 (2022). https://doi.org/10.55927/fjmr.v1i1.411
2. Saw, Y.Q., Dissanayake, D., Ali, F., Bentotage, T.: Passenger satisfaction towards metro infrastructures, facilities and services. In: Transportation Research Procedia, pp. 3980–3995. Elsevier B.V. (2020)
3. Duxbury, N.: Cultural sustainability, tourism, and development. Cult. Sustain. Tourism Dev. 1–18 (2021). https://doi.org/10.4324/9780367201777-1
4. Sonuç, N.: Culture, tourism and sustainability (cultural heritage and sustainable tourism, social sustainability of tourism, socio-cultural sustainability of tourism). Encyclopedia Sustain. Manag. 1–7 (2020). https://doi.org/10.1007/978-3-030-02006-4_457-1
5. Richards, G.: Making places through creative tourism? Cult. Sustain. Tourism Dev. 36–48 (2021). https://doi.org/10.4324/9780367201777-4
6. Burgers, G.-J.: Tourism, leisure and cultural heritage: the challenge of participatory planning and design. New Front. Regional Sci.: Asian Perspect. 71–85 (2021). https://doi.org/10.1007/978-981-16-3623-3_5
7. Sosa-Larrainzar, E., Niño-Ramos, S., Perdomo-Abud, M.I.: Cultural-natural heritage of Texmelucan towards sustainable tourism. Sustainable Tourism. Puebla and other entities, pp. 65–80 (2023). https://doi.org/10.35429/p.2023.3.65.80
8. Li, H., Lüthje, M., Björn, E., Miettinen, S.: Fostering stakeholder engagement in sustainable cultural tourism development in nature-based sites. In: The Routledge Handbook of Nature Based Tourism Development, pp. 183–200 (2023). https://doi.org/10.4324/9781003230748-15
9. Upadhyay, P.: Promoting employment and preserving cultural heritage: a study of handicraft products tourism in Pokhara. Nepal. J. Tourism Advent. **3**, 1–19 (2020). https://doi.org/10.3126/jota.v3i1.31354
10. Sustainable, A., Results, C.: Tourism Destination Management Achieving Sustainable and Sustainable Tourism: International Cooperation for Development

11. Shahid, M., Ullah, K., Imran, K., Masroor, N., Sajid, M.B.: Economic and environmental analysis of green transport penetration in Pakistan. Energy Policy **166** (2022). https://doi.org/10.1016/j.enpol.2022.113040
12. Gregory-Smith, D., Wells, V.K., Manika, D., McElroy, D.J.: An environmental social marketing intervention in cultural heritage tourism: a realist evaluation. Market. Sustain. Tourism. 174–191 (2019). https://doi.org/10.4324/9780203711668-11
13. Cheng, Z., Chen, X.: The effect of tourism experience on tourists' environmentally responsible behavior at cultural heritage sites: the mediating role of cultural attachment. Sustainability **14**, 565 (2022). https://doi.org/10.3390/su14010565
14. Wulandari, M.P., Agustina, I.F.: Infrastructure facilities in tourism village management. Indonesian J. Cult. Community Dev. **11** (2022). https://doi.org/10.21070/ijccd2022769
15. Kasztovszky, Z.: Large facilities and cultural heritage research. In: Handbook of Cultural Heritage Analysis, pp. 13–21 (2022). https://doi.org/10.1007/978-3-030-60016-7_2
16. Mandić, A., Mrnjavac, Ž., Kordić, L.: Tourism infrastructure, recreational facilities and tourism development. Tourism Hospital. Manag. **24**, 41–62 (2018). https://doi.org/10.20867/thm.24.1.12
17. Prihartanto, E.: Pengembangan Wilayah Pesisir Timur Kota Tarakan Berdasarkan Sarana Dan Prasarana (2020)
18. Wang, H., Luo, S., Furuya, K.: What attracts tourists to press the shutter in cultural heritage tourism? An analysis of visitor-employed photography and visual attributes: a case study on Japan's Kairakuen Garden. Tourism Recreation Res. 1–17 (2022). https://doi.org/10.1080/02508281.2022.2153993
19. Gamage, K.P.: Activity economy and cultural heritage tourism in Sri Lanka(Special Reference in Galle and Kandy). TRIVALENT ත්‍රිසංයුජ: J. Archaeol. Tourism Anthropol. **3**, 54–70 (2022). https://doi.org/10.4038/tjata.v3i1.59
20. Ratten, V.: Cultural conditions and heritage tourism in Southeast Asia. Driving Entrepreneurship Southeast Asia 31–43 (2022). https://doi.org/10.4324/9781003260783-3
21. Engmann, R.A.A.: Coups, castles, and cultural heritage: conversations with Flight Lieutenant Jerry John Rawlings, former President of Ghana. J. Heritage Tourism **16**, 722–737 (2020). https://doi.org/10.1080/1743873x.2020.1817929
22. Khalil, N., Kozmal, H.: The role of heritage hotels in the development of creative cultural tourism: the case study of Egypt. J. Assoc. Arab Univers. Tourism Hospital. **18**, 97–133 (2020). https://doi.org/10.21608/jaauth.2020.40470.1065
23. Toker, B., Rezapouraghdam, H.: Intangible cultural heritage and management of educational tourism. In: Advances in Hospitality, Tourism, and the Services Industry, pp. 199–216 (2021). https://doi.org/10.4018/978-1-7998-4318-4.ch010
24. Jimura, T.: Tourism development at World Heritage Sites. In: World Heritage Sites: Tourism, Local Communities and Conservation Activities, pp. 34–48 (2018). https://doi.org/10.1079/9781786392688.0034
25. Black, R.: Improving engagement between tourists and staff at natural and cultural heritage tourism sites: exploring the concept of interpretive conversations. Tourism Recreation Res. **43**, 82–90 (2017). https://doi.org/10.1080/02508281.2017.1382981
26. Jigyasu, N.: Authentic versus staged identity – problematizing tourism interventions for authenticity and integrity in historic cities. J. Cult. Heritage Manag. Sustain. Dev. (2022). https://doi.org/10.1108/jchmsd-07-2022-0116
27. 7 Commodification and Integrity. Intangible Cultural Heritage and Tourism in China, pp. 126–149 (2023). https://doi.org/10.21832/9781845418649-010
28. Portalés, C., Casas, S., Alonso-Monasterio, P., Viñals, M.J.: Multi-dimensional acquisition, representation, and interaction of cultural heritage tangible assets. Adv. Hospital. Tourism Serv. Ind. 72–95 (2018). https://doi.org/10.4018/978-1-5225-2927-9.ch004

29. Aldhanhani, H.M.O.R., Zainudin, M.Z.: Willingness of residents to participate in UAE heritage and culture tourism industry. Int. J. Sustain. Construct. Eng. Technol. **13** (2022). https://doi.org/10.30880/ijscet.2022.13.04.011
30. Sulaiman, M.Z., Wilson, R.: Tackling the tourism translation challenge: a cultural-conceptual approach. Transl. Tourism. 49–66 (2019). https://doi.org/10.1007/978-981-13-6343-6_4
31. Marchegiani, L.: From Mecenatism to crowdfunding: engagement and identification in cultural-creative projects. J. Heritage Tourism **13**, 143–151 (2017). https://doi.org/10.1080/1743873x.2017.1337777
32. Zhang, H., Leung, X.Y., Bai, B.: Cultural attractiveness index for sustainable cities: Tourism Agenda 2030. Tourism Rev. **78**, 411–426 (2023). https://doi.org/10.1108/tr-05-2022-0255
33. den Dekker, T.: A Practical framework for creating cultural tourism experiences. Creat. Manag. Exp. Cult. Tourism. 59–70 (2019). https://doi.org/10.1142/9789813233683_0004
34. Mele, E., Kerkhof, P., Cantoni, L.: Cultural localization in online heritage promotion. J. Heritage Tourism **16**, 300–316 (2020). https://doi.org/10.1080/1743873x.2020.1779731
35. Carson, S., Pennings, M.: Performing Cultural Tourism: Communities, Tourists and Creative Practices. books.google.com (2017)

Evaluating and Optimizing Ferry Port Performance for Enhanced Service and Economic Growth

Deska Uliyani[1], Wisnu Handoko[1], Ahyani[2(✉)], Sri Sarjana[3], and An Naufal Gintara[1]

[1] Land Transportation Department, Politeknik Transportasi Darat Indonesia - STTD, Bekasi, Indonesia

[2] Road Transportation Management Department, Politeknik Transportasi Darat Indonesia - STTD, Bekasi, Indonesia
ahyani@ptdistt.ac.id

[3] Marketing Innovation and Technology Department, Politeknik Transportasi Darat Indonesia - STTD, Bekasi, Indonesia

Abstract. Optimizing the performance of ferry ports plays an important role in supporting the economy, safety, efficiency, and regional development as a long-term investment that provides economic benefits to the community. This study aims to evaluate the performance of the port as an effort to improve performance in providing optimal services to passengers and vehicles in current and future conditions. A quantitative approach is developed in this research method. The analysis is applied referring to the performance evaluation, modal interaction matrix, trip segment analysis, analysis of the level of importance and service performance, analysis of facility needs, analysis of the planned number of future requests using the smallest quadrant method, analysis of the zoning system. The results of study stated that the current port performance reached a value of 58.86 (moderate category), modal interaction matrix of -133.3 (sufficient category), trip segment analysis pedestrians using motorbikes have longer travel times than cars and there are 11 quadrant I service attributes and 5 quadrant III attributes that need improvement so that efforts are made by providing descriptive recommendations, fulfilling facility and service needs, improving pedestrian facilities, arranging passenger counter areas and adjacent waiting rooms, providing information facilities and implementing a zoning system. The five-year facility fulfilment plan with a predicted number of 104,591 passengers and 14,891 mixed vehicles as required by the waiting room of 413 m^2, canteen area of 62 m^2, administration area of 62 m^2, utility room of 134 m^2, public space of 67 m^2, ready-to-load parking lot of 1,285 m^2 and drop-off parking lot of 481 m^2. For that, the findings of this study can be implemented so that ferry ports can improve their operational performance by providing better services to users and supporting economic growth and regional development.

Keywords: Ferry port · Port Performance · Performance Evaluation · Importance Performance Analysis

M. Awang (Ed.): ICCAEET 2024, PTT, pp. 145–162, 2025.
https://doi.org/10.1007/978-3-031-94347-8_13

1 Introduction

1.1 Theoretical Phenomena

The current development of ports is known as the sixth generation port model (6GP) which adapts and transforms from the first generation port model (1GP) which prioritizes the evolution of port operations, technological advances, port resilience to natural disasters and supply disruptions [1]. Maritime ports have a vital contribution to global trade as an effort to support the supply network system [2], create a key link in the trade chain and become a driver of development in the maritime economic sector [3]. Ports play an important role as vital intersections connecting land and sea where connectivity and spatial integration serve as competitiveness indicators that determine performance, which can be optimized through the establishment of coordinated collaborative mechanisms and provide valuable insights in port management decision making [4]. Port efficiency can be managed well through the arrangement of arrival distribution, terminal capacity, and supply offered by shipping companies by considering yard area saturation and average ship load factor referring to tactical policies including the introduction of flexible time and alternatively by introducing additional boat services during peak hours [5]. The quality function deployment (QFD) approach needs to be applied in the port environment to formulate problems and determine priority scales [6]. Port entry rules for foreign ships need to be applied equally in various countries, not to be applied weakly in developed countries, but the rules are applied more strongly in developing countries [7]. It is important for port authorities to enforce the rules by acting firmly and imposing sanctions on ship operators who do not comply with applicable rules and regulations [8]. Diagnostic instruments are needed for evaluating and monitoring port performance to meet various requirements flexibly [9]. Some factors that influence the choice of port include the distance between ports, waiting times, fees for port services, ship stopover schedules, and the destinations offered by the port [10]. Ports are a means of providing services and economic development that provide connecting routes between land and sea modes of transportation [11]. Transportation is related to living needs for supporting the movement of people and goods, accessibility to some locations and settlement, supplied goods and service for consumption [12]. Transportation play role as the catalyst to support economy growth and regional development [13]. One of the important element of transportation is the port [14].

1.2 Research Problems

According to government regulation number 61 in 2009 concerning ports, a port is a place consisting of land and surrounding waters with certain boundaries as a place for government activities and economic activities that are used as a place for ships to lean, anchor, load and unload passengers and or load and unload goods equipped with shipping safety facilities and port supporting activities as well as a place for inter and inter-modal transportation. Kuala Tungkal Ferry Port has been operating since 2017, connects Jambi Province and Kepulauan Riau Province, which operational services are from Kuala Tungkal – Telaga Punggur with a travel time of ± 14 h served by 3 ships and Kuala Tungkal – Dabo with a travel time of ± 9 h served by 1 ship. After reviewing the data from the last 5 years, there has been an increase in the demand passengers and vehicles, where in 2019 there were 10,824 passengers and 2,444 vehicles, 14,964 passengers and 4,375 vehicles in 2020, 20,305 passengers and 4,835 vehicles in 2021, 42,070 passengers and 7,126 vehicles in 2022, the last one is 51,961 passengers and 7,894 vehicles in 2023, so that the percentage increase in number of passenger from 2019 until 2023 is 380.1% and an average 5 years increase is 51.16% and the percentage increase in number of vehicle is 223% and an average 5 years increase is 36.92%.

The increasing of number passengers and vehicles hasn't been balanced with the availability of facilities so that it can affect performance of a port. Maximum performance of a port can be achieved if the port is supported by adequate facilities, professional human resources, and a good management system [15]. In the current condition, the performance of Kuala Tungkal Ferry Port wasn't supported by the availability of facilities according to the condition of the waiting room passenger with an area of 42 m^2 which is equipped with 20 chairs. This condition makes passengers to prefer to wait in around the terminal yard and gangway area. Apart from that, there is also a build-up in the ready-to-load parking area. In other condition, ticket counter for passengers and vehicles are still being combined, passengers and vehicle inspection area are also still combined. Two-wheeled vehicles parked anywhere, and the weight of the vehicles boarded in the ship is not known because Kuala Tungkal Ferry Port does not have weight-bridge which means it does not fulfil regulations of the Minister of Transportation 103 in 2017 which required every port to provide weight-bridge and portal facilities. Therefore, it is necessary to make improvements that are proposed as solutions to improve the performance of port including improving passenger and vehicle facilities, arranging zoning systems, improving pedestrian facilities and provision of better service information.

The performance of service for passengers and vehicles is not optimal, this condition is indicated by unavailability of evacuation route guidance facilities, evacuation points, wheelchairs, disabled facilities, nursery room, information site-plan or layout area, onward transportation, route network maps for transport service, incomplete first aid kit (P3K), the condition of the fire extinguisher (APAR) which has expired and the condition of the toilet which has not been keep clean. The research method was carried out by comparing existing conditions with existing regulatory standards, the service standards are benchmarks used as guidelines for providing service and as a reference for assessing quality, responsibility and promises to the community to provide quality, fast, easy, affordable, and measurable services which include safety, security, comfort, convenience or affordability, reliability or regularity and equality. This condition will affect performance values, so the performance improvement of Kuala Tungkal Ferry Port is needed. The purpose of this study is to evaluate the performance of the port as an effort to improve performance in providing optimal services to passengers and vehicles in current and future conditions.

2 Method

This study implements a combination approach in the form of quantitative method. The results of the quantitative approach are easier to replicate because the procedures are systematic and standardized [16]. Quantitative data is often used in decision-making in various fields because it provides a strong basis for making policies [17], quantitative data collection can be faster and cheaper [18]. This study uses the primary data and secondary data processed by using analysis techniques that are in accordance with the plan to improve the performance of passenger and vehicle service. Primary data collection allows researchers to use a variety of approaches that are appropriate to the purpose such as surveys, interviews, observations, or experiments [19]. The data collections are in the form of observation and interviews in the form of questioners and survey forms. Primary data used for supporting the research are assessment of passenger, vehicles and ships operators satisfaction, maneuver time, number of vehicles boarding or disembarking to or from the ship, time for vehicle boarding or disembarking time to or from the ship, vehicle queuing time to board the ship, inventory of the availability of basic and supporting facilities along with the implementation of service standards, assessment of modal interaction matrix, distance between facilities service data, parking generation data in drop-off/pick-up area, assessment of passengers and vehicles service levels while secondary data includes regional study condition, the number of passengers, vehicles and ship in the last 5 years, from 2019 until 2023, dock operational time in day, number of days ready for dock operational (excluding maintenance and repair time), availability of operational procedure system (SOP) documents, information on the availability of human resources at ports, port site-plan or layout, characteristics of infrastructure and ships data in the port.

Several analytical techniques applied include performance analysis based on performance evaluation of river, lake and crossing port consisting of aspects of service performance, operational performance and supporting performance. Service performance analysis allows ports to identify how well services are being delivered to passengers and vehicles such as waiting times, accessibility, comfort and safety. Operational performance evaluation is important to ensure that port activities are carried out efficiently which includes the management of vehicle and passenger traffic, space distribution and workflow in the port. Supporting performance analysis needs to involve infrastructure, facilities and technology that support port operations such as parking areas, waiting rooms and other facilities needed to handle larger volumes. Apart from that analytical techniques are modal interaction matrix, trip segment analysis, and importance performance analysis. Prediction of the number of passenger and vehicles demand in the next 5 years used least quadrant method, calculation of the needed facilities which consist of passenger terminal facilities and parking areas. Service performance analysis is used to measure how well the services are being provided to port users, helping to identify areas for improvement to improve the user experience. The modal interaction matrix attempts to identify how different modes of transport connect and interact, helping to better plan for improved connectivity and ease of movement of passengers and goods.

3 Result and Discussion

Port performance evaluation related to service performance, operational performance and supporting performance are in accordance with the provisions of Director general of land transportation regulations number 539 in 2022. The result of the performance assessment based on service performance indicator, operational performance indicator and supporting performance indicator. By conducting performance evaluation, ports can identify bottlenecks and inefficiencies in daily operations that allow port managers to implement improvements and increase operational efficiency. Performance evaluation allows ports to assess and improve the quality of services provided to users, including speed of service, comfort, and security that have a positive impact on customer satisfaction. Performance evaluation provides accurate and relevant data and information for better decision making where port managers can use the information for strategic and operational planning. Through evaluation, ports can assess the use of resources such as manpower, facilities, and infrastructure to help optimize resource use and reduce waste. In addition, performance evaluation also helps ensure that ports comply with all applicable regulations and standards because regulations are important to maintain legality and safe operations. Thus, port performance evaluation is an important tool for managing and developing ports effectively and efficiently and ensuring that ports can meet the needs and expectations of all stakeholders.

Table 1. The result of the performance assessment

No	Indicator	Indicator Value	Quality	Performance Value
1	Development of the Numbers of vehicles	100	0.035	3.50
2	Development of the Numbers of passengers	100	0.035	3.50
3	Development of the Numbers of ships	100	0.007	0.7
4	Customer satisfaction of vehicles carrier	79.78	0.126	10.05
5	Customer satisfaction of pedestrian passengers	76.00	0.126	9.58
6	Customer satisfaction of ship operator	76.97	0.126	9.70
7	Maneuver time	40	0.152	6.08
8	Average time for vehicle boarding time	60	0.038	2.28
9	Average time for vehicle disembarks	60	0.038	2.28

(continued)

Table 1. (*continued*)

No	Indicator	Indicator Value	Quality	Performance Value
10	Time to queue for vehicle to board the ship	20	0.076	1.52
11	Berth Occupancy Ratio	40	0.152	6.08
12	Availability of technical plan document for port development and port licensing	7.14	0.009	0.06
13	Implementation of the action plan contained in the planning document	0	0.009	0.00
14	System and procedure service	50	0.009	0.45
15	Implementation of the zoning system	50	0.009	0.45
16	Basic Facilities	84.09	0.015	1.26
17	Supporting Facilities	45.83	0.003	0.14
18	Passengers service standards	40.44	0.006	0.24
19	Vehicles service standards	53.57	0.006	0.32
20	Ships service standards	45.00	0.006	0.27
21	Availability of human resources	29.27	0.009	0.26
22	Human resource competency	14.63	0.009	0.13
Total Performance Value			**1.00**	**58.86**

Performance value of the Kuala Tungkal Ferry Port is currently reaching value of 58.86 (medium) refer to Table 1. The performance of the port should be supported by the availability of facilities to improve the convenience of users especially passengers to move between facilities, then an evaluation is carried out related to the interaction of facilities and the convenience of passengers in moving from one place to the other. Performance evaluation helps identify the needs of port users, especially in terms of convenience when moving between facilities whereby understanding the interactions between facilities and how users move from one place to another, the port can make significant improvements to improve convenience. Port managers can plan and implement facility improvements, including the addition of passenger facilities, parking areas, and other supporting infrastructure to meet user needs. Performance evaluation provides insight into how port operations can be optimized including rearranging facilities, improving traffic management systems, and providing additional services that can facilitate the movement of passengers and vehicles. Performance evaluation helps ensure that the port meets applicable standards and regulations to ensure safe and efficient operations. Ports that have good performance and adequate facilities will be more competitive in attracting service users who play an important role in the sustainability and economic growth of the region. Performance evaluation also includes an assessment of security and safety aspects so that the port can identify and address potential risks that can endanger users and port operations. Overall, the performance evaluation of the Kuala Tungkal Ferry Port is an important step to ensure that the port can function optimally, provide adequate services, and meet the needs and expectations of users well.

Dropoff/pickup Parking Area													
Motorcycle Parking Area	8	8 0											
Checking Ticket Area	5	7 -2	5	6 -1									
Waiting Room	4	7 -3	4	8 -4	9	9 0							
Canteen	5	6 -1	5	7 -2	8	8 0	8	9 -1					
Toilets	5	6 -1	5	6 -1	8	8 0	8	9 -1	9	9 0			
Passenger Counter	5	7 -2	5	7 -2	6	7 -1	6	9 -3	7	9 -2	6	7 -1	
Sum of Negative Value		-9		-10		-1		-5		-2		-1	-28
Modal Interaction Matrix	Dropoff /pickup Parking Area		Motorcycle Parking Area		Checking Ticket Area		Waiting Room		Canteen		Toilets		Total

Fig. 1. Modal Interaction Matrix

The interaction between of facilities used by pedestrians based on Fig. 1 that the total negative value is -28 so that the normalized score of the interaction between the facilities is -133.3. This result show that the interaction between facilities at Kuala Tungkal Ferry Port is in the sufficient category. To determine the level on convenience of passengers for moving between facilities at the port so that calculate a trip segment analysis with the division of dis-utility segments. Disutility segment is determined based one the passengers who will depart starting from the gate to the inside of the shop and getting off starting from the ship to the exit gate using car and motorcycle modes. Based on the result of the analysis, the time taken by passengers using a car to enter the port is 34.35 min which the distance of 253 m and exit 26.16 min with the distance of 194.5 m while the time taken by passengers using a motorbike to enter is 35,02 min with the distance of 259 m and exit 29.02 min with the distance of 231.5 m. The time taken by passengers using who enter and exit is influenced by the existing obstacles as well as the layout of facilities and service condition in the Kuala Tungkal Ferry Port. Regarding the condition of service to passengers and vehicle, an analysis was carried out using the importance performance analysis method based on passenger and vehicle service indicator.

The assessment of the level value important and performance service based on the result of the quadrant in Fig. 2, it is show the attributes that need to be improved the main priority (Quadrant I) are Emergency Rescue Equipment (fire extinguisher (APAR), evacuation route guidance, evacuation points, emergency phone number), health facilities, security facilities (complaint stickers), toilets, waiting room area, weight-bridge, weight-bridge officers, drop-off or pick-up parking area, vehicle weight counting device, number of counters and vehicle separation lanes. Attributes that need to be maintained

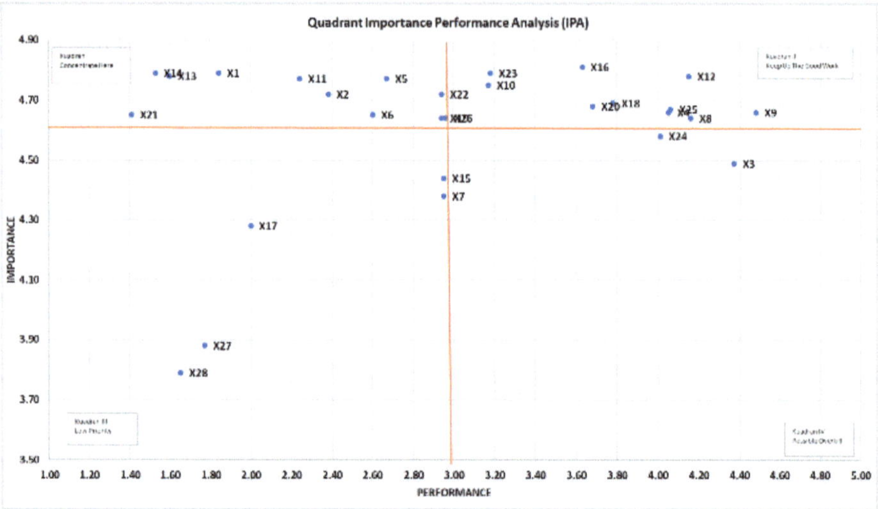

Fig. 2. Result of the Importance Performance Analysis

(Quadrant II) are security facilities (CCTV), canteens, lighting lights, arrival or departure passenger lanes, temperature control facilities, arrival information, departures and travel disruptions, passenger information service facilities, passenger baggage service or porter officers, ticket queue services, vehicle loading queue time. Attributes that need to be improved but are in low priority (Quadrant III) are prayers rooms, service information (site-plan or layout and tariffs), advanced transportation information, disable facilities and nursery room while attributes that have excess performance (Quadrant IV) are uniformed security officers and the identify and availability of officers.

Based on the port performance evaluation assessment in Tabel 1, there are several indicators that need to be improve so that efforts to improve performance carried out, one of which is by providing descriptive for all aspect of performance assessment. Port performance improvement is a crucial aspect that has a direct impact on various sectors including the economy, logistics and public welfare. An efficient port minimizes waiting time for ships, passengers, and goods to reduce logistics costs, improve supply chain efficiency and promote smoother trade. A well-performing port attracts more trade and investment to improve the economic competitiveness of a region or country, create jobs and drive economic growth. Port performance improvement improves service quality, which contributes to user satisfaction, satisfied users are more likely to return to port services and provide positive recommendations. High-performing ports tend to have better safety and security standards, reducing the risk of accidents and incidents to protect passengers, crew, and goods from harm. Higher operational efficiency is usually accompanied by more environmentally friendly practices, such as reduced emissions and waste. Port performance improvement allows for more optimal use of existing resources, including manpower, infrastructure, and technology to reduce waste and increase productivity. Performance improvement ensures that ports comply with all regulations and standards set by relevant authorities to maintain operational legality and gain the trust of stakeholders. Overall, improving port performance is key to creating a more efficient, safe, and sustainable transport and logistics system that not only benefits the port itself but also supports broader economic growth and societal well-being.

Table 2. Recommendations for Performance Improvement

No	Indicator	Recommendations	Description
1	Development of the number of vehicles	• Improve the services of passengers and vehicles • Improve the facilities of safety and security • Full fitment the facilities of port	To anticipate increased demand
2	Development of the number of passengers		
3	Development of the number of ships	• Identify of the user services demand potential • Optimize the ship operational services	Ship needs are influenced by number of productivities, ship trips, and ship operational time
4	Customer satisfaction of vehicle carrier	• Improve the quality of services • Full fitment of the facilities • Arrangement the zonation system • Optimize the operational of regulation	Customer satisfaction can affect the increase demand, it's guided by the aspect of service quality
5	Customer satisfaction of pedestrian passengers		
6	Customer satisfaction of ship operator		
7	Maneuver time	• Coordinate between ship operators with officers in the port • Carry out the supervision of load or unload activity in the port • Optimize the services regulation of arrival and departure ships • Evaluate of the ships needed • Prepare of the operational schedule • Provide information to service users if there are delays or problem with the ship services	The preparation of the operational schedule is adjusted to the performance of ship operational services which can guided by regulation director general of sea transportation number HK103/2/2/DJPL-17 concerning the calculation of the service performance which includes Waiting Time, Postpone Time, Approach Time, Berth Time, Effective time, Idle Time, Not Operation Time, ET/BT Ratio, Turn Round Time, BOR
8	Average time for vehicle boarding time		
9	Average time for vehicle disembarks		
10	Time to queue for vehicle to board the ship		
11	Berth Occupancy Ratio	• Optimize the use of the dock • Arrange the operational schedule of ships • Carry out the supervision of services ship	Kuala Tungkal Ferry Port have one dock facilities which the value of BOR is 26,5%, it's still in accordance with the provisions of the BOR recommended by UNCTAD
12	Availability of technical plan document for port development and port licensing	• Carry out the compilation of plan document in accordance with the direction of port development policy • Coordinate port development with the budget provided by the local government (Pemda)	The technical plan document must be adjusted to port condition and in implementing it need to be coordinated with other related agencies such as BPTD
13	Implementation of the action plan contained in the planning document		

(*continued*)

Table 2. (*continued*)

No	Indicator	Recommendations	Description
14	System and procedure service	• Prepare standard operational procedure (SOP) documents that are not yet available • Implementation standard operational procedure (SOP) documents in accordance with regulations • Carry out the supervision and evaluation of implementation standard operational procedure (SOP) documents which has been set	The preparation of the standard operational procedure (SOP) is guided by regulation director general of land transportation number SK.2681/AP.005.DRJD/2006 concerning the operational of ferry port
15	Implementation of the zoning system	• Prepare and propose documents for the implementation of the zoning system documents to the Land Transportation Management Center (BPTD) to carry out the zoning system • Carry out operational arrangements and controls at the port in accordance with the provisions • Improve and Arrangement of the facilities	The implementation of system zoning is an effort to improve safety, security, smoothness, comfort, and order. The provisions for implementation are in accordance with the regulation of the Ministry of Transportation number 91 in 2021
16	Basic facilities	• Carry out routine and periodic checks and maintenance relate to the availability and condition of the port facilities • Improve the facilities of port, there are not yet available	The improving of facilities to accommodate transportation demands, support service customer satisfaction and improve safety, security, smoothness, comfort
17	Supporting facilities		
18	Passengers service standards	• Implementation of the service standards according to provisions • Carry out routine and periodic checks and maintenance of the facilities • Complete the service facilities there are not yet available in the port	The condition of service facilities that are not functioning is repaired or replaced, and service standards are applied in accordance with regulation director general of land transportation number KP.5062/AP.005/DRJD/2020
19	Vehicles service standards		
20	Ships service standards		
21	Availability of human resources	• Identify the duties and functions of each position and the human resource needs at the port • Identify of accordance the duties and functions at the port • Facilitate the improvement of competency by participating in training and or continuing education	Human resources needs can be adjusted with the institution position structure map and institution need
22	Human resource competency		

Based in the Tabel 2 above, the plan for improving performance in the research is related to passenger and vehicles services which include the availability of facilities, the implementation of service standards, the arrangement of the system zoning so that if the performance improve recommendations are implemented, it will affect the increase in the performance value of the port. In addition, the recommendations proposed as an effort to improve performance are technical recommendations related to the fulfilment of the availability of passenger and vehicles facilities which include passenger terminal areas, including waiting rooms, ready-to-load parking areas, drop-off, or pick-up parking areas. The need for facilities is calculated based on the current ideal conditions and the condition of the next 5 years plan.

The port performance improvement plan that is carried out focuses on passenger and vehicle services by considering the availability of facilities, the application of service standards and the arrangement of system zoning. The implementation of performance improvement recommendations is expected to increase the port′s performance value. The technical recommendations submitted include the fulfilment of facilities for passengers and vehicles, including areas passenger terminal, waiting room, loading and unloading parking area, and drop-off or pickup parking area. Facility needs are calculated based on current conditions and 5-year predictions, using the least squares method to produce a linear trend equation in predicting passenger numbers and vehicle demand.

Table 3. Passenger and Vehicle Model Equations

No	Category	Equations	No	Category	Equations
1	Adults	Y = 26,125.8 + 10,758.2 X	8	Class V.A vehicle	Y = 18.8 + 0.5 X
2	Children	Y = 1,839.2 + 120 X	9	Class V.B vehicle	Y = 2,475 + 688.4 X
3	Class I vehicle	Y = 10.4 + 3.6 X	10	Class VI.A vehicle	Y = 3.6 + 2.3 X
4	Class II vehicle	Y = 879 + 256.9 X	11	Class VI.B vehicle	Y = 48.4 + 33.1 X
5	Class III vehicle	Y = 9 + 4.4 X	12	Class VII vehicle	Y = 18.6 + 15.5 X
6	Class IV.A vehicle	Y = 1,491.6 + 331.5 X	13	Class VIII vehicle	Y = 0.4 + 0.2 X
7	Class IV.B vehicle	Y = 379.8 + 28.7 X	14	Class IX vehicle	Y = 0.2 + $8.77(10^{-8})$ X

Based on the equations of the linear trend model of passenger and vehicle demand in Table 3 where the equations model is projected with X variable used, namely time. It's known that the projection of passenger and vehicle demand for the next 5 years shows an increasing trend with the prediction of the number of passengers (adults and children) in 2024 is 60,839 an increase of 14.6% from 2023. Prediction in 2028 is 104,591 Passengers meanwhile, the prediction data for the number of vehicles in 2024 is 9,430 mixed vehicles with the number of 4-wheel vehicle (R4) is 7,737. This condition increases by 16.3% from 2023 and prediction the number of vehicles in 2028 is 14,891 mixed vehicle which include 12,138 4-wheel vehicle (R4) vehicle and 2,753 R2 vehicle so that the daily average 4-wheel vehicle (R4) in 2024 is predicted to be 21 vehicles and in 2028 is 33 vehicles. Based on existing condition data and prediction results, calculations were made to calculate the needs of passenger terminal facilities, ready-to-load parking areas, dropout/pick-up parking area and weight-bridge placement plan in ideal conditions and the next 5-year plan.

Table 4. The Result of Calculation of Facility needs in Ideal Conditions and Next 5-Year Plan Conditions

No	Description	Existing Condition (m^2)	Ideal Condition (m^2)	Plan Condition (m^2)	Difference (m^2)
1	Passenger Terminal				
	a. Waiting Room Area (A1)	42	317	413	+371
	b. Canteen (A2)	30	48	62	+32
	c. Administration Room (A3)	26	48	62	+36
	d. Utility Room (A4)	73.25	103	134	+61
	e. Public Area (A5)	163.5	51	67	−96
2	Ready to Load Parking Area	200	836	1,285	+1,085
3	Dropoff/Pickup Parking Area	332.5	226	481	+149
4	Weight Bridge	Not yet available	Available	Available	

The calculation of facility requirements in ideal condition and plan condition is guided by Minister of Transportation Regulation (KM) 52 in 2004. In addition, it needs to add weight-bridge facilities as the provisions of Minister of Transportation Regulation (PM) 103 in 2017. Refer to Table 4 presents the different between existing condition, ideal condition, and plan next 5-year condition so that need add facilities for accommodate passengers and vehicles demands for increase passengers and vehicles services. The Weight-bridge placement plan is guided by Minister of Transportation Regulation 103 in 2017, every port is required to have a portal facilities and weight-bridge which placed before the tollgate. In addition, for the weight and supervision of good transportation in weight-bridge area need to be equipped with competent officers. By calculating the needs based on ideal conditions and 5-year projections, the port can plan and develop adequate facilities to accommodate the increasing number of passengers and vehicles. Evaluation of existing conditions, ideal conditions, and the 5-year plan helps identify deficiencies in current facilities and plan for improvements needed to improve passenger and vehicle services. Calculation of facility needs helps in providing adequate facilities such as passenger terminals, waiting rooms, parking areas, and weighbridges, all of which are essential for smooth operation.

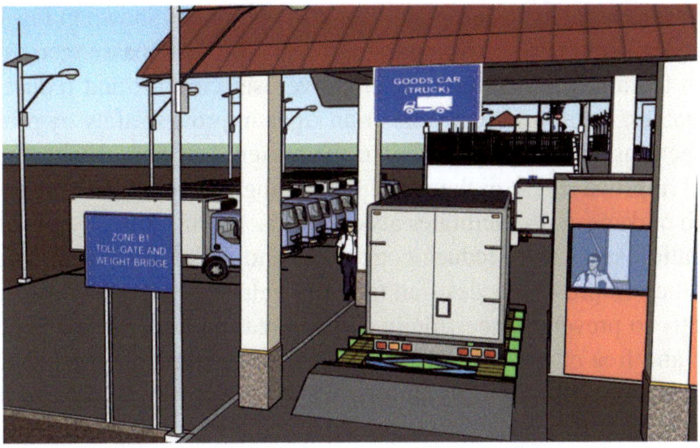

Fig. 3. Weight-bridge placement plan

To improve the quality of service to passengers and vehicles, efforts can be made to improve service performance as recommended based on the results of the assessment of the level of importance and performance in Fig. 3 to achieve aspects of safety, comfort, ease of access, reliability, and regularity of the implementation of service standards. Improving the quality of service to passengers and vehicles is very important and involves several performance improvement efforts that must be carried out based on the results of the assessment of the level of importance and performance. To achieve convenience or affordability for passenger movement in ideal facility conditions, improved pedestrian facilities by providing protective facilities in the form of roofs on existing pedestrian paths, installing pedestrian signs and arranging motorbike and drop-off or pick-up parking area. Improving pedestrian facilities including providing roofs, installing signs, and arranging parking areas, are important steps to create a more comfortable, safe, and efficient environment at the port that not only enhances the user experience but also supports smoother and more effective port operations.

Fig. 4. Pedestrian Facilities and drop-off or pick-up parking area plan in Ideal Condition

In addition, convenience or affordability passengers make transfer to minimize passenger movement time by arranging the passenger counters and waiting room are connected close to each other as well as providing service information facilities. Providing easy or accessible passenger movement in ideal facility conditions is essential to creating a good user experience and ensuring efficient port operations shown in Fig. 4 and Fig. 5. Providing shelter such as roofs on pedestrian paths protects passengers from extreme weather, such as rain or hot sunlight to improve user comfort and reduce disruptions caused by weather. Installing clear pedestrian signs improves safety by providing easy-to-follow directions and alerting drivers to the presence of pedestrians to help prevent accidents and improve safety in the port area. Arranging motorcycle parking and drop-off or pick-up parking areas facilitates access for users who come by private vehicle or use transportation services to reduce congestion and make movement around the port facility smoother. Improving pedestrian facilities helps organize the flow of user movement efficiently to prevent congestion in certain areas and ensure that pedestrian paths remain clean and free of obstacles. Good pedestrian facilities improve the overall user experience by making travel more comfortable, safe, and easy which can increase user satisfaction and loyalty. Ensuring that pedestrian facilities comply with applicable standards and regulations, including safety and accessibility guidelines, ensures that the port meets legal and regulatory requirements. The implementation of effective pedestrian facilities in the passenger terminal area improves safety, comfort, and efficiency, as well as providing a better user experience and supporting smoother port operations.

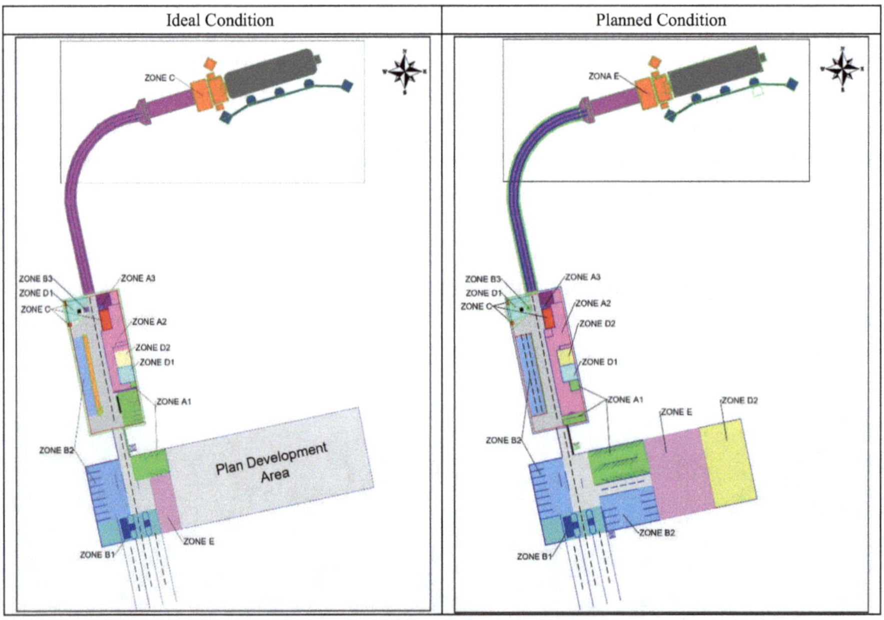

Fig. 5. Ideal and Planned Condition

The arrangement of the zoning system based on regulations includes zone A which is intended for People consisting of Zone A1 as a counter placement area and parking for delivery/pick-up vehicles, Zone A2 as a passenger waiting room area, Zone A3 as a passenger ticket inspection area. Zone B intended for vehicles consists of zone B1 as a tollgate and weight-bridge area, zone B2 as a ready-to-load parking lot, Zone B3 as a vehicle ticket inspection area. Zone C for vital port facilities. Zone C includes the dock, fuel, freshwater facilities, and other facilities. Zone D is designated as a restricted area consisting of zone D1 for offices and zone D2 for commercial areas. Zone E serves as a parking area for vehicles that have tickets but are not yet scheduled to enter the ferry port or as a parking lot. Based on the ideal conditions and the next five-year plan in Fig. 6, there are differences where there is an increase in the availability of facilities in the passenger zone (A1), vehicle zone (B2), commercial zone (D2), and parking lot zone (E). The arrangement of facilities at the port in the next 5 years will utilize the available development area by expanding the passenger terminal area, loading-ready parking lot, and drop-off or pick-up parking lot to accommodate the needs of passengers and vehicles. In the implementation of the zoning system, supporting facilities such as the installation of signs are needed to regulate the port, there are by enhancing safety, security, smoothness, comfort, and order, as recommended in the zoning system signage arrangement in Fig. 8 below.

The zoning system arrangement facilitates the management and organization of the port area according to the function and purpose of each zone which helps in ensuring that each area has a clear designation and reduces confusion. Zone A for the public (counters, waiting rooms, ticket checks), zone B for vehicles (toll gates, ready-to-load parking, vehicle ticket checks), zone C for vital port facilities (piers, fuel, clean water facilities), zone D for restricted areas (offices and commercial areas), and zone E for vehicle parking areas. Evaluation of ideal conditions and the 5-year plan shows the need to improve facilities in the passenger, vehicle, commercial, and parking zones aimed at meeting increasing demand and accommodating the growth of users and vehicles. Utilizing available development land to expand port facilities in accordance with the 5-year plan that ensures that the port can accommodate growing demand and continue to function efficiently. Good port facility arrangement requires the installation of signs to regulate the movement and use of areas that are important for improving safety, security, smoothness, comfort, and order at the port. Proper arrangement and installation of signs contribute to improving safety and security in the port to prevent accidents and ensure that the port area is well managed. Overall, proper zoning system arrangement and facility improvement in accordance with the 5-year plan are the keys to managing the port effectively, meeting user needs, and ensuring safe and efficient operations.

The installation of supporting facilities in the form of signs serves to improve safety, security, and smoothness, comfort, and order of the port. The signs installed include guide signs, command signs, and prohibition signs. The guide signs installed explain the instructions in words, traffic control guide signs, and public utility location guide signs shown in Fig. 7. Guide signs with words- The word describes the zone area, which consists of the passenger area A1 indicator sign placed at the passenger zone entrance, the zone A2 indicator sign placed in the waiting room area, the zone A3 indicator sign placed in the passenger ticket check area, the zone B1 indicator sign placed in the passenger ticket check area. Placed in the vehicle toll gate area and weighbridge, zone B3 signposts placed in the vehicle ticket inspection area, zone C signposts placed in the vital port facility area, zone D1 signposts placed in the office area, and zone D2 signposts placed in the office area. Placed in a commercial area. In addition to signs explaining the location of the zone area, signs are also installed indicating the location

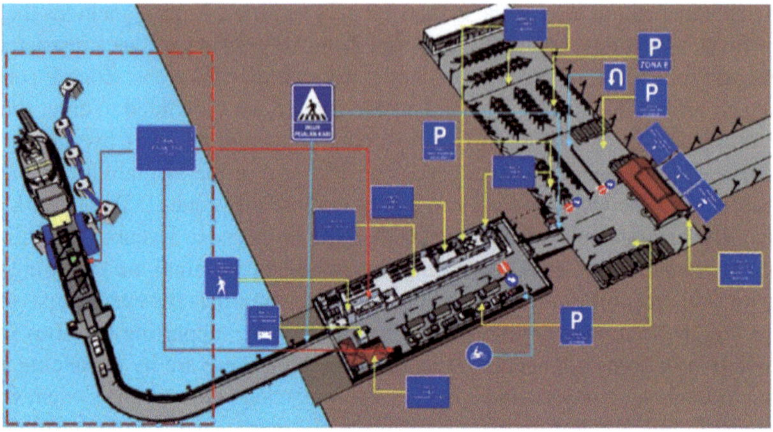

Fig. 6. Sign Arrangement Scheme

of public utilities, such as signs indicating the location of parking facilities installed in the drop-off/pick-up parking area, ready-to-unload parking area (zone B2), parking pocket area (zone E), and pedestrian path signs installed in gangway or corridor areas and on passenger-only roads. Traffic control signs installed include U-turn location signs installed in ready-to-unload parking areas. Command signs installed include entry signs to the certain roads installed at the entrance to the ready-to-unload parking lot and at the entrance to the drop-off/pick-up parking lot. The installed command signs also include signs directing the use of special motorcycle traffic lanes installed at the ready-to-unload parking lot for two-wheeled vehicles. In addition, signs in the form of words explaining the special lanes for goods vehicles, passenger cars and buses are also installed, as well as exit lane signs installed at the toll gates for vehicles according to the lanes for each vehicle. The prohibition signs installed include signs prohibiting entry by motorized vehicles and non-motorized vehicles which are placed in the parking area ready for loading and unloading and the parking area for dropping off or picking up.

To strengthen the depth of this study, it can be compared with Merak Ferry Port which also serves ferry transportation routes with high passenger and vehicle volumes. This port has some similarities in terms of strategic roles, but there are significant differences in performance and facilities. Merak has a much larger capacity in terms of the number of passengers and vehicles served per day. This port has undergone major improvements in terms of parking facilities, passenger waiting rooms and traffic management systems within the port. Merak has implemented stricter zoning to manage passenger and vehicle flows, including the provision of adequate parking areas, organized drop-off areas and well-protected pedestrian facilities. Merak Port has an automatic weighbridge integrated with the toll system, thereby speeding up the inspection process and reducing waiting times for heavy vehicles. Practical and immediate recommendations that can be taken to address the problems at the proposed Kuala Tungkal Ferry Port based on potential impacts and feasibility, for example, building a special drop-off and pick-up parking area near the port entrance to facilitate passengers and vehicles. In addition, it is necessary to install a protective roof along the existing pedestrian path to provide protection from the weather. In addition, installing clearer signs and directions to regulate the movement of vehicles and passengers, as well as placing information boards at strategic points. It is also important to expand the passenger waiting area by adding seats and providing more comfortable areas such as air-conditioned waiting rooms or outdoor areas with weather protection. Based on the existing trend and analysis using the least squares method, the projection of passenger and vehicle growth at Kuala Tungkal Ferry Port in the next five years shows a significant increase. To accommodate the increase in the number of passengers and vehicles, several urgent facility improvements include the development of waiting rooms and passenger service facilities must be expanded, the current parking area needs to be enlarged to accommodate the increase in incoming vehicles, the addition of vehicle monitoring and weighing facilities must be implemented more quickly to maintain operational safety and efficiency.

4 Conclusion

This study states that the current performance of Kuala Tungkal Ferry Port is in the moderate category with a value of 58.86. Interaction between facilities is classified as sufficient, with a value of -133.3. The entry and exit times of motorized vehicles and pedestrians indicate a need for increased efficiency. Improvement efforts include the provision of passenger and motorized vehicle facilities such as passenger terminals, parking areas, toll gates, and weighbridges. Based on the quadrant analysis, facilities such as waiting rooms, canteens, administration areas, utility rooms, and parking areas need to be improved. The plan for the provision of facilities for the next five years, including a 413 m^2 waiting room and a 1,285 m^2 parking area, is expected to meet the increasing number of passengers and vehicles. The arrangement of the zone system and the improvement of pedestrian facilities, such as protective roofs and service information, are also important to support user accessibility and comfort. The implementation of service standards will improve security, smoothness, comfort, and order at the port.

References

1. Song, Z.Y., Lin, C.W., Feng, X., Lee, P.T.W.: An empirical study of the performance of the sixth generation ports model with smart ports with reference to major container ports in mainland China. Transp. Res. Part E Logist. Transp. Rev. **184**, 103460 (2024). https://doi.org/10.1016/j.tre.2024.103460
2. Fahim, P.B.M., Rezaei, J., Montreuil, B., Tavasszy, L.: Port performance evaluation and selection in the physical internet. Transp. Policy **124**, 83–94 (2022). https://doi.org/10.1016/j.tranpol.2021.07.013
3. Nong, T.N.M.: Performance efficiency assessment of vietnamese ports: an application of delphi with kamet principles and DEA model. Asian J. Shipp. Logist. **39**, 1–12 (2023). https://doi.org/10.1016/j.ajsl.2022.10.002
4. He, Y., Xu, X.: Spatial correlation network structure of port performance and its drivers: a case study of Chinese coastal ports. Ocean Coast. Manag. **244**, 106780 (2023). https://doi.org/10.1016/j.ocecoaman.2023.106780
5. Belcore, O.M., Polimeni, A., Di Gangi, M.: Case studies on transport policy performance analysis for a maritime port with high-frequency services : an Italian case study. Case Stud. Transp. Policy **17**(101263), 1–11 (2024). https://doi.org/10.1016/j.cstp.2024.101263

6. Duru, O., Galvao, C.B., Mileski, J., Robles, L.T., Gharehgozli, A.: Developing a comprehensive approach to port performance assessment. Asian J. Shipp. Logist. **36**, 169–180 (2020). https://doi.org/10.1016/j.ajsl.2020.03.001
7. Hosch, G., Miller, N.A., Yvergniaux, Y., Young, E., Huntington, T.: IUU safe havens or PSMA ports: a global assessment of port State performance and risk. Mar. Policy **155**(105751), 1–20 (2023). https://doi.org/10.1016/j.marpol.2023.105751
8. Tampubolon, F., Seno, W., Sarjana, S.: Analysis of lashing gear in ship loading spaces to meet safety transportation. E3S Web Conf. **519**, 1–13 (2024). https://doi.org/10.1051/e3sconf/202451903024
9. Ha, M.H., Yang, Z., Seo, Y.J.: Port performance measurement from a multistakeholder perspective. Int. Encycl. Transp. 396–405 (2021). https://doi.org/10.1016/B978-0-08-102671-7.10506-8
10. Talley, W.K., Ng, M.: Port congestion probability: port performance evaluation using cargo port choice equilibrium. Transp. Res. Part E Logist. Transp. Rev. **185**, 103488 (2024). https://doi.org/10.1016/j.tre.2024.103488
11. Nanyam, V.P.S.N., Kumar Jha, N.: Modeling challenges affecting the performance of major ports of India. Asian J. Shipp. Logist. **39**, 26–38 (2023). https://doi.org/10.1016/j.ajsl.2023.06.002
12. Hidayat, A., Hijriah, H., Setiowati, N.O.: Kelayakan jaringan transportasi sungai kabupaten mimika feasibility of mimika regency river transportation network. J. Urban Reg. Spat. **3**(2), 217–224 (2023). https://doi.org/10.35965/jups.v3i2.338
13. Lengkono, D.Y., Aksà, K., Arief, R., Rofaidah, A.: Transportation infrastructure in supporting the development of Oyama beach tourism objects in Lokotoy village, North Banggai district, Banggai Laut regency. J. Urban Plan. Stud. **3**(2), 129–136 (2023). https://doi.org/10.35965/jups.v3i2.348
14. Yasin, F., Wahid Kamma, A., Baharuddin, I.: Implementasi standar pelayanan penumpang transportasi laut di pelabuhan speed boad loleo rute loleo-tidore pada dinas perhubungan kota tidore kepulauan. J. Ilm. Wahana Pendidik. **9**(7), 260–267 (2023). https://doi.org/10.5281/zenodo.7813539
15. Zurkiyah, S.A.: Analisis tingkat pelayanan dermaga pelabuhan penumpang teluk nibung asahan, tanjung balai sumatera utara. Semin. Nas. Tek. UISU 248–252 (2021)
16. Amruddin, et al.: Metodologi penelitian kuantitatif dan kualitatif. penerbit: Media Sains Indonesi (2022)
17. Sekaran, U.: Research methods for business, 4th Edition. Jakarta: Salemba Empat (2006)
18. Kusumastuti, H., Pranita, D., Viendyasari, M., Rasul, M.S., Sarjana, S.: Leveraging local value in post-smart tourism village to encourage sustainable tourism. Sustainability **16**(873), 1–26 (2023). https://doi.org/10.3390/su16020873
19. Fachri, H., Sarjana, S.: Performance evaluation through the effectiveness of resources and reputation: a case study of hospitals in Indonesia. Probl. Perspect. Manag. **20**(3), 72–82 (2022). https://doi.org/10.21511/ppm.20(3).2022.06

Enhancing Traffic Safety for Motorcyclists and Pedestrians

Rifat Arief Setyawan[1], Rianto Rili Prihatmantyo[2], and Sri Sarjana[3](✉)

[1] Student in Land Transportation Department, Politeknik Transportasi Darat Indonesia - STTD, Bekasi, Indonesia
[2] Railway Transportation Management Department, Politeknik Transportasi Darat Indonesia - STTD, Bekasi, Indonesia
[3] Marketing, Innovation and Technology Department, Politeknik Transportasi Darat Indonesia - STTD, Bekasi, Indonesia
srisarjana@gmail.com

Abstract. It is hoped that optimal support for traffic safety will create a safer and more comfortable traffic environment for all road users, especially motorcyclists and pedestrians. This study aims to improve traffic safety for drivers and pedestrians on Cipatat Road segment 7 and segment 8, West Bandung Regency. Quantitative methods are applied in this study with analysis techniques using road safety inspection to determine road infrastructure and provide recommendations for road infrastructure proposals, the gross output method to determine accident costs, and roadside hazards to determine the risks arising from roadside hazards. The results of the study indicate that segments 7 and 8 from the results of spot speed analysis and 85th percentile it is known that the highest speed in the direction of entry of segment 7 is 58 km/h which is categorized as exceeding the set speed limit, while for the exit direction with a speed of 62.8 km/h is also categorized as exceeding the set speed limit. Then in segment 8 the highest speed in the direction of entry of 58 km/h has exceeded the set speed limit, the speed for the exit direction of 66.1 km/h has exceeded the set speed limit. The high number of accidents on Cipatat Road segment 7 and segment 8 is mostly caused by human factors and poor road infrastructure. Therefore, it is necessary to install speed limit signs, install caution signs, install guide signs, crossing facilities, add zebra crossings and rumble strips to protect the safety and comfort of road users.

Keywords: Traffic Safety · Motorcyclist · Pedestrians · Road Safety Inspections · Road Infrastructure

1 Introduction

Road safety must be an integral part of every comprehensive and integrated traffic system improvement process [1]. Components traffic safety essentials include engineering, law enforcement, and education, where engineering focuses on the design and construction of safer transportation systems that exist on roads and vehicles [2]. Traffic law enforcement is understood to be able to reduce the rate of traffic accidents globally, although

M. Awang (Ed.): ICCAEET 2024, PTT, pp. 163–178, 2025.
https://doi.org/10.1007/978-3-031-94347-8_14

there are differences in treatment in each region regarding its effectiveness [3]. Road users' knowledge of understanding traffic safety regulations makes an important contribution to reducing the level of traffic accidents [4]. Complex traffic situations related to density and speed require in-depth knowledge in the field of traffic safety management to develop strategies to prevent accidents [5]. In addition, it is necessary understand trip characteristics to consider zone-based traffic safety estimates to compare spatial distributions referring to mobility and transportation conditions, and need to consider route choices by considering accessibility analysis aimed at avoiding overestimation [6]. Speed management is a safe way to drive traffic on the road with the main target of eliminating fatalities and serious injuries due to road crashes by reducing risk exposure, reducing risk levels, and increasing protection from harmful impact forces, so that reducing traffic speeds can have the opportunity to avoid crashes and reduce the impact level of crashes so that injuries are less severe [7]. One effort to improve traffic safety is adding visual guidance facilities to actively guide driving safely [8]. Regulatory implementation is needed to lower the threshold vehicle speed, especially on roads that do not have traffic separation in order to increase traffic safety and improve the quality of accessibility [9]. For this reason, it is necessary to carry out road traffic safety simulations, which currently receive little attention from various stakeholders, including the authorities in charge of transportation safety [10].

Road safety should be an integral part of any comprehensive and integrated traffic system improvement effort through an approach that combines engineering, law enforcement, and education as the key to creating a safer traffic environment. Traffic engineering, which involves the design and construction of safer transportation systems, is essential to reducing the risk of crashes. Good road infrastructure and vehicles designed with safety in mind are the foundation of a safe traffic system. Traffic law enforcement plays a vital role in reducing crash rates, although its effectiveness can vary from region to region, consistent enforcement can increase compliance with traffic regulations and ultimately reduce the number of crashes. Road users' knowledge of traffic safety regulations makes a significant contribution to reducing crash rates. The implementation of safety regulations has improved overall road safety, which has an important message to encourage increased public awareness and the quality of driver education in strengthening a safety culture, including through efforts to understand the risks of dangerous driving behaviour [11]. Many factors cause traffic accidents, including weather, vehicles, road conditions and driver habits. To reduce traffic accidents, research needs to be conducted in areas with high accident rates. According to government regulation no 37 of 2017, traffic safety is a state of avoiding everyone from the risk of accidents during traffic caused by humans, vehicles, roads or the environment. The high rate of traffic accidents in Indonesia makes traffic accidents ranked 6th as the main cause of premature or early death in Indonesia according to the Institute of Health Metrics and Evaluation. The number of traffic accidents is expected to continue to increase along with the increasing growth of motor vehicle users in Indonesia. The growth of motorcycle users in Indonesia has the highest number compared to other vehicles. Every year the growth rate of motorcycles in Indonesia reaches 19% to 37% according to the Indonesian Motorcycle Industry Association. The Bandung Road Safety Annual Report explains that around 68 incidents of traffic accidents in Indonesia involved motorcycles, this indicates that motorcycles are the largest contributor to the number of accidents in Indonesia. Third highest mortality rate in Indonesia is due to the large number of traffic accidents that occur, where Indonesia's position as a country with the highest ratio of deaths due to traffic accidents.

Safety plays an important role in all aspects [12], including road safety which provides comfort for traffic users [13]. The importance of road safety is a shared concern in order to reduce the risk of traffic accidents [14]. Attitudes and behaviour that consider traffic safety have an influence on perceptions of traffic safety risks [15]. There are several factors that must be considered in traffic accidents including road infrastructure factors and road facility factors where the vehicles used are in accordance with roadworthiness, human factors due to negligence, or natural factors that can occur suddenly at any time which can cause traffic accidents. The Cipatat Highway Section based on Road Status is a National Road and based on Road Function it is an arterial road. Based on the 2023 West Bandung Regency PKL Team General Report, it is known that the Cipatat Highway Section is ranked 2nd in the Accidents Prone Area (DRK) ranking in West Bandung Regency. Based on data obtained from Cimahi Police for Cipatat Highway Section in the last 5 years, there have been 143 accidents with 190 victims, 23 deaths, 1 serious injury, 166 minor injuries. The condition of Cipatat Highway Section related to road equipment facilities is available but the condition is not good and there is still a lack of road equipment facilities such as speed limit signs, warning signs, guide signs, caution signs, public street lighting and there are several road conditions where the pavement and road markings are damaged, which can cause traffic accidents. Lack of awareness for road users to be disciplined such as driving at high speeds causes accidents [16, 17].

Data source from the West Bandung Police Resort which explains that the number of accidents that occurred on the Cipatat Road Section at blackspot points segment 7 (Km 3 - Km 3.5) and segment 8 (Km 3.5 - Km 4) West Bandung Regency over the past five years has decreased, although in some years there has been an increase in incidents and caused the number of fatalities to increase where in 2023 there were 31 accidents with a total number of accident victims of 39 fatalities including 7 people died, 1 person was seriously injured, 31 were slightly injured. The condition of traffic flow and speed on the Cipatat Road section segments 7 and 8 is quite high because of its function as an arterial road and national road. Although there has been a decrease in the number of accidents over the past five years, there are still fluctuations indicating that accidents still occur. The expected goal in this study is to find various efforts to improve traffic safety for drivers and pedestrians on Cipatat Road segment 7 and segment 8, West Bandung Regency.

2 Method

Quantitative approach is applied in this study. Quantitative research is a type of research that is systematic, planned, and structured [18], research that is based on assumptions, then variables are determined and then analyzed using valid research methods [19], process starts from theory, hypothesis, research design, selecting subjects, collecting data, processing data, and analyzing data [20]. Identification of problems that have been known from observations in the field with the limitations of the problems determined so that the problems raised do not deviate from the discussion. The sequence of steps applied in this study begins with conducting field observations which are one of the preliminary studies carried out at the study location, namely the Cipatat road section, to get a general picture of field conditions. Problem identification is carried out to find out what problems are related to safety on the Cipatat road section. Collection of secondary data from related agencies and primary data, namely data obtained from direct surveys at the research location.

Cipatat road section is a national road based on road status and is an arterial road based on road function, with a 2/2 undivided road type, has 2 directions, a total road width of 7 m, a width per lane of 3.5 m, and the Cipatat highway section has a length of 4.4 km shown in Fig. 1. The condition of Cipatat road can be said to be good with asphalt road pavement. Cipatat road is a road section that is often used by the people of West Bandung Regency or other areas around it, because Cipatat road is a connecting road section between West Bandung Regency and Cianjur Regency and there is mix traffic between goods vehicles and private vehicles. Land use around Cipatat road is in the form of shops, green areas.

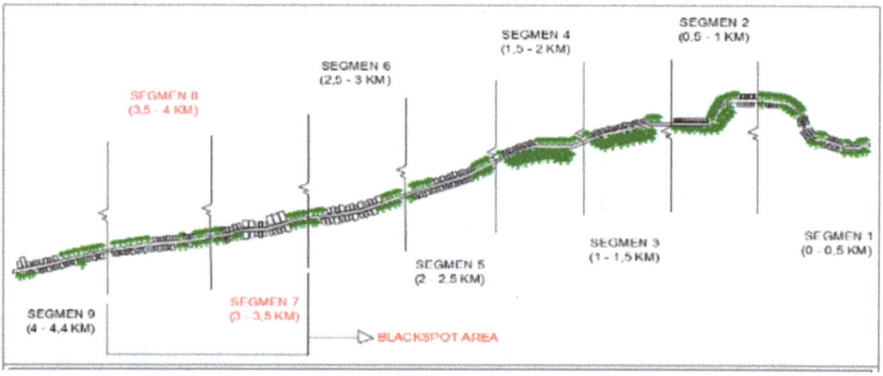

Fig. 1. Cipatat road section

Data sources were obtained by conducting a survey to obtain primary data including road inventory data, instantaneous speed data, driver behavior data on road sections, pedestrian data. Meanwhile, secondary data needed to conduct this study came from the Cipatat Police, West Bandung Regency and from the general transportation pattern report in West Bandung Regency 2023 in the form of road network map data and traffic accident data from the Cipatat Police Traffic Unit, West Bandung Regency and the chronology of accidents on the cipatat highway. Accident characteristics analysis was obtained from accident data from the Cipatat Police, West Bandung Regency from 2018–2022. The severity of accident victims is divided into three categories, death, serious injury and minor injury. Accident data analysis using the upper control limit (UCL) method. Driver behavior analysis begins by conducting a driver behavior survey by observing and recording how road users behave when crossing.

3 Result and Discussion

Characteristic analysis is expected to determine the accidents that occurred on the Cipatat road segment 7 and 8 based on the number of accidents, type, time, number of victims and collision diagrams to describe the visualization of the accidents that occurred. Traffic accident data for 5 years was taken on the Cipatat road segment, then separated into segments 7 and 8 which were discussed in this study to determine the number of accidents that occurred. Determination of the number of accidents is based on the chronology of accidents on the Cipatat road segment where there were 143 accidents on the road segment. Based on the number of accidents, the highest accident rate occurred on segments 7 and 8, so segments 7 and 8 were the subjects of this study. Data obtained on the Cipatat road shows that the highest number of accidents occurred in 2018, 12 accidents and the lowest occurred in 2021, amount 3 accidents.

Analysis of traffic accident characteristics can help identify accident patterns and trends on the road sections studied. By knowing when and how accidents occur, more effective preventive measures can be designed. Collision diagrams that illustrate accident visualization can facilitate understanding of accident distribution, both geographically and temporally. The visualization presented facilitates identification of areas or times that are most prone to accidents, which can be used as a basis for decision making. Determining road sections based on the highest accident rates, on segments 7 and 8, allows the study to focus on areas that need the most intervention, allowing for more effective resource allocation in an effort to reduce accidents. Accident data for 5 years provides a strong basis for evaluating changes in accident rates over time.

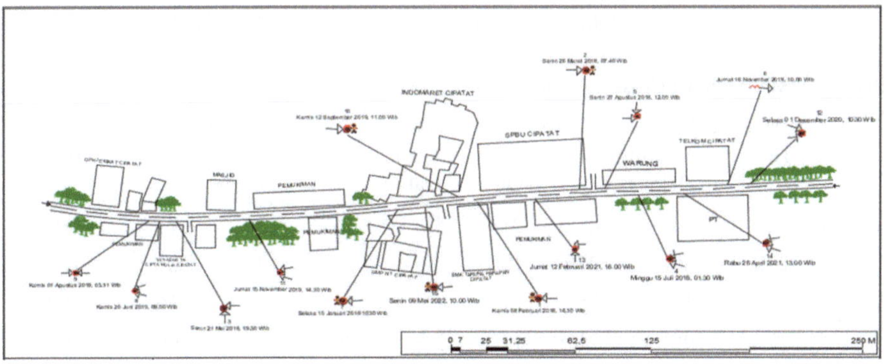

Fig. 2. Collision Diagram for Cipatat Road Segment 7

The collision diagram of segment 7 and segment 8 shown in Fig. 2 and Fig. 3 explains the accidents that occurred in both segments, where in segment 7 there were 15 accidents that occurred including 5 side-side collisions, 3 front-side collisions, 1 front-front collision, 5 human collisions, 1 single collision. While in segment 8 there were 16 accidents that occurred including 4 side-side collisions, 5 front-side collisions, 3 front-front collisions, 3 human collisions, 1 pile-up. Collision diagrams provide in-depth details of accidents that occur at a location, including the type of collision and

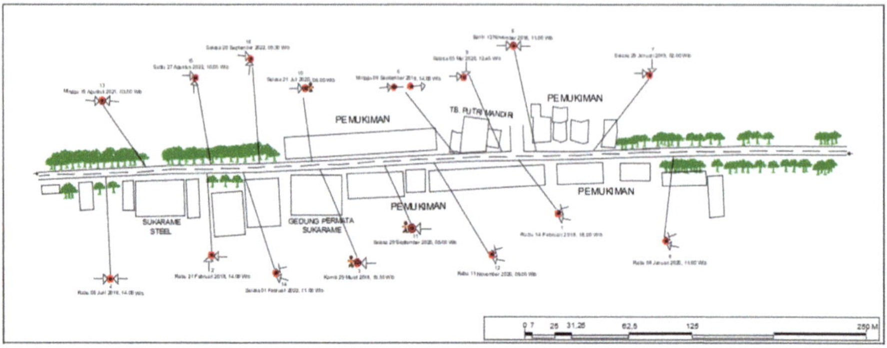

Fig. 3. Collision Diagram for Cipatat Road Segment 8

the causal factors, making it important to identify recurring patterns and understand the root causes of accidents so that more appropriate preventive measures can be taken. Collision diagrams include important information such as the type of vehicles involved, the maneuvers of the vehicles before the collision, road conditions, and the influence of alcohol. These details allow for a comprehensive analysis of every aspect of the accident which helps in understanding the overall situation. The visual depiction through collision diagrams allows for an easier and more intuitive understanding of how accidents occur, helping experts, policy makers, and other stakeholders to quickly identify problems and plan appropriate solutions.

Table 1. Motorcyclist Behavior Segment 7 and Segment 8

Black Spot	Motorcyclist Crossing Black Spot								
	Wearing Helmet	Not Wearing Helmet	Lights On	Lights Not On	Going Against Traffic	Zig-zag	Playing Mobile Phone	Chatting	Total Incidents
Segment 7	42	14	48	21	5	15	21	14	180
	23%	8%	27%	12%	3%	8%	12%	8%	100%
Segment 8	45	16	35	21	4	13	19	16	169
	27%	9%	21%	12%	2%	8%	11%	9%	100%

Driver behavior analysis was conducted to determine the behavior of drivers crossing the Cipatat road segment 7 and 8 which affect the possibility of traffic accidents. In the Table 1, it can be seen that the percentage of motorcycle driver behavior is highest in segment 7, the percentage of drivers turning on the lights 27%, followed by wearing helmets 23% then the lights are not on and playing with cellphones 12%. Based on this, it can be known that there are drivers who obey traffic laws and those who do not, where the behavior of drivers playing with cellphones, chatting provides the possibility of accidents due to driver negligence in driving. While the behavior of motorcycle drivers is highest in segment 8, namely drivers wearing helmets 27%, followed by lights on 21% then lights not on 12% and playing with cellphones 11%. Based on this data, it can be seen that there are drivers who obey traffic laws which is marked by the behavior of drivers playing with cellphones, chatting provides the possibility of accidents due to driver negligence.

This study allows the identification of risky driver behavior, such as playing with a mobile phone or chatting while driving. These behaviors can significantly increase the likelihood of a traffic accident due to driver negligence. By analyzing driver behavior, it is possible to reveal the level of compliance with traffic regulations, such as wearing a helmet or turning on vehicle lights. Knowing how many drivers comply with or violate the rules can help in formulating more effective safety policies. Driver behavior data provides a basis for determining which areas or behaviors require more attention in accident prevention efforts, for example if the level of helmet use is low, interventions such as safety campaigns or law enforcement can be focused on that aspect.

Table 2. Car Driver Behavior Segment 7 and Segment 8

Black Spot	Car Driving Behaviour Crossing							
	Wearing Seat Belt		Not Wearing Seat Belt		Zig-zag	Playing Mobile Phone	Chatting	Total Incidents
	Driver	Passenger	Driver	Passenger				
Segment 7	32	25	36	42	9	12	11	167
	19%	15%	22%	25%	5%	7%	7%	100%
Segment 8	36	27	32	41	7	13	14	170
	21%	16%	19%	24%	4%	8%	8%	100%

In the Table 2, it can be seen that the highest percentage of car driver behavior in segment 7 is that passengers do not wear seat belts 25%, followed by drivers 22%. This indicates that there is car passenger behavior that obeys traffic rules and some do not obey traffic rules, where the behavior of passengers who do not wear seat belts provides the opportunity for fatalities with a fairly high level of severity if an accident occurs. While the highest car driver behavior in segment 8 is that passengers do not wear seat belts 24% followed by drivers wearing seat belts 21%. This shows that there is car passenger behavior that obeys traffic rules and some do not obey traffic rules, where the behavior of passengers who do not wear seat belts provides the opportunity for fatalities with a fairly high level of severity.

Comparing segment 7 and segment 8, this study shows that playing with mobile phones is the most common behavior in both segments, with a higher percentage in segment 8 (63%) compared to segment 7 (57%). This suggests that segment 8 has a higher risk of pedestrian accidents due to negligence. Segment 8 shows a higher percentage of risky behaviors such as playing with mobile phones, this could be a priority area for safety interventions such as increased supervision or more intensive education campaigns in that segment. Therefore, it is necessary to evaluate whether there is a difference in the effectiveness of safety policies or infrastructure between the two segments. If segment 8 has a higher level of risky behaviors, there may be a need for policy adjustments or specific interventions to reduce the risk in that segment.

Table 3. Data Recapitulation of Incoming Direction - Outgoing Direction in Segment 7

Vehicle Type	Minimal Speed (Km/h)		Maximum Speed (Km/h)		Average Speed (Km/h)		85 Percentile (P85)	
	Incoming direction	Outgoing direction	Incoming direction	Outgoing direction	Incoming direction	Outgoing direction	Incoming direction	Outgoing direction
Motorcycle	30	31	68	80	42.9	50	49	62.8
Car	27	24	64	63	42.8	41	50.4	49
Public passanger car	20	28	60	54	36.8	37	45	42.1
Pick up	29	22	76	54	42.2	38	53.4	49.7
Small Truck	24	20	60	60	42.1	39	49.3	52.85
Medium Truck	27	25	65	60	41.2	38	47	47
Large Truck	31	21	56	56	40.9	39	47	50.9
Bus	39	29	61	62	44.1	45	58	60.8

To determine the maximum speed limit based on technical data and traffic data, the 85% percentile (P85) analysis is used, where 85% of vehicles travel at or below that speed on the road segment as seen in the table of 85% percentile speed limits for vehicles obtained from the instantaneous speed survey on the Cipatat road segment 7. The existing speed from the results of the vehicle survey with the number of samples taken based on the average daily traffic of 138 motorcycle vehicle samples, cars, public passenger cars (MPU), pick-ups, small trucks as many as 160 vehicle samples, 8 medium trucks as many as 4 large trucks, and 2 buses as many as 2 vehicle samples. Recapitulation of incoming motorcycle data at P85 speed reaches 49 km/h, cars 50.4 km/h, MPU 45 km/h, Pick Up 53.4 km/h, small trucks 49.3 km/h, medium trucks 47 km/h, large trucks 47 km/h, Bus 58 km/h. Based on the table above, it can be seen that the speed of vehicles in this segment has exceeded the planned speed limit of 40 km/h, therefore it is necessary to install speed limit signs in this segment. The existing speed from the results of a vehicle survey with the number of samples taken based on the average daily traffic of motorcycles as many as 138 vehicle samples, cars, MPUs, pick-ups, small trucks as many as 160 vehicle samples, medium trucks as many as 8 vehicle samples, large trucks as many as 4 vehicles, and buses as many as 2 vehicle samples. Recapitulation of motorcycle data heading out at P85 speed reaches 62.8 km/h, cars 49 km/h, MPUs 42.1 km/h, Pick ups 49.7 km/h, small trucks 52.9 km/h, medium trucks 47 km/h, large trucks 50.9 km/h, Buses 60.8 km/h. Based on the Table 3, it can be seen that the speed of vehicles in this segment has exceeded the planned speed limit of 40 km/h, therefore it is necessary to install speed limit signs in this segment.

To determine the maximum speed limit based on technical data and traffic data, an 85% percentile analysis is used, where 85% of vehicles travel at or below that speed

Table 4. Data Recapitulation of Incoming Direction - Outgoing Direction in Segment 8

Vehicle Type	Minimal Speed (Km/h)		Maximum Speed (Km/h)		Average Speed (Km/h)		85 Percentile (P85)	
	Incoming direction	Outgoing direction	Incoming direction	Outgoing direction	Incoming direction	Outgoing direction	Incoming direction	Outgoing direction
Motorcycle	30	32	68	82	43.3	49.5	50.8	66.1
Car	27	26	64	64	42.5	41.1	50.4	48.4
Public passanger car	20	27	61	49	36.7	36.9	45	42.6
Pick up	29	21	62	54	41.7	37.5	53.4	49.7
Small Truck	24	20	60	60	42.1	38.8	49.3	52.9
Medium Truck	27	24	65	63	41.2	38.1	47	47.5
Large Truck	31	23	56	57	40.5	38.6	47	49.2
Bus	39	25	61	61	44.1	43.9	58	59.6

on the road segment. This can be seen in the table of the 85% percentile speed limit of vehicles obtained from the instantaneous speed survey on Jalan Cipatat segment 8. The existing speed from the results of the vehicle survey with the number of samples taken based on the average daily traffic of 138 motorcycle vehicle samples, cars, MPUs, pick-up, small trucks as many as 160 vehicle samples, 8 medium trucks as many as 4 large trucks, and 2 buses as many as 2 vehicle samples. Recapitulation of motorcycle data entering the P85 speed reaches 50.8 km/h, cars 50.4 km/h, MPUs 45 km/h, pick up 53.4 km/h, small trucks 49.3 km/h, medium trucks 47 km/h, large trucks 47 km/h, buses 58 km/h. Based on the table above, it can be seen that the speed of vehicles in this segment has exceeded the planned speed limit of 40 km/h, therefore it is necessary to install speed limit signs in this segment. The existing speed from the results of the vehicle survey with the number of samples taken based on the average daily traffic of motorcycles as many as 138 vehicle samples, cars, MPUs, pick-ups, small trucks as many as 160 vehicle samples, medium trucks as many as 8 vehicle samples, large trucks as many as 4 vehicles, and buses as many as 2 vehicle samples. Recapitulation of motorcycle data heading out at P85 speed reaches 66.1 km/h, cars 48.4 km/h, MPU 42.6 km/h, pick up 49.7 km/h, small trucks 52.9 km/h, medium trucks 47.5 km/h, large trucks 49.2 km/h, buses 59.6 km/h. Based on the Table 4, it can be seen that the speed of vehicles in this segment has exceeded the planned speed limit of 40 km/h, therefore it is necessary to install speed limit signs in this segment.

Several causes of accidents on Cipatat road segments 7 and 8 after analysis there are two factors causing accidents, road infrastructure and humans. Efforts that need to be made to improve road safety include improving road infrastructure. Analysis of infrastructure conditions found that accidents were caused by poor infrastructure conditions such as poor sign conditions, no signs, some street lights were out, no zebra crossings for crossing facilities.

In the Table 5, there are several proposals for installing warning signs, vehicle speed limit prohibitions or road markings and rumble strips at certain points to warn vehicles

Table 5. Image of Proposed Segment 7

No	Existing Image	Proposed Image	Description
1	 -6.828282600909434, 107.37650870947289	 -6.828282600909434, 107.37650870947289	Installation of no parking signs
2	 -6.828268892859811, 107.37649583648671	 -6.828268892859811, 107.37649583648671	Installation of convex road glass
3	 -6.828446595528106, 107.37640500056798	 -6.828446595528106, 107.37640500056798	Installation of 40 Km/h Vehicle Speed Limit Signs. To limit vehicle speeds that exceed the safe limit.
4	 -6.828513248735814, 107.37632067872302	 -6.828513248735814, 107.37632067872302	Installation of school location signs

of geometric changes in the road to be more careful or to limit vehicles from exceeding the specified speed limit and pedestrian facilities to make it easier for pedestrians to cross. Some of the treatments in this segment include the installation of no parking signs, installation of convex road glass, installation of 40 km/h vehicle speed limit signs. To limit vehicle speeds that exceed safe limits. Installation of school location signs, installation of pedestrian facility signs and zebra crossings and rumble strips to reduce accidents in this area, where there are accidents that hit people, installation of caution signs before the right turn warning sign, to warn drivers about the bend to be more careful. Installation of mosque location signs to indicate the presence of a mosque in this area and installation of church location signs to indicate the presence of a church in this area.

The installation of warning and prohibition signs and rumble strips can help drivers become more aware of changes in road geometry such as sharp bends, dips or inclines and areas with lower speed limits to help prevent accidents by providing drivers with timely information. Clearly marked and posted speed limits on the road, supported by signs and rumble strips, can help control vehicle speeds, especially in high-risk areas such as near schools, pedestrian facilities or areas with sharp bends to reduce the risk of fatal accidents. The installation of pedestrian facilities such as zebra crossings and signs indicating the presence of crosswalks or pedestrian facilities is essential to protect pedestrians to help ensure that pedestrians have a safe place to cross, which reduces the likelihood of accidents involving pedestrians. The installation of no parking signs in accident-prone or congested areas helps prevent vehicles from parking in places that could disrupt traffic flow or block the view of other drivers to improve traffic flow and reduce the risk of accidents. Clear and visible signs in strategic locations facilitate enforcement of speed limits, parking restrictions and other traffic regulations in increasing driver compliance with traffic regulations, ultimately contributing to road safety. The proposed installation of signs, road markings and other safety facilities is essential in creating a safer road environment, reducing the risk of accidents and protecting all road users, including drivers and pedestrians.

In the Table 6, it can be seen that there are several proposals for installing speed limit signs or road markings and warning tapes and road barriers at certain points to limit vehicles from exceeding the specified speed limit and pedestrian facilities to make it easier for pedestrians to cross and road barriers installed so that vehicles do not fall to the left side of the road which looks quite deep. Some of the handling in this segment, for example, the installation of speed limit signs for vehicles of 40 km/h. To limit the speed of vehicles that exceed the safe limit, installation of convex road glass, Installation of pedestrian facility signs and zebra crossings and warning tapes to reduce accidents in this area, where there are accidents that hit humans, Installation of Speed Limit Signs for vehicles of 40 km/h. To limit the speed of vehicles that exceed the safe limit, and road barriers on the left side of the road. Based on the analysis of the causes of accidents due to human factors, for example, driver behavior and pedestrian behavior. The chronology of the accident was caused by the driver not mastering his vehicle where the steps that can be taken to overcome this problem are to tighten the issuance of driving licenses so that road users are truly proficient in driving their vehicles.

Roadside hazards are features or objects beside the road that may affect safety in the roadside area as fixed objects measuring 100 mm or more. Roadside hazards include other features such as rocks or steep slopes that can contribute to the severity of a collision, causing serious injury to vehicles that leave the road, so roadside hazard management is needed. Things that can be done to improve roadside hazard safety include guidelines for handling accident-prone locations that can be done, including guardrails and safety fences. 2. Accident blackspot investigation unit guidelines that can be done include

Table 6. Image of Proposed Segment 8

No	Existing Image	Proposed Image	Description
1	 -6.830053765596521, 107.37207715557948	 -6.830053765596521, 107.37207715557948	Installation of speed limit signs for vehicles 40 km/hour. To limit the speed of vehicles that exceed the safe limit
2	 -6.830023320459103, 107.37177295360563	 -6.830023320459103, 107.37177295360563	Installation of convex road glass
3	 -6.83037088127728, 107.37061338611005	 -6.83037088127728, 107.37061338611005	Installation of pedestrian facility signs and zebra crossings and warning tapes to reduce accidents in this area, where there are accidents that hit humans.
4	 -6.830444893518353, 107.3702859746413	 -6.830444893518353, 107.3702859746413	Installation of speed limit signs for vehicles 40 km/hour. To limit the speed of vehicles that exceed the safe limit. As well as a road barrier on the left side of the road.

painting and providing reflectors on certain objects, making road markings to guide around disturbances, re-placing no-road barriers for pedestrians (in rural areas), repairing or installing lighting, safety fences, crash pads, installing speed limit signs and installing cat's eye signs on asphalt. The five steps of roadside hazard management strategies can be done by keeping vehicles on the road, eliminating hazards, relocating hazards, modifying hazards and closing hazards.

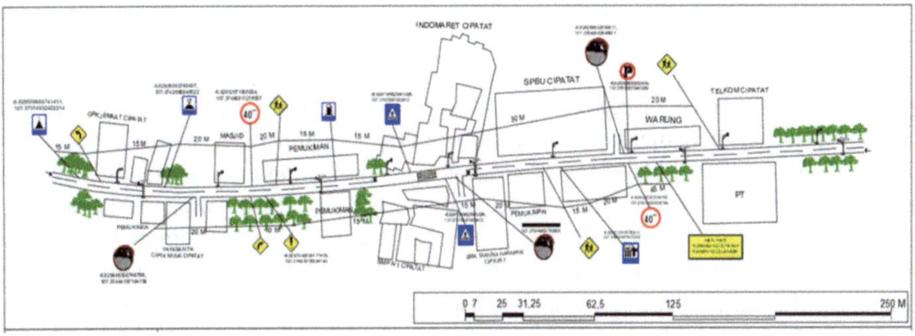

Fig. 4. Proposed Road Infrastructure Handling on Cipatat Road Segment 7

Based on the results of the analysis that has been carried out on Cipatat road segment 7, road safety inspection analysis to determine recommendations for the installation, addition or repair of road signs and markings, spot speed analysis that has been carried out to determine the speed limit set on this road section, and roadside hazards to determine the dangers of roadside risks, can be seen in Fig. 4 proposed recommendations for handling road infrastructure, there are several additional signs and road markings including the installation of a 40 km speed limit sign from the results of the instantaneous speed analysis, it is known that on the Cipatat road segment 7 in and out directions there is a speed that exceeds the speed limit. Installation of convex road glass from the collision diagram analysis and roadside hazards that there was an accident from the direction of the intersection so that the installation of the glass is necessary. Then the installation of signs indicating crossing facilities, zebra crossings and rumble strips in school areas to facilitate pedestrians who are mostly school students in order to reduce accidents in the future. Installation of signs indicating gas station facilities, mosques and churches to indicate that there are places of worship or gas stations at that location. As well as repairs to several points of public street lighting locations that sometimes often go out and endanger drivers at night.

The results of the analysis on segment 8, road safety inspection analysis to determine recommendations for the installation, addition or repair of road signs and markings, spot speed analysis that has been carried out to determine the speed limit set on the road section, and roadside hazards to determine the dangers of roadside risks, can be seen in Fig. 5 in the form of proposed recommendations for handling road infrastructure and there are several additional road signs and markings including the installation of a 40 km speed limit sign from the results of the instantaneous speed analysis, it is known that on the Cipatat road segment 8, both in and out, there is a speed that exceeds the

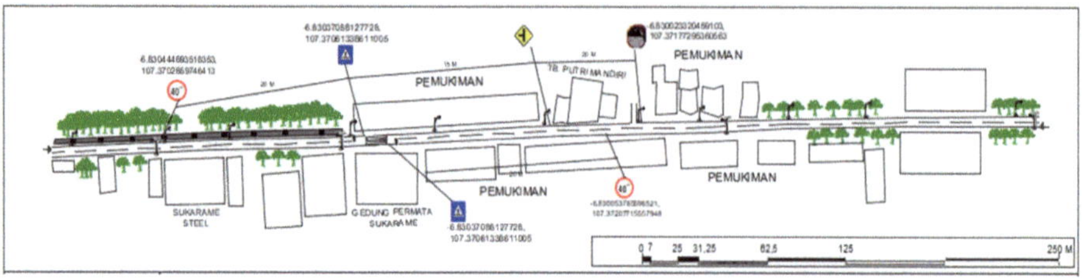

Fig. 5. Proposed Road Infrastructure Handling on Cipatat Road Segment 8

speed limit. Installation of convex road glass from the collision diagram analysis and roadside hazards where there is an accident from the direction of the intersection so that it is necessary to install glass. Installation of signs indicating crossing facilities, zebra crossings and rumble strips to reduce accidents in the future. Installation of road barriers because on segment 8 on the right side of the road there are trees and empty land whose land is lower than the road so that it is necessary to install fences to prevent drivers from falling into the right side of this road. There are improvements to several points of public street lighting locations that sometimes often go out and endanger drivers at night.

Implementation of the concept of safe roads on Cipatat road segments 7 and 8 from the aspects of self-explanatory, forgiving road users and self-regulation roads that explain road conditions to drivers with the help of warning signs so that they can reduce the risk of accidents, and from the regulatory side, whether they have met the rules of safe roads or not. In the existing one, it seems that it does not meet the standards of safe roads because there are several signs and road markings, noise tape to reduce speed that should be there to provide warnings regarding speed limits or warnings have not been seen in the existing conditions, From the analysis that has been carried out, it can be seen that there is a very clear comparison related to the concept of safe roads before the recommendation was proposed. Where before the recommendation was proposed, it was considered not to meet the standards in segments 7 and 8 for road safety, due to the lack of existing road infrastructure. After the analysis and recommendations were carried out, it was seen that the concept of safe roads had met the safety standards that should have existed, where there were speed limit signs, noise tape, convex road glass to reduce the risk of accidents in the future.

4 Conclusion

The infrastructure conditions on Cipatat road segments 7 and 8 in this study were concluded to be inadequate with several faded road markings, incomplete and damaged signs, and dead street lights. The high number of accidents in these segments was mostly caused by infrastructure and human factors, such as vehicle speeds exceeding the limit, violations of traffic rules, and pedestrian behavior that paid little attention to safety. In segment 7, there were many side-by-side accidents and collisions with humans, while in segment 8, there were more front-side and front-front collisions. Analysis of driver and pedestrian behavior also showed many violations, such as drivers not turning on lights, not wearing helmets, playing with mobile phones, and passengers and drivers not wearing seat belts. In addition, the planned speed limit of 40 km/h was often violated, with vehicle speeds exceeding the limit, both in segments 7 and 8. Danger factors on the edge of the road, especially in segment 7, also contributed to the high risk of accidents.

References

1. Almatar, K.M., Minhas, K.S., Butt, F.M.: Combining road safety and traffic operations: an integrated approach to implement system-wide transportation improvements. Transp. Res. Interdiscip. Perspect. **24**(101091), 1 (2024). https://doi.org/10.1016/j.trip.2024.101091
2. Yasanthi, R.G.N., Wickens, C.M., Jonah, B., Mehran, B., Suggett, B.: Determinants of traffic safety enforcement behaviour among police officers: a narrative review. Case Stud. Transp. Policy **16**(101206), 1 (2024). https://doi.org/10.1016/j.cstp.2024.101206
3. Morimoto, A.: History and prospects of traffic enforcement for traffic safety in Japan. IATSS Res. **48**, 129–135 (2024). https://doi.org/10.1016/j.iatssr.2024.02.004
4. Zhang, Z., Li, H., Hu, H., Chen, T., Ren, G. : Do non-motorists understand the traffic safety laws protecting them? results from a Chinese survey. Travel Behav. Soc. **36**(100779) (2024). https://doi.org/10.1016/j.tbs.2024.100779
5. Ma, Q., et al.: A big data analytics method for the evaluation of maritime traffic safety using automatic identification system data. Ocean Coast. Manage. **251**(107077), 1–21 (2024). https://doi.org/10.1016/j.ocecoaman.2024.107077
6. Asadi, M., Ulak, M.B., Geurs, K.T., Weijermars, W.: A methodological framework to conduct joint zone-based analysis of traffic safety and accessibility. J. Transp. Geogr. **118**(103949), 1–14 (2024). https://doi.org/10.1016/j.jtrangeo.2024.103949
7. Vadeby, A., Howard, C.: Spot speed cameras in a series - effects on speed and traffic safety. Accid. Anal. Prev. **199**(107525), 1–12 (2024). https://doi.org/10.1016/j.aap.2024.107525
8. Jiao, F., Du, Z., Shi, Z., Li, P., Yan, Y., Sun, F.: Traffic safety comprehensive evaluation of urban tunnel visual guiding system based on extension matter-element model: a case study in tunnel curves. Traffic Inj. Prev. (2024). https://doi.org/10.1080/15389588.2024.2359629
9. Kristoffersson, I., Große, C., Olsson, L.: Traffic safety versus accessibility: Investigating resistance against speed limit reductions. Transp. Res. Part A Policy Pract. **179**(103920), 1–13 (2024). https://doi.org/10.1016/j.tra.2023.103920
10. Leich, A., Nippold, R., Schadschneider, A., Wagner, P.: Physical models of traffic safety at crossing streams. Phys. A Stat. Mech. its Appl. **640**(129669), 1 (2024). https://doi.org/10.1016/j.physa.2024.129669
11. Saeidi, T., Mishra, S., Mehran, B.: Influencing factors on drivers' support for traffic safety laws in Canada. Case Stud. Transp. Policy **15**(101150), 1 (2024). https://doi.org/10.1016/j.cstp.2024.101150
12. Gao, M., Zhu, X., Cheng, X. : Safety – Premise for play: Exploring how characteristics of outdoor play spaces in urban residential areas influence children's perceived safety. Cities **152**(105236) (2024). https://doi.org/10.1016/j.cities.2024.105236
13. Galal, A., Ghizzawi, F., Donmez, B., Roorda, M.J. : Human factors affecting truck – vulnerable road user safety: a scoping review. Transp. Rev. 2379905 (2024). https://doi.org/10.1080/01441647.2024.2379905
14. Chen, D.Y., et al.: Risk factors and predictive model of cerebral edema after road traffic accidents-related traumatic brain injury. Chinese J. Traumatol. **27**, 153–162 (2024). https://doi.org/10.1016/j.cjtee.2024.02.001
15. Duy Dinh, D., Hoai Nam, V. : Examining the determinants of electric bike riders' attitudes towards traffic safety and risky driving behaviours amongst high-school students in Vietnam. Transp. Res. Interdiscip. Perspect. **21**(100889), 1–11 (2023). https://doi.org/10.1016/j.trip.2023.100889

16. Thorslund, B., Thellman, S., Nyberg, V., Selander, H.: Simulator-based driving test prescreening as a complement to driver testing – toward safer and more risk-aware drivers. Accid. Anal. Prev. **194**(107335), 1–8 (2024). https://doi.org/10.1016/j.aap.2023.107335
17. Ren, W., Yu, B., Chen, Y., Bao, S., Gao, K., Kong, Y. : Self-explaining analysis of facility environments on 2-lane rural roads with an improved lightweight CNN considering drivers' visual perception. Int. J. Transp. Sci. Technol. 1–18 (2024). https://doi.org/10.1016/j.ijtst.2024.08.002
18. Pranita, D., Sarjana, S., Musthofa, B.M., Kusumastuti, H., Rasul, M.S.: Blockchain technology to enhance integrated blue economy : a case study in strengthening sustainable tourism on smart Islands. Sustainability **15**(5342), 1–24 (2023). https://doi.org/10.3390/su15065342
19. Mustikaningsih, D., Cahyandito, M.F., Kaltum, U., Sarjana, S.: Building business performance through partnership strategy model : evidence from renewable energy industry in Indonesia. Int. J. Energy Econ. Policy **9**(5), 297–307 (2019)
20. Pranita, D., Sarjana, S., Musthofa, B.M.: Mediating role of sustainable tourism and creative economy to improve community wellbeing. Afr. J. Hosp. Tour. Leis. **11**(2), 781–794 (2022). https://doi.org/10.46222/ajhtl.19770720.257

Design Economic Creative Center in Lamongan District with Vernacular Architecture Approach

Ahmad Nijam Qoirun Naser(✉), Yayuk Sri Rahayu, Afin Ulul Azmi, and M. Mukhdif Al-Afghoni

Universitas Islam Darul 'Ulum Lamongan, Jl. Airlangga No.03, Merjoyo, Sukodadi, Kec. Sukodadi, Kabupaten Lamongan, Jawa Timur 62253, Indonesia
ahmadnijam.2020@mhs.unisda.ac.id

Abstract. The Economic Creative Center is a place for empowerment in the creative industries in the lamongan district by prioritizing intellectuals as wealth that creates money, employment opportunities, income, and welfare. The description of this creative industry is not only focused on production and distribution but also involves social and culture, Identifying from the economic and social values, the application of Vernacular Architecture to the design is very suitable. The design concept of the Economic Creative Center also considers integration with activities, facilities, comfort, and circulation. The method used in this design is to conduct field observations, namely by directly reviewing field conditions and collecting data by interviewing related parties. in addition, literature studies and comparative studies of objects related to technical factors and building requirements that affect the design design. Then identify aspects that can be applied in the design concerning technical aspects, design aspects that are in accordance with the design criteria to be achieved. From the problems obtained, it is then analyzed to produce a design concept with the theme of Vernacular Architecture. As a final result, the design can accommodate all creative industry activities and produce designs that can collaborate locality and creativity.

Keywords: Creative Economy · Economic Creative Center · Lamongan · Vernacular Architecture

1 Introduction

Urban The growth of the creative economy in Indonesia has a rapid impact on the national economy, this factor is based on the fact that the creative economy is a form of a new economic concept that intensifies information and creativity by relying on ideas and knowledge from human resources as the main production factor [1]. The contribution of the creative economy to Gross Domestic Product (GDP) in 2018 reached around 7.38% [2]. This economic concept is very suitable for industry 4.0, a digital industry where all parts collaborate and communicate in real time anywhere and anytime with the use of information technology [3]. In addition to the digitalization factor in the creative industry, the human resource factor is also very influential in developing the

M. Awang (Ed.): ICCAEET 2024, PTT, pp. 179–187, 2025.
https://doi.org/10.1007/978-3-031-94347-8_15

creative economy, namely the social and culture of Indonesian society, because social and culture should be a potential source of ideas that cannot be exhausted [4]. From this potential, the government has made the development of the creative economy part of the national priority agenda, among others there are 16 creative economy sectors that are being developed and included as creative economy subsectors, namely (1) application and game developers, (2) architecture, (3) interior design, (4) visual communication design, (5) product design, (6) fashion, (7) film, animation, and video, (8) photography, (9) craft, (10) culinary, (11) music, (12) publishing, (13) advertising, (14) performing arts, (15) fine arts, and (16) television and radio [5].

One form of national creative economy development is through the National Creative Economy Movement (GEKRAFS) program, as an organization that helps oversee the development of the creative economy ecosystem in Indonesia through the Minister of Tourism and Creative Economy (Manparekraf) [6]. One of the Gekrafs programs that has been held in several regions, namely Lamongan Regency through DPC Gekrafs Lamongan to become a partner of the local government in ensuring that the regional economy continues to grow. Lamongan will implement three important things, namely Gekrafs as a place to learn and share knowledge, synergize and collaborate to create a new idea, and a place to empower the community and open up the widest possible employment opportunities [7]. Through the Instagram social media account Gekrafs Lamongan has carried out many activities such as mural competitions, barista competitions, theater workhsop, band festivals and many more, as a form of development in the lamongan creative economy, but there are still many creative industry players in Lamongan who have not been accommodated due to limited space and facilities to develop, hone, and collaborate in one place.

Referring to the three important things that will be implemented in the lamongan creative economy, namely learning and sharing knowledge, synergizing and collaborating, and a place to empower the community for employment, then from the potential and problems that already exist. The design of the Economic Creative Center as a form of solution in accommodating creative industry players in the lamongan district to get facilities in learning, showing, collaborating, and selling products, the Economic Creative Center can become the City branding of the lamongan district in the creative industry in lamongan. By lifting the creativity and culture of the existing locality, it is hoped that the users of the Economic Creative Center can reach all circles of society, especially the younger generation of Lamongan in order to channel creativity, culture, forming an economic chain at the Economic Creative Center.

Lamongan district has also created its own creative economy through existing communities and creative craftsmen, such as the digital art community who bring elements of lamongan locality to their personal branding designs, for example t-shirts with lamongan cultural designs, from culinary t-shirts of soto lamongan, nasi boran, and so on. The batik craftsmen of sedang duwur who bring elements of the lamongan catfish icon in their batik drawings. And from lamongan arts activists who bring issues that occur in lamongan through their performances and also lamongan documentary films to capture existing cultural moments. Even so, problems such as a place that accommodates their creativity and a system that can collaborate between culture and creativity still need to be studied to form an Economic Creative Center design that can be integrated in the economy and elements of locality, so the approach of vernacular architecture in the design of the Economic Creative Center can be a solution. Because vernacular architecture can be interpreted as an original culture owned by a community that grows from social conditions and is still simple (Humble), refers to the work of humans/ordinary people (under priviledged, common people), adopted continuously for several generations, which includes architecture, language, art and music [8]. So the outline of the vernacular architecture approach to the design of the Economic Creative Center in the lamongan district can attract interest in creative industry players with the resources they already have through existing creativity and socio-culture.

2 Methods

The design method used in the design uses the Linear analysis technique (Reekie R. Frasier, 1972) which is adapted to the design approach. The stages in the linear analysis technique are known by theacronym "B-A-S-I-C" (Fig. 1).

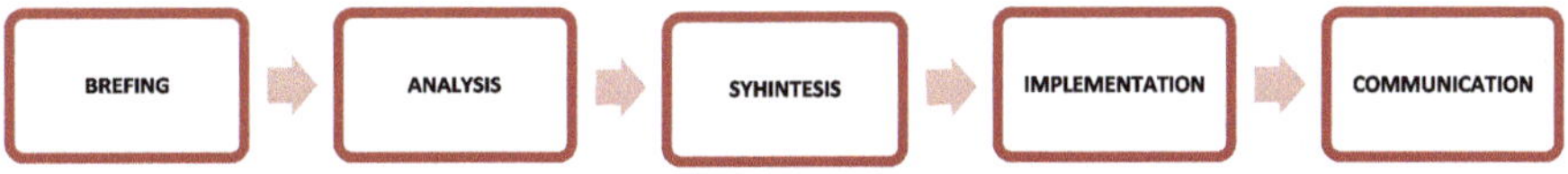

Fig. 1. Linear Analysis Technique Diagram

1. **Briefing,** which is the stage of collecting & processing data for design;
2. **Analysis,** is a process that contains responses and data that has been obtained and analyzed in detail and the responses are in the form of images and adverbs;
3. **Synthesis,** is the concept of the conclusion of the analysis that is adjusted to the principles of the design approach;
4. **Implementation**, is the final result and concept that has been obtained in the previous stage, in the form of architectural drawings and design work drawings.
5. **Communication,** is the presentation stage and the results of the design, poster presentations, mockups, etc.

3 Desing Resu

3.1 Basic Concept

Creative industry in Lamongan district, where this building requires a design that can collaborate between creativity and culture to help develop the creative economic growth of creative industry players. So the application of vernacular architecture in the design of the Creative Economic Center to be able to collaborate between creativity and culture will apply the principles implemented in buildings such as local styles, designs that place the situation in the location, shapes that depict local traditions, which will underlie this design (Fig. 2).

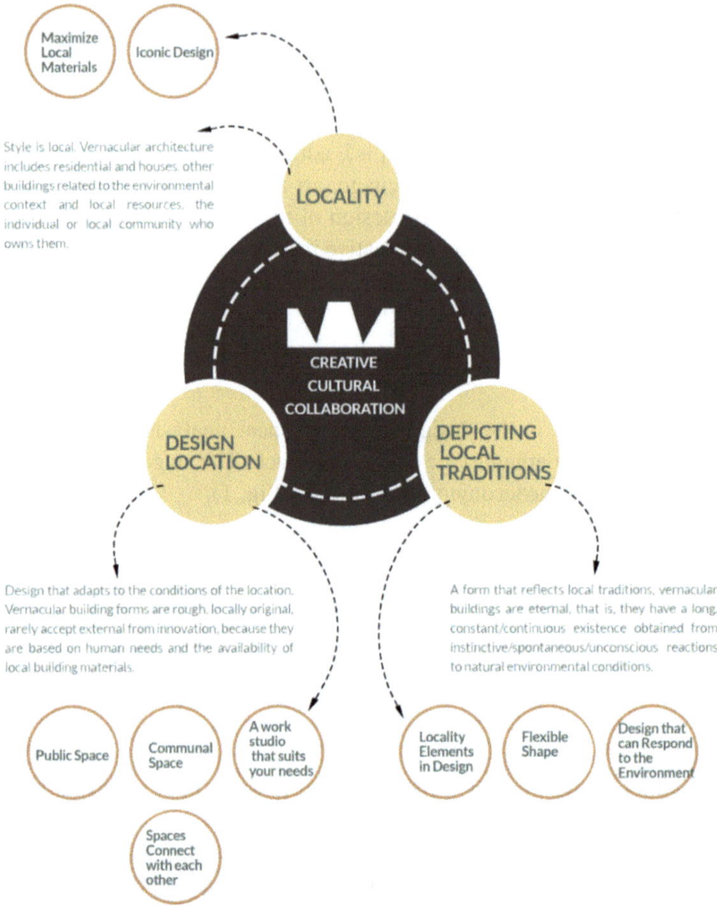

Fig. 2. Basic Concept

From the principles of vernacular architecture and the needs of the Economic Creative Center, the Creative Culture Collaboration Tagline is formed, which is based on the need for creative actors to be able to collaborate with the creativity of innovation with culture or excellence that already exists in the lamongan district, so the Creative Culture Collaboration Tagline becomes the basis of the concept in the design of the Economic Creative center.

3.2 Site Concept

The concept of the Economic Creative Center site applies three principles of Vernacular Architecture in it, namely:

Locality, The concept on the site illustrates like a temple area in Indonesia which is divided into sacred and non-sacred areas surrounded by petirtaan.

Design Placing Location, The shape of the site adapts to the circumstances of the existing site where the concept of the site transforms from the shape of rice fields.

Describing Local Traditions, The shape of the site is surrounded by a water area which takes the concept of petirtaan and the path uses a rock arrangement. Locality, The concept on the site illustrates like a temple area in Indonesia which is divided into sacred and non-sacred areas surrounded by petirtaan (Fig. 3).

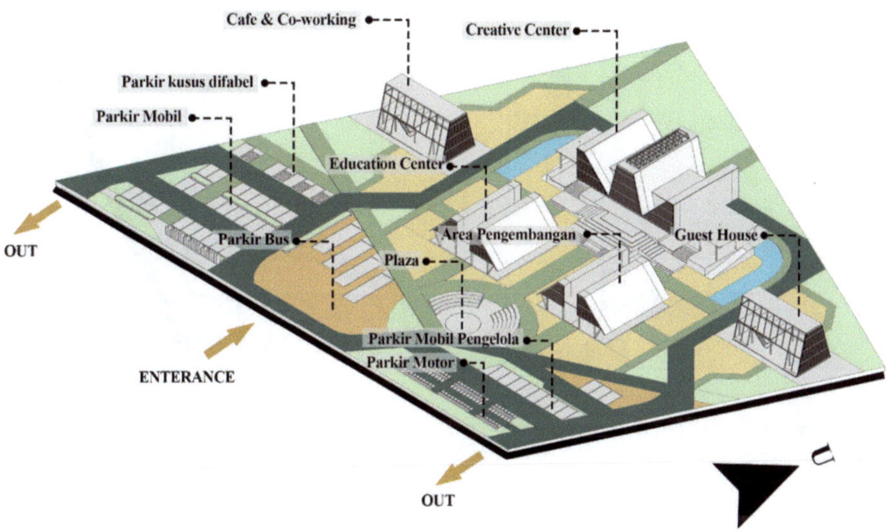

Fig. 3. Site Concept

3.3 Site Circulation Concept

(Figure 4)

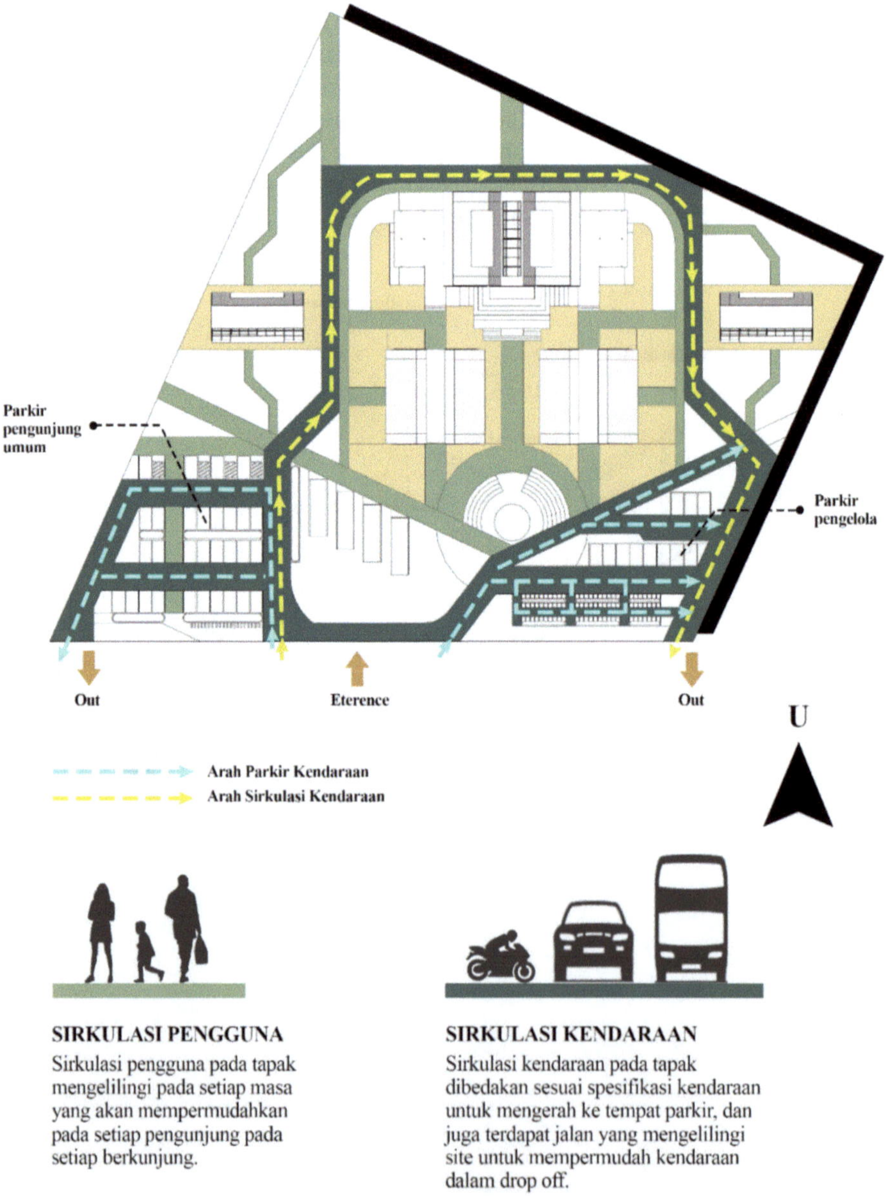

Fig. 4. Site Circulation Concept

3.4 Space Concept

The concept of interconnected spaces between interior and exterior and providing flexible space for activities is an application of the principle of harmony with the environment from Vernacular Architecture (Fig. 5).

Fig. 5. Space Concept

3.5 Design Result of Interior Space

1. Pay attention to the basic pattern of shapes from the functions and needs of thebuilding which form the mass formation. First SecondaryTertiary
2. The direction of circulation leads to paths at each time that are continuous with each other, Circulation
3. The form at each time is a vernacular implementation of the locality elements of Lamongan district, which transforms the shape of the paduraksa gate, Sendang Duwur tomb and the Bentar style gate which has become the iconic Lamongan gate. In this design, the roof is transformed into a form that can function as a space within it (Fig. 6).

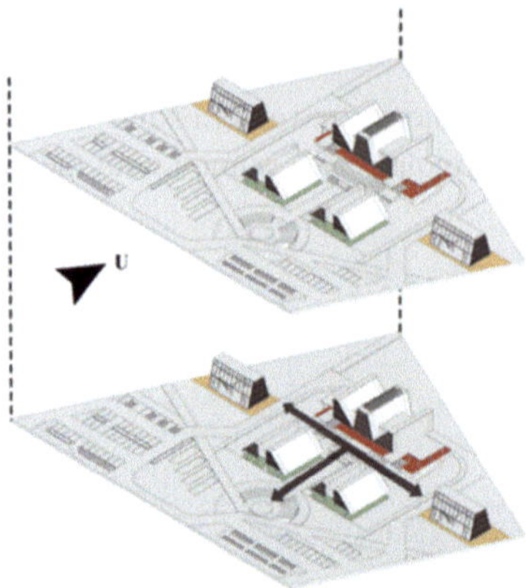

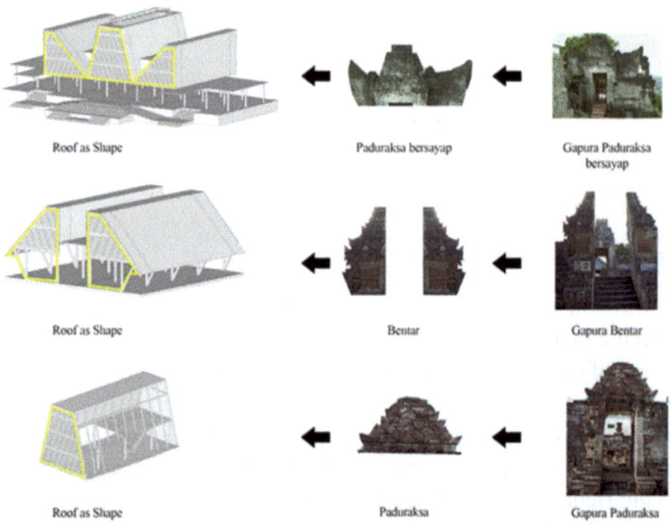

Fig. 6. Shape Concept

4 Conclusion

The vernacular architecture approach to the design of the Economic Creative Center is very suitable in implementing vernacular principles in design. Which vernacular is very concerned about the existing on the site and also has elements of locality in its design. This principle is very helpful for creative industry players, especially in the lamongan district where personal branding on each product takes elements of cultural locality and local wisdom from the lamongan district, the final output of the results of this design is a collaboration between culture and creativity.

By integrating natural elements into the workspace, we not only create a healthier and more pleasant environment, but also encourage innovation and collaboration. Investing in human- and nature-oriented co-working space design is a strategic step to prepare Indonesia for the challenges of the future. By creating workspaces that support creativity, productivity and well-being, we can empower the younger generation to be the driving force of Indonesia's digital economy.

The results of this study can illustrate the vernacular architectural approach which is often considered ancient and traditional and is less applicable today, but in the design of the Economic Creative Center it is very suitable to be applied with the concept of collaborating culture and creativity.

References

1. (Paramita dkk, 2021:vii). Ratna, Paramita, dkk. Manajemen Industri Kreatif Bahan Ajar Perkuliahan Bagi Mahasiswa. Lumajang: WIDYA GAMA PRESS ITB WIDYA GAMA LUMAJANG (2021)
2. (Pahlevi dkk, 2018:ii). Andreas, Pahlevi, dkk. Kolase Pemikiran Ekonomi Kreatif Nasional. Indonesia: CV. Oxy Consultant (2018)
3. (Purba 2021)
4. (Pangestu, 2008:206). Mari, Pangestu. 2008. Pengembangan Industri Kreatif Menuju Ekonomi Kreatif Indonesia 2025. Jakarta: RENCANA PENGEMBANGAN 14 SUBSEKTOR INDUSTRI KREATIF 2009–2015 2008 © Departemen Perdagangan RI. Diakses pada ......
5. (Pahlevi 2018:i)
6. (Hendriyani, 2024). I Gusti, Hendriyani, Siaran Pers: Menparekraf Apresiasi Gekrafs Turut Mengawal Pengembangan Ekonomi Kreatif Indonesia (2024). https://kemenparekraf.go.id/berita/siaran-pers-menparekraf-apresiasi-gekrafs-turut-mengawal-pengembangan-ekonomi-kreatif-indonesia. Diakses Tanggal 1 Mei 2024

The Effect of Video on Safety and Work Knowledge of Vocational Students in Surabaya

Feri Harianto(✉), Jenny Caroline, Anggita Maharani, Salsabila Putri, and R. Deby Ismayu

Department of Civil Engineering, Faculty of Civil Engineering and Planning, Institut Teknologi Adhi Tama Surabaya, Surabaya, Indonesia
jennycar@itats.ac.id

Abstract. Serious work accidents in the construction industry require efforts to be controlled to minimize the number of accidents. The occurrence of work accidents causes physical injury and lowers the mental spirit of workers. The leading cause of work accidents is human factors, including a lack of knowledge about work safety. Vocational high school students must be equipped with work safety knowledge while still in school. Understanding work safety knowledge in students requires suitable media to convey it. This research aims to determine the effect of video as a learning medium on work safety attitudes and expertise. This study uses a quasi-experimental method with video treatment. The respondents of this study were 30 students of SMKN 3 Surabaya. Data collection used a questionnaire. The material in the video is related to the definition of K3, the use of personal protective equipment, and K3 motivation. The analysis used a t-test. The results of the analysis showed that video learning media had a positive effect on the attitudes and knowledge of vocational high school students regarding work safety. The results of this study indicate that video is an easy-to-understand learning media to provide skills and shape work safety attitudes to prospective students in construction projects. Workers who know about work safety can reduce work accidents in the construction sector.

Keywords: Personal protective equipment · learning videos · occupational safety

1 Introduction

Implementing construction projects has a high risk of work accidents [1]. Serious work accidents in the implementation of construction projects have long been attempted to be reduced in number. Prevention efforts have not provided the expected results; therefore, a breakthrough must be made. The leading cause of work accidents is unsafe behaviour when doing work. One of the causes of work accidents is the lack of awareness and knowledge of worker safety and the inability to take appropriate action to protect themselves [4]. The causes of unsafe behaviour of construction workers can be divided into human error and violations when doing work. The type of human error is based on skills, rules, and knowledge. Worker errors when doing work are closely related to mentality [3].

M. Awang (Ed.): ICCAEET 2024, PTT, pp. 188–193, 2025.
https://doi.org/10.1007/978-3-031-94347-8_16

Research related to work safety in construction projects has been widely conducted, for example, mobile-based work safety communication through films [5], Virtual Reality (VR) [8], Augmented Reality [9], work safety and health games [1, 10], video [2], and Naturalistic Injuries Simulation (NIS). Various work safety communication media that provide pretty good results are videos [2]. The material from the video in the previous study consisted of general and specific materials. The general material contained occupational safety and health regulations, first aid for work accidents, and how to report procedures in the event of a work accident. In contrast, the specific material talked about dangerous behaviour in 12 jobs on construction projects. In addition, the video in the previous study was applied to construction workers. Therefore, in this study, communication media was developed through videos with personal protective equipment material and applied to students of building construction vocational schools. Effective occupational safety communication will increase occupational safety knowledge [11]. This study aims to improve the understanding of occupational safety in the implementation of construction projects among students of building construction vocational schools using video media.

Workers and the new generation need to understand occupational safety and health and who will become the next generation in an industry [12]. Therefore, it is necessary to carry out an understanding of K3 for vocational high school students. Education in Vocational High Schools is a secondary school that educates and equips its students with a piece of good knowledge with the hope of producing graduates who have skills so that they become graduates who are ready to enter the world of work as determined by the business world and can work according to their talents and interests.

2 Research Methodology

The research method uses a quasi-experimental design, and this experimental design was chosen considering the difficulty of controlling external factors that affect work safety attitudes, especially when using personal protective equipment. The object of this study was vocational students of SMK 3 class X Surabaya majoring in Building Drawing Engineering. The number of respondents was 30 people taken from 2 classes [17]. Data collection will be done using a longitudinal design of a questionnaire. A longitudinal design is a re-measurement of the same student's attitude every 15 days and is carried out three times before treatment and three times after treatment. Considerations are made three times to see the stability of students' work safety attitudes. The treatment in this study used a video that provided information about the meaning of work safety, personal protective equipment, and work safety motivation. The duration of the video is 4 min and 2 s. Six audio-visual media experts and three material experts validated the video. The assessment for audio-visual media experts is learning design with video and visual communication. The evaluation of material experts is reviewed regarding learning objectives, learning materials, learning methods, learning resources, and learning activities. The dependent variable is the occupational safety attitude of personal protective equipment, and the independent variable is the video treatment [7, 14]. The questionnaire grid used to measure the occupational safety attitude variable, especially on the importance of personal protective equipment, is as in Table 1. The questionnaire assessment uses the criteria very poor (score 1–1.79), poor (score 1.8 – 2.59), sufficient (score 2.6 – 3.39), good (score 3.4 – 4.19), and very good (score 4.2 – 5). The analysis uses the student t-test to measure occupational safety attitudes.

Hypothesis in the study:

Ho: There is no difference in understanding of K3 material for students before and after being taught with video learning.

H_1: There is a difference in understanding of K3 material for students before and after being taught with video learning

Table 1. Variables of occupational safety attitudes regarding the use of personal protective equipment

Indicators	Reference
Prevention of work accidents	[1]
Understanding work safety	[1]
gloves	[18, 19]
Safety shoes	[18, 19]
Safety Clothing	[18, 19]
Helmets	[18, 19]
Goggles	[18, 19]
Respiratory protective equipment	[18, 19]
Safety Shirts	[18, 19]
Implementation of work safety standards	[1]

3 Result and Discussion

The questionnaire used to measure occupational safety attitudes must meet the validity and reliability tests. For the validity test, the correlation value of each question item to the total question item value r count > 0.361 so that the questions in the questionnaire are declared valid. For the reliability of the questionnaire, the value of Cronbach's alpha (0.941) > 0.6 [15], so the questionnaire is declared reliable. The occupational safety attitude variables were measured three times before and after treatment to see respondents' stability in answering questions. The average before and after treatment results are as in Fig. 1.

Testing before treatment was carried out three times with a 15-day interval. Before conducting a paired mean difference test (t-test) on each test or 1st to 6th. Then, it is necessary to perform a normality test. The results of the normality test show that the significance value is 0.22, and the value is greater than 0.05, so the data has a normal distribution pattern. In Table 2, the t-test results explain that in measuring the work safety attitude variable before treatment, the average value in each test did not differ

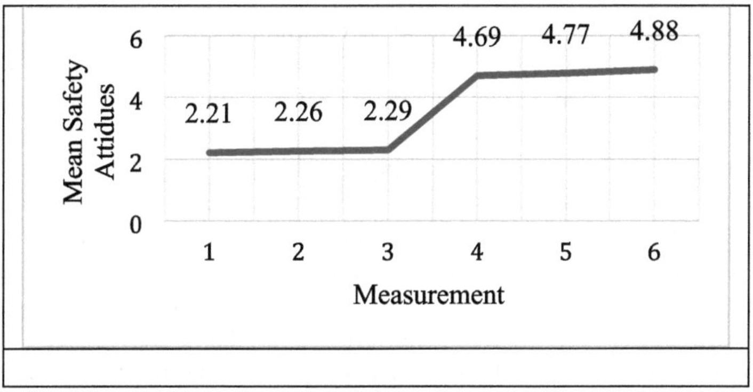

Fig. 1. Mean Safety Attitudes

significantly (sig > 0.05). After the 3rd test, respondents were given treatment, then in the 3rd and 4th tests after the treatment in the form of a video, the average value of the work safety attitude was significantly different (sig. 0.00 < 0.05), meaning that the provision of treatment in the form of a video to respondents had a good impact related to work safety attitudes. The average value of the measurement after therapy in the form of a video on the 4th and 5th measurements the 5th and 6th measurements showed no significant difference in the average (sig > 0.05).

Table 2. Mean Test Between Measurements

Mean Test	Significant Value
Test 1 – Test 2	0.660
Test 2 – Test 3	0.766
Test 3 – Test 4	0.000
Test 4 – Test 5	0.475
Test 5 – Test 6	0.590

Learning through video is an interactive process that is easy to understand and interesting for individual and group learning. The characteristics of video are that it has elements of images, sound, movement, and colour that have concise language to strengthen the learning material. Images in videos help strengthen memory in the brain. Sound can enhance students' focus. Colours in videos have a positive impact, encouraging a joyful learning spirit and creating a cheerful atmosphere. Additionally, colours help emphasize important material and categorize information. Movement in videos increases student engagement in the learning process. Therefore, videos serve as an effective learning medium to create an engaging learning experience and motivate students. The development of learning programs through learning videos can be delivered in a way that is easy to understand and can develop students' knowledge since they were in school. In addition, computer-based learning videos make students feel happy and interested so that their development aspects can increase. K3 learning videos can be published online and used repeatedly to facilitate teachers, parents, and the community [16]. The use of learning videos in the process of understanding K3 material can be accessed anywhere and anytime to foster a culture of awareness of the importance of taking care of yourself and being aware of the importance of complying with the K3 regulations that have been determined to prevent unwanted risks for vocational students [16, 19] concluded that audio-visual media in the form of video is a concise, flexible medium and is considered very good as a medium for teaching and learning processes both in terms of delivering information and in efforts to increase motivation and development of student knowledge [6]. The video used as the treatment in this study has a duration of 4 min and 5 s, which is in line with research findings that state videos shorter than 5 min provide a good level of understanding for participants [20]. Furthermore, a duration of less than 5 min helps keep participants engaged in watching it [21]. The existence of a work safety training program through video media motivates students and can improve control of work accident risks [13].

4 Conclusions

Based on the results and discussions, the conclusion is that video is one of the learning media that positively impacts vocational students regarding work safety attitudes. Vocational students are young prospective construction workers who must understand work safety attitudes. In contrast, vocational students are young prospective construction workers and newcomers at high risk of accidents. A limitation of this study is the necessity to expand the sample size of vocational students in building construction across different regions, taking demographic factors into account to enhance the validation of video-based learning. This study contributes to the identification of the need for a video-based occupational safety training program for construction workers prior to commencing their tasks.

References

1. Harianto, F., Anwar, N., Wiguna, I.P.A., Suryani, E.: Conceptual system model dynamic OSH performance improvement of building construction projects. In: Lie, H.A., Sutrisna, M., Prasetijo, J., Hadikusumo, B.H., Putranto, L.S. (eds.) Proceedings of the Second International Conference of Construction, Infrastructure, and Materials. Lecture Notes in Civil Engineering, vol. 216, pp. 441–448. Springer, Singapore (2022). https://doi.org/10.1007/978-981-16-7949-0_39
2. Harianto, F., Anwar, N., Wiguna, I.P.A., Suryani, E.: Experiment the effect of providing monetary incentives and safety patrols on work safety behavior in construction project implementation. In: Lie, H.A., Sutrisna, M., Prasetijo, J., Hadikusumo, B.H., Putranto, L.S. (eds.) Proceedings of the Second International Conference of Construction, Infrastructure, and Materials. Lecture Notes in Civil Engineering, vol. 216, pp. 449–457. Springer, Singapore (2022). https://doi.org/10.1007/978-981-16-7949-0_40
3. Karimi, H., Taghaddos, H.: The influence of craft workers' educational attainment and experience level in fatal injuries prevention in construction project. Saf. Sci. **117**, 417–427 (2019)

4. Breslin, F.C.: Educational status and work injury among young people. Can. J. Public Health **99**(2), 121–124 (2008)
5. Endroyo, B., Yuwono, B.E., Mardapi, D.: Model of learning/training of occupational safety and health (OSH) based on industry in the construction industry. Proceedia Eng. **125**, 83–88 (2015)
6. Huang, Y.H., Yang, T.R.: Exploring on-site safety knowledge transfer in the construction industry. Sustainability **11**(22), 6426 (2019)
7. Edirisinghe, R., Lingard, H., Broadhurst, D.: Exploring the potential for using video to communicate safety information to construction worker: case studies of organizational use. Constr. Manage. Econ. **34**(6), 1–11 (2016)
8. Strzałkowski, P., Bęś, P., Szóstak, M., Napiórkowski, M.: Application of virtual reality (VR) technology in mining and civil engineering sustainability **16**(6) (2024)
9. Liu, T.Y., Tan, T.H., Chu, Y.L.: QR code and augmented reality- supported mobile english learning system. Lect. Notes Comput. Sci. **5960**, 37–52 (2010)
10. Greuter, S., et al.: Designing a game for occupational health and safety in the construction industry. In: Proceedings of 8th Australian Conference on Interactive Entertainment Playing the System IE, vol. 12 (2012)
11. Alsamadani, R., Hallowell, M., Javernick-Will, A.N.: Measuring and modelling safety communication in small work crews in the US using social network analysis. Constr. Manage. Econ. **13**(6), 568–579 (2013)
12. Hussain, R., Pedro, A., Lee, D.Y., Pham, H.C., Park, C.S.: Impact of safety training and implementation on training-transfer: targeting migrant construction workers. Int. J. Occup. Saf. Ergon. **26**(2), 1–24 (2018)
13. Hubbard, B., McGlothlin, J., Hubbard, S.: Case study: safety training issues for student interns. J. Occup. Environ. Hyg. **6**(8), 35–40 (2009)
14. Zebua, Y.: Media PowerPoint Berbasis Video Terhadap Hasil Belajar Materi Keselamatan dan Kesehatan kerja (K3) Pada mahasiswa Pendidikan Teknik Bangunan IKIP Gunungsitoli. Jurnal Rev. Pendidikan dan pengajaran **3**(2), 469–475 (2020)
15. Ghozali, I.: Analisis Multivariate dengan SPSS. 3nd, Universitas Diponegoro, Semarang (2005)
16. Sya'bania, N., Anwar, M., Wijaya, M: Pengembangan Media Pembelajaran Berbasis Video Animasi dengan Model Pembelajaran Inkuiri Terbimbing untuk Meningkatkan Motivasi dan Hasil Belajar Peserta Didik. Chem. Educ. Rev. (CER), **4**(2), 34–44 (2020)
17. Alwi, I.: Kriteria Empirik Dalam Menentukan Ukuran Sampel Pada Pengujian Hipotesis Statistika dan Analisis Butir Jurnal Formatif. **2**(2), 140–148 (2012)
18. Guan, L.R., Xian, G.J., Rajendran, S.D., Wahab, S.N.: A Study on the effectiveness of personal protective equipment (PPE) on building construction workers. In: E3S Web of Conference, vol. 136 (2019)
19. Schenker, M.B., Orenstein, M.R., Samuels, S.J.: Use of protective equipment among California farmers. Am. J. Ind. Med. **42**, 455–464 (2002)
20. Yu, Z., Gao, M.: Effect of video length on a flipped English classroom. Sage J. **12**(1), 1–14 (2022)
21. Guo, P.J., Kim, J., Rubin, R.: How video production affect student engagement: an empirical study of MOOC Videos. In: Proceedings of The First ACM Conference on Learning (2014)

Sustainable Playground Design Guided by Play Patterns of the Alpha Generation in Urban Environments: A Case Study in Tangerang, Indonesia

J. F. Bobby Saragih[1(✉)], G. Suharjanto[1], and Hendra Wahyu Cahyaka[2]

[1] Architecture Departement, Faculty of Engineering, Bina Nusantara University, Jakarta, Indonesia
bsaragih@binus.edu

[2] Civil Engineering Departement, Faculty of Engineering, Surabaya State University, Surabaya, Indonesia

Abstract. Play is an essential aspect of children's development, yet it is often overlooked by planners when designing playgrounds. Research shows three key principles that children prefer playgrounds, providing freedom, excitement, and surprises. Nevertheless, many existing playgrounds fail to embrace these principles, resulting in suboptimal use or even abandonment. With the generation shift, particularly the emergence of the Alpha generation, it becomes important to understand these children's play patterns and preferences to create sustainable playgrounds—an aim of this research. This research is exploratory, with qualitative methodology through pictures and interviews, meanwhile, quantitative data was employed to observe trends. Based on play patterns, children draw their ideal playground, and interpretation of the images is carried out as a basis for analysis. In-depth interviews were conducted with random student to verify the consistency between their drawings and verbal responses. Carried out in an urban context, in Tangerang City, Indonesia, involving 72 fifth and sixth-grade students from Sekolah Tiga Penuai Indonesia, a private school. These findings suggest that sustainable playgrounds can be achieved by combining three design elements: 1) a nature-inspired playground featuring trees and water; 2) traditional play structures such as swings and slides; and 3) a design that offers challenges and stimulates adrenaline through more adventurous activities.

Keywords: sustainable · playground · design · children · behavior

1 Introduction

Interesting story of one morning, 3 male students from Pemuai Indonesia School, representing the Alpha Generation, wearing school uniforms, discussing in one corner of the school. Listening to what they were talking about, they continued the online discussion they had yesterday, which they initiated independently using the Google Meet application. Technological progress is closely related to them, and this phenomenon has never happened a few years before.

M. Awang (Ed.): ICCAEET 2024, PTT, pp. 194–208, 2025.
https://doi.org/10.1007/978-3-031-94347-8_17

Traditional and natural public playgrounds may become threatened and even extinct due to the emergence of digital and virtual playgrounds. The impact of technology on Children's Play and how technology has changed how and where children play could be explored. The phenomenon shows that children's interest in playing on the traditional and natural playground is decreasing, there are three obstacles experienced by children in visiting it: distance, transportation, and awareness [1]. Even though playing provides several positive benefits for children, a part in recreation and physical activity [2], escaping stress, building competence, and forming supportive social groups [3]. The facts prove that some children had varying experiences, encountering accidents while playing [4], conflict between peers, seniors, and juniors, etc. [5]. Unpleasant incidents also occurred at various public playgrounds in New Z [6]. Conflicts arise due to spatial competition among children discovered In Japanese public playgrounds [7], which changes children's play behavior [8]. Australian playgrounds note the potential for such conflicts [9]. Research shows that conflicts and accidents in playgrounds are often caused by improper design of equipment [10] and mistakes in spatial design [11]. These conditions can potentially trigger the unsustainability of playgrounds, which are problematic, inadequate, or boring [12].

Regarding his research, Maxey questioned the significance of ensuring sustainable playgrounds in residential areas [13]. The question appears relevant in current conditions, particularly during the Generation Alpha children, who have limited playtime [14] and tend to prefer playing with gadgets or smartphones [15]. If playgrounds are important for children's development, it's crucial to understand their design preferences based on behavior and character. This design preference is interesting to understand, complementing research on playground design based on the Alpha generation, which is still minimal [16] This is a component of the plan to achieve a sustainable playground.

2 Research Problem and Methodology

This research aims to understand the playing behavior of children in the Alpha Generation. This understanding is quite important when children are faced with technology-based digital and virtual playgrounds. The existence of traditional and natural playgrounds seems to be experiencing degradation along with advances in technology, however, several studies still consider that traditional and natural playgrounds are still needed, and rely on the physical and social, expression of human development in childhood, therefore, the question that needs to be understood is: How the idea design of a sustainable playground by play patterns of Alpha Generation?

This research is exploratory research aimed at understanding issues or phenomena that are not yet clearly defined or are still minimally understood. In this case, it focuses on the design of play spaces based on the play patterns of the Alpha generation. The appropriate method for exploratory research must be flexible, open-ended, and capable of deeply extracting information without being constrained by specific hypotheses. Common data collection methods include interviews, documentation, and content analysis. This research was carried out in 2022 at Penuai Indonesia School, Tangerang – Indonesia, a private school with diverse students from ethnicities and religions. Using a grounded theory approach to 72 Students, Grade 5 (21 Female Students and 15 Male Students) and Grade 6 (18 Female Students and 18 Male Students). The students range in age from 9 to 11 years; some come from lower socioeconomic status (15.3%), and the majority come from medium socioeconomic (84.7%), as evidenced by their house. Due to licensing limitations, interviews could only be carried out with 30 students (41%), around 28 students said that they had played in modern playgrounds.

Using the qualitative method, efforts to understand students' perspectives are carried out in several ways: 1. Drawing, through sketch it is hoped that they will be able to express imagination, imagine desires, and imagine actions [17], exploring students' perceptions and experiences regarding their dream playground, the informants were given the task of spontaneously "drawing a dream playground" on a piece of paper provided by the researcher, several studies support this method, drawings are an effective means to reflect their observations, experiences, problems or ideas [18]. Becomes a marker and meaningful semiotic space for children's thoughts [19] to obtain the information needed [20]; 2. Apart from drawing, interviews are also carried out, drawings should be supported by interviews with the children [21]. Conducted in a structured manner (using an interview guide) many important things that were not represented in the pictures were obtained through the interviews. Analysis was carried out by coding the images created and comparing them with the results of the interviews.

3 Literature Review

3.1 Alpha Generation Play Patterns

Alpha Generation, born between 2010–2025 is used to dealing with digital playgrounds and virtual playgrounds [22], they are the screenagers, digital natives, and connected or wired generations [23] and have now, entered a new era of play space that is realized through the metaverse [24]. All of this tends to be done independently and does not rely on physical activity, what they do is contrary to the basic theory of play, which relies on the physical and social, expression of human development in childhood [25]. Research proves that children's play has more opinions, and some take sides that play is an intrinsically universal activity, regardless of cultural differences [26], controversial behavior, play is perceived as an unclear, difficult-to-understand [27], what it is for, how it originated, and how it developed [28]. In the end, research still leaves a discourse opinion on the idea of play, and others agree as a physical activity, but, others tend to be a metaphysical activity [29], with play more as a contributor to children's mood and well-being [30] and reduce their stress [29] Regarding Alpha Generation play behavior, the idea of play as part of the metaphysical activity seems more relevant to their ecology, decreased social skills, increased screen time, and decreased physical activity. Research proves Limited playtime [14] and the limited play space pushed them to finally deal with digital play.

3.2 Imagination and Children's Preferences Design

Several studies explain well the imagination, which is a mental capacity, the process of thinking and deciding about what might happen, planning and anticipating the future, creating creative ideas, and considering alternatives that are far and near to actual experiences that transcend space and time [31]. Recent theories on developmental psychology explain that children can imagine by explaining and predicting mental concepts through 1. Imagining desires and 2. Imagining Actions [17]. As a mental process, imagination is a creative result of combining pieces of past information obtained through a combination of systematic stimulation and varied sensory input [32], with children. The more sensory stimulation given, the more open the potential for imagination, the process of visualizing imagination is a stimulated recall process [33]. Research by Kan-Kılıç, D. (2021) About My Dream Playground with the participatory method which involves children aged 0–12 years proves that: children appreciate Playgrounds that can be used by various ages, which can meet the needs of climbing, sliding, jumping, playing on the ground [34]. Another study with the same aim involving Low SES and High SES children using the drawing method also proved that playgrounds preference was derived in the order of "Openness", "Safety", and "Diversity", and in terms of morphological properties, the playground must be able to meet the needs "Adventure", "Five senses elements", "Visual elements" and "Interactivity" [35]. Children's imagination and children's preference for playground design can be seen from the research above.

3.3 Children Playground

Designing a sustainable and child-friendly public playground in various cities is challenging. However, by utilizing ideas from multiple studies, the playground is carefully crafted to include natural elements [14] and provide safe and comfortable gaming equipment [36]. Efforts to provide safe and comfortable playgrounds for children continue to be carried out, as implementation of standards, and preventing falls and fall-related fractures. Apart from being related to dimensions, several other standards that are also important to understand are: having high standards of safety and durability [37], the principle of inclusion, and providing equal access [38], security and a sense of belonging, social affiliation, and creative expression and exploration [39]. Discourse between standard and non-standard play spaces continues, research has shown that children want a playground that follows standards, respects children's wishes, and looks original and creative [40]. Meanwhile, several other studies found different facts, standard playgrounds are not desirable, and children prefer playing in non-standard playground configurations [41].

Research believes that, in the context of playgrounds, children have an innate biological tendency known as biophilia, which allows them to come into direct contact with nature and experience nature and has the potential to create longer and more complex play [42]. Meanwhile, from a typical perspective, playgrounds are known as traditional equipment playgrounds, contemporary-design playgrounds, and adventure playgrounds (which supply play material, not play equipment) [43]. The equipment includes climbers, monkey bars, slides, swings, teeter-totters (seesaws), merry-go-round, playground climber and some add to it with a Natural Playground which is marked with Tree/grass, Water, Sand Pit, Sun, and Beasts). A traditional playground is a playground equipped with manufactured equipment made of metal or brightly colored plastic and a Digital playground. Children's digital playgrounds have evolved from being a series of convergent digital media consisting of social media platforms, mobile applications, and the internet of toys [44].

4 Result

4.1 Playground Preference

Drawing is the best way for children to express intertextual and creative ideas, this process can build imagination and narratives of their design [45] Grugeon, E., & Harding, L. (2012.) and as a form of children's thinking [46] (Jo Deguara & Cathy Nutbrown (2018). The drawing process is carried out spontaneously between learning activities; it is hoped that it will be able to explore information related to hopes and dreams about the playground. This research involved the Alpha Generation, in the 9–11-year age range, who were born after 2010 [22]. Aligned with data collection techniques typically employed in exploratory research, such as documentation, the researcher requested that respondents create drawings depicting their ideal and dream playground.

"Please draw the playground you dream of"
Researcher

Seventy-two students were involved and carried out the drawing process; it was proven that two students (2.7%) drew a gaming room as their dream playground, and 70 students (97.2%) of children preferred to draw a swing, a monkey bar-like traditional playground and drawing trees and water as a symbol of a natural playground, as their dream playground with several different variations. From several available theories, digital playgrounds are identified in several ways: interactive whiteboards, iPads, and other tablet devices are computing devices. The choice to sketch the gaming room is an expression of their daily playing behavior that is familiar to children's daily lives. Several digital-based applications are used, and digital has become part of their playing life [47]. They tend to use technology as part of their playing life [23, 48], Fig. 1 shows the results of a child's drawing, which shows that he prefers to play using electronic media to fulfill his playing needs. The pattern of playing with electronic media is often found, especially for children who come from medium and up socioeconomic status.

Those who choose to play digitally, because they have good technological literacy, can use technology as part of their play behavior. In contrast to the playing behavior of previous generations, currently, the use of technology is a real thing [22] There is a logical reason for the 2.7% of students who want a digital playground: most of every day they play and sometimes on certain days, they spend more of their playing time with the digital playground. To validate their designs and consider limitations in time and space, interviews were conducted randomly, allowing children to articulate additional insights that were not captured in the drawings, below are the answers provided by the children:

"... play VR (Virtual Reality) ... when playing in TZ (modern playground) ... "
Informant No. 31: Girl, Grade 5

"... playing online games.... on Saturday and Sunday "
Informant No. 9: Boy, Grade 5

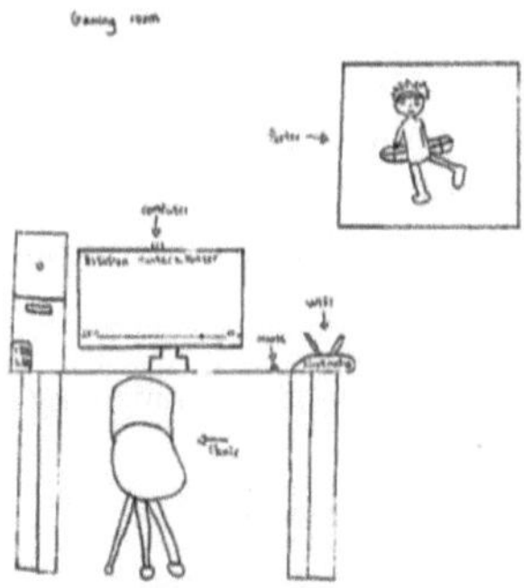

Fig. 1. Gaming room. Source: drawn by informant no.9, 2022

When they play digital-based games they tend to do it themselves, many researchers say that the Alpha generation tends to be individual, less interactive with each other, and engrossed in their world. This research proves that the presence of friends is very important in their playing activities, friends have an influence on playing behavior, and social interaction with peers is important in playing behavior [49], especially friends they know well, so it is not wrong when they play at school, they tend to play associative and cooperative games, different when they are in the community or neighborhood playground where tend to play solitary and parallel games, children -children are not comfortable playing with children they do not know [50]. All students have used digital to meet their playing needs, compared with digital, interviews conducted with 97.2% of students, who preferred to play on traditional playgrounds, below are the answers provided by the children:

Why choose to draw *a swing, a monkey bar, like traditional playgrounds?*
Researcher

"...playing games is interesting, but sometimes boring..."
Informant No. 15: Boy, Grade 6

"...online games are not interesting...rarely to play..."
Informant No. 27: Girl, Grade 5

Apart from that, they chose to play at the Traditional Playground because they hoped it could be done with their peers.

"... if we play football... bike... we can meet friends..."
Informant No. 27: Girl, Grade 5

The statement above proves that digital playgrounds are something they often use, but when the choice of playing is given, it turns out that children from the alpha generation still prefer to do playing activities in traditional playgrounds.

4.2 Playground Apparatus

From the student's drawing, each child drew one or more playground apparatus on A4 paper. Table 1 explains the variety of playground equipment sketched by children, which was very diverse, students sketched 1–7 pieces of equipment.

Table 1. Variety of Playground Equipment by Gender (n = 72)

Gender	Variety of Playground Equipment						
	1	2	3	4	5	6	7
Boys	9	7	10	8	4	1	0
Girls	6	7	9	6	3	0	2

Source: Researcher, 2022

Table 2. Frequency of Traditional and Natural Playground Equipment Drawing by Grade

Class	Playground Equipment			
	Traditional Playground		Natural Playground	
	Swing	Slide	Tree	Water
Grade 5	11	12	13	9
Grade 6	19	19	13	18
Total	30	31	26	27

Source: Researcher, 2022

Table 2 proves that for Traditional Playgrounds there are only 2 dominant pieces of equipment, swing, and slide. The sketch describes the swing and slide as challenging equipment. By age, the data reveals that the age factor is very determining, grade 6 students are more interested in swings and slides than grade 5 students. Meanwhile, Natural Playground also shows the same phenomenon, grade 6 is more interested in trees and water than grade 5 students. Meanwhile, several other supporting equipment, tables, chairs, and fences are also part of the picture, and this is in line with other research, the provision of supporting equipment, chairs, fences, and the softness of the surface are also things that children pay attention to [51]. By Gender, research using a design approach proves that there are real differences between school playground designs chosen by male and female students [52]. In line with this research, this research also proves gender differences in the choice of playground element, apart from that the age factor also influences this choice, 9–11-year-olds have different playing behavior compared to those aged below, and physical changes make them prefer to do play activities which are more of a challenge. Table 3 explains two games on traditional playgrounds that are challenging. Playground element swings and slides are more popular with Grade 5 and Grade 6 girls. Boys generally have more interest in playing sports and natural equipment, the option to play in nature is possible because natural elements provide a more challenging game and invite adrenaline, below are the answers provided by the children:

"...traditional playground that is more challenging..."
Informant No. 25, Girl, Grade 6

"...requires adrenaline..."
Informant No. 15: Boy, Grade 6

Challenging and inviting adrenaline is possible because nature playgrounds always come with changes and present many possible variations in playing [29]. The element of the game that invites challenge and adrenaline can be seen in Fig. 2, where informant number 25 describes this clearly, through the image of a roller coaster.

Table 3. Frequency of Traditional Playground Equipment Drawing by Gender

Class	Playground Equipment			
	Swing		Slide	
	Girls	Boys	Girls	Boys
Grade 5	7	4	7	5
Grade 6	11	8	10	9
Total	18	12	17	14

Source: Researcher, 2022

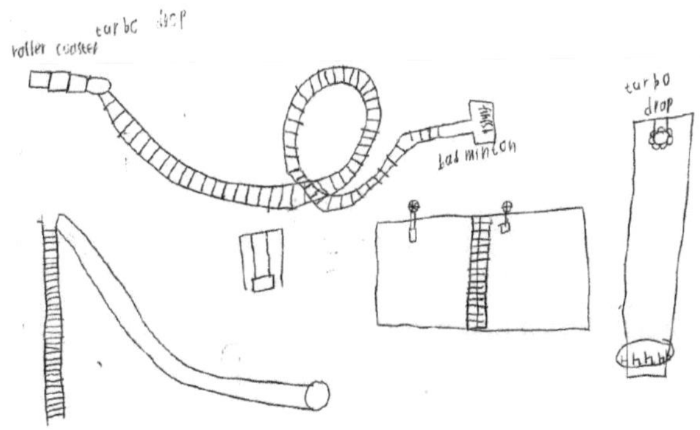

Fig. 2. More challenging Playground Apparatus. Source: Sketch by informant no.25, Girl, 2022

Meanwhile, sketches made by several students show the phenomenon of play by including natural elements in the playground, sometimes even being more important elements than other equipment. Table 4 shows the tendency for boys to play more dominantly with natural types of equipment. The preference design approach proves that boys students prefer to play in areas that are connected to nature (Fig. 3), while girls students prefer the presence of a sandbox [53], this research also proves the same thing.

Table 4. Nature Playground Sketches by Gender

Class	Playground Equipment			
	Tree		Water	
	Girls	Boys	Girls	Boys
Grade 5	5	8	4	5
Grade 6	5	8	6	12
Total	10	16	10	17

Source: Researcher, 2022

For children from lower socioeconomic, engaging with electronic media is a luxury. They are more likely to rely on designated play areas available in their communities or on materials they find in their surroundings. Below are the answers provided by the children

"...rarely play games.... rather than playing games...it is healthier to play in nature..."
Informant no. 4, Boy, Grade 6,

:

Fig. 3. Nature Playground. Source: Sketch by informant no.4, Boy, 2022

4.3 Playground Area

Cognitively, researchers classify play behavior with terms such as functional play, dramatic play, or constructive play [49]. Meanwhile, in terms of equipment completeness, several studies have classified them as traditional playgrounds and natural playgrounds. Looking at children's pictures, 97.2% of children prefer traditional playgrounds as their dream playground. Efforts to provide a playground that is equipped with playground elements and natural materials certainly require a certain area. From the overall sketch, a clear pattern of arrangement of playground elements and diversity of playground elements can be seen. This diversity is important to respond to the needs of many different children [54]. Data shows that Females display more diversity in equipment compared to males (Picture 4). Play element seems to have a strong influence, to maximize physical activity in children, playgrounds must be designed with lots and variety of play elements [28].

The more sensory stimulation given, the more open the potential for imagination, the process of visualizing imagination is a stimulated recall process [55, 33]. Meanwhile, paying attention to the drawings produced by children, children tend to need a playground area that is open and outdoor, some children drew the sun as a symbol of natural lighting and several clouds surrounding the playground area. When the drawings are examined, it can be observed that they include problems environment that they can observe in their immediate environment [21].

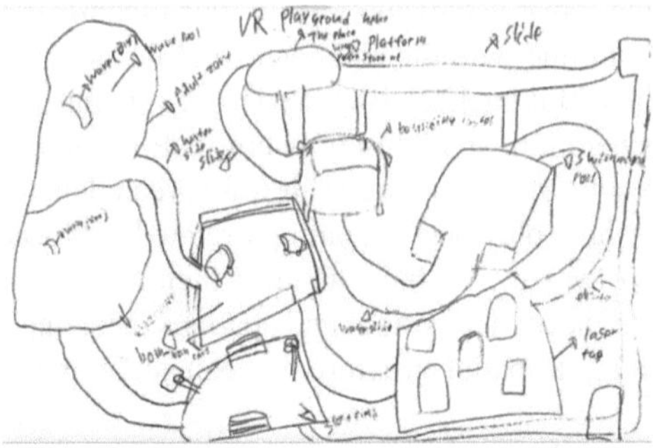

Picture 4. The Variety of Games. Source: Drawn by informant no. 4, Girl, 2022

5 Discussion

Children are part of the community and playing is part of children's daily lives. Research has proven that playing is a strategy for releasing children's high energy, positive self-development (ego), exploration skills, children's self-esteem, and success, as well as cognitive, physical, and psychological development. Children [33]. Spatially, the playground is considered an ideal place to achieve this goal. It is designed with lots and a variety of play equipment to maximize physical activity in children [28]. Understanding of traditional playgrounds equipped with play equipment includes climbers, monkey bars, slides, swings, and teeter-totters (seesaws), merry go round, playground climbers, all play This equipment seems to have similarities in almost all countries in the world, both in developed and developing countries.

Especially for the Alpha generation, it turns out that traditional playgrounds are still part of their dreams. The images presented further confirm that traditional playgrounds are an interesting and challenging space for children. If it is related to the imagination where children are allowed to display their dream playground, then it seems that children are more focused on the activities they have seen and used, the sensory stimulation that they have felt so far is very visible from their drawings, the images produced by children are part of the stimulated process. Recall process [55, 33] even tends to visualize fragments of past information [32], representing the play activities they have engaged in. For children aged 9–11 years, the imagination expressed in an image has the aim of achieving cognitive and physical satisfaction according to their stage of development, "...*a more challenging playground...requires adrenaline...* " and their social development " "... *if you play ball... Bike... You can meet friends...*" This is very proven, with traditional playgrounds children can imagine what they will get from the existing equipment, in line with research which says that children can imagine desires and imagine action, [17].

However, it cannot be denied that alpha generation children have good digital literacy so they are quite adept at using technological games that can be easily accessed via digital devices. This research involved the majority of children from Medium Socio-Economic Status who had great opportunities to use technology-based games but did not choose this as part of fulfilling their play needs, boredom was their main reason for putting aside these games, "...*playing those games interesting, but sometimes boring...*", they need dynamic play activities in line with the character of alpha generation children who are creative and dynamism in thinking [22]. If viewed from a spatial perspective, for children especially the alpha generation, spatial play is not that important, if it can fulfill their play needs. From the results of the interviews, it was revealed that some of them enjoyed their solitude more, for them play was not entirely related to physical activity, "...*if you like to read story books at home...*". Understanding their drawings also does not provide any indication regarding the area, position, and spatial shape, one thing they hope for is the possibility of obtaining a spatial structure that can accommodate a diversity of play equipment.

For policymakers and urban planners, this study helps identify play equipment preferences for school and community playgrounds. Interview data shows that children prefer the play activities available in community playgrounds. Although adults select and purchase play equipment when building a new playground, the consumers of play equipment in these environments are children. A child's input into product selection can lead to better child outcomes in a variety of areas. This area of research may be worth exploring in the future [50]. How Do the Children Play? The Influence of Playground Type on Children's Play Styles. Research proves that children's involvement and receiving input from children in the design process is important [57], and urgent [58], top-down processes must be abandoned [59]. This form of children's participation in design occurs when children are actively involved and hear their voices.

6 Conclusion

Planning and designing a playground are not easy, and children's involvement in this process is often neglected. This research wants to fill this void, specifically involving the alpha generation as users of existing playgrounds. Even though there have been attacks from the Digital Playground through various technological media, alpha generation children still have a great desire to be able to play in traditional playgrounds and natural playgrounds. From the children's drawings and interviews carried out, there are differences in the need for playground design ideas. Based on research, it has been proven that sustainable playgrounds can be realized if 1. Provide natural playground elements, trees, and water, 2. Prefer equipment that provides traditional playgrounds, swings, and slides. 3. Another need of the alpha generation children is a playground that is more challenging and requires adrenaline, this is in line with the idea of a playground that is free, exciting, and surprising [29]. Efforts to fulfill these needs are expected to be able to create a sustainable playground and not make children become bored and leave the playground and move to another place. The results of this study are expected to shift the prevailing paradigm of playrooms, which tend to focus heavily on physical elements. The ideal playground, as perceived by children and their play patterns, is simpler than current designs. The inclusion of a "third point" element is key to the successful and sustainable utilization of playgrounds.

This research focused on children of the Alpha generation predominantly from lower socioeconomic backgrounds, which may differ if the study were conducted with children from middle to upper socioeconomic statuses. Further research is necessary to enhance the findings of this study, creating opportunities to establish sustainable playgrounds for the Alpha generation. Such spaces can be developed using data-driven guidelines that planners can implement.

References

1. Ramsay, G., Dodds, R., Furtado, D., Mykhayletska, Y., Kirichenko, A., Majedian, M.: The barriers to millennials visiting rouge urban national park. Sustainability **9**(6), 904 (2017)
2. Hyndman, B.: Recommendations for future school playground strategy research. In: Hyndman, B. (eds.) Contemporary School Playground Strategies for Healthy Students. Springer, Singapore (2017). https://doi.org/10.1007/978-981-10-4738-1_16
3. Chawla, L., Keena, K., Pevec, I., Stanley, E.: Green schoolyards as havens from stress and resources for resilience in childhood and adolescence. Health Place **28**, 1–13 (2014)
4. Naeini, H.S., Lindqvist, K., Jafari, H.R., Mirlohi, A.H., Dalal, K.: Playground injuries in children. Open Access J. Sports Med. **2**, 61 (2011)
5. Doll, B., Sarver, S.C.: Playground conflicts: everyday opportunities for children to manage conflict. In: Justice, Conflict, and Wellbeing, pp. 337–354. Springer, New York, NY (2014)
6. Boyle, D.E., Marshall, N.L., Robeson, W.W.: Gender at play: fourth-grade girls and boys on the playground. Am. Behav. Sci. **46**(10), 1326–1345 (2003). https://doi.org/10.1177/0002764203046010004
7. Mayeza, E.: 'Girls don't play soccer: children policing gender on the playground in a township primary school in South Africa. Gend. Educ. **29**(4), 476–494 (2017). https://doi.org/10.1080/09540253.2016.1187262
8. Imai, N., Shikano, A., Kidokoro, T., Noi, S.: Risky play and social behaviors among japanese preschoolers: direct observation method. Int. J. Environ. Res. Public Health **19**(13), 7889 (2022). https://doi.org/10.3390/ijerph19137889
9. Aminpour, F.: From the child's perspective: How the layout of outdoor school environments shapes conflict between children's self-directed play. J. Environ. Psychol. **79**, 101727 (2022)
10. Oostakhan, M., Babaei, A.: Safety assessment in the urban park environment in Alborz Province Iran. Int. J. Injury Control Safety Promot. **20**(1), 79–84 (2013)
11. Olsen, H., Kennedy, E.: Safety of school playgrounds: field analysis from a randomized sample. J. Sch. Nurs. **36**(5), 369–375 (2020)
12. Evans, J.: In search of peaceful playgrounds. Educ. Res. Perspect. **28**(1), 45–56 (2001)

13. Maxey, I.: Playgrounds: from oppressive spaces to sustainable places? Built Environ. (1978-) , 18–24 (1999)
14. Carr, V., Luken, E.: Playscapes: a pedagogical paradigm for play and learning. Int. J. Play **3**(1), 69–83 (2014)
15. Elias, N., Lemish, D., Dalyot, S., Floegel, D.: "Where are you?" an observational exploration of parental technoference in public places in the US and Israel. J. Child. Media **15**(3), 376–388 (2021)
16. Jha, A.K.: Understanding generation alpha (2020). https://doi.org/10.31219/osf.io/d2e8g
17. Kushnir, T.: Imagination and social cognition in childhood. Wiley Interdiscipl. Rev. Cogn. Sci. e1603 (2022)
18. Chang, N.: Children's drawings; science inquiry and beyond. Contemp. Issues Early Child. **6**(1), 104–106 (2005)
19. Deguara, J., Nutbrown, C.: Signs, symbols, and schemas: understanding meaning in a child's drawings. Int. J. Early Years Educ. **26**(1), 4–23 (2018). https://doi.org/10.1080/09669760.2017.1369398
20. Mart, M., Simsar, A., Uyanik, G.: The playground perception of syrian refugee children. Child Ind. Res. **15**, 349–372 (2022). https://doi.org/10.1007/s12187-021-09869-2
21. Günindi, Y.: The environment in my point of view: analysis of the perceptions of the environment of the children attending kindergarten through the pictures they draw. Procedia Soc. Behav. Sci. **55**, 594–603 (2012)
22. dos Reis, T.A.: Study on the alpha generation and the reflections of its behavior in the organizational environment. Quest J. J. Res. Humanit. Soc. Sci. **6**, 9–19 (2018)
23. Tootell, H., Freeman, M., Freeman, A.: Generation Alpha at the intersection of technology, play, and motivation. In: 2014 47th Hawaii International Conference on System Sciences, pp. 82–90. IEEE (2014)
24. Han, J., Heo, J., You, E.: Analysis of metaverse platform as a new play culture: focusing on roblox and Zepeto. In: International Conference on Human-centered Artificial Intelligence, pp. 1–10 (2021)
25. Early Education, The British Association for Early Childhood Education. Friedrich Froebel (2020). https://www.earlyeducation.org.uk/about-froebel
26. Haight, W.L., Wang, X.-L., Fung, H.H., Williams, K., Mintz, J.: Universal, developmental, and variable aspects of young children's play: a cross-cultural comparison of pretending at home. Child Dev. **70**, 1477–1488 (1999)
27. Ridge, M.: Play and games: an opinionated introduction. Philos Compass **14**(4), e12573 (2019)
28. Furley, P.: The nature and culture of physical activity play: towards a biocultural model of play, practice, and teaching in sports. Evol. Behav. Sci. **15**(2), 208 (2021)
29. Saragih, J.F.B., Tedja, M.: How do children create their space for play? In: IOP Conference Series: Earth and Environmental Science, vol. 109, no. 1, p. 012045. IOP Publishing (2017)
30. Kourti, A., et al.: Play behaviors in children during the COVID-19 pandemic: a review of the literature. Children **8**(8), 706 (2021)
31. Taylor, M.: Imagination. In: Zelazo, P.D., (ed.), The Oxford Handbook of Developmental Psychology, Body and Mind, vol. 1, pp. 791–831. Oxford University Press (2013)
32. Pelaprat, E., Cole, M.: "Minding the gap": Imagination, creativity, and human cognition. Integr. Psychol. Behav. Sci. **45**(4), 397–418 (2011)
33. Jonsson, S.: Hearing the Forest for the Trees: The Designing of a Soundscape with the Theoretical and Methodological Starting Points of World Building & Soundscape Ecology (2022)
34. Kan-Kılıç, D.: My dream playground: children's playground design. J. Curr. Res. Soc. Sci. **11**(2), 221–252 (2021)

35. Han, H.S.: Analysis of preferred element in constructing creative play space design. J. Digit. Convergence **18**(10), 389–396 (2020)
36. Raja Mohd Rozuan, R.M.A.A., Ismail, K., Wan Omar, W.R.: Safety factors for playground equipment to promote social sustainability (2019)
37. Ruth, L.C.: Playground design and equipment (2008). Web : http://www.wbdg.org
38. Burke, J.: Just for the fun of it: making playgrounds accessible to all children. World Leisure J. **55**(1), 83–95 (2013)
39. Koller, D., Farley, M.: Examining elements of children's place attachment. Children's Geographies **17**(4), 491–500 (2019). https://doi.org/10.1080/14733285.2019.1574336
40. Lundman, R.: Coping with standards: integrating art and safety into the design of a creative playground. J. Plan. Educ. Res. (2021). https://doi.org/10.1177/0739456X21994650
41. Sporrel, K., Caljouw, S.R., Withagen, R.: Children prefer a nonstandardized to a standardized jumping stone configuration: Playing time and judgments. J. Environ. Psychol. **53**, 131–137 (2017)
42. Cengiz, C., Boz, A.Z.: Biophilic playgrounds as playscapes in child-nature interaction. Int. J. Sci. Technol. Res. **5**, 216–226 (2019)
43. Hayward, D.G., Rothenberg, M., Beasley, R.R.: Children's play and urban playground environments: a comparison of traditional, contemporary, and adventure playground types. Environ. Behav. **6**(2), 131–168 (1974). https://doi.org/10.1177/001391657400600201
44. Smith, K.L., Shade, L.R.: Children's digital playgrounds as data assemblages: problematics of privacy, personalization, and promotional culture. Big Data Soc. **5**(2) (2018). https://doi.org/10.1177/2053951718805214
45. Grugeon, E., Harding, L.: Discovering creativity on the playground. In: Literacy Through Creativity, pp. 42–50. David Fulton Publishers (2012)
46. Deguara, J., Nutbrown, C.: Signs, symbols, and schemas: understanding meaning in a child's drawings. Int. J. Early Years Educ. **26**(1), 4–23 (2018)
47. Has-Tokarz, A.: Alpha generation in children's libraries in Poland. "Digital Kids–children's digital playground" is an example of a project. Інформація, комунікація, суспільство 2022, **19**, 75 (2022)
48. Kaplan-Berkley, S.: Digital tools and streaming media converge to inspire social interactions of generation alpha. Int. J. Early Childhood **54**(2), 185–201 (2022)
49. Wilson, E.O.: Sociobiology: The new synthesis. Harvard University Press (2000)
50. Stanton-Chapman, T.L., Schmidt, E.L.: How do the children play? the influence of playground type on children's play styles. Front. Psychol. **12**, 703940 (2021)
51. Nasar, J.L., Holloman, C.H.: Playground characteristics to encourage children to visit and play. J. Phys. Act. Health **10**(8), 1201–1208 (2013)
52. Khan, M., Bell, S., McGeown, S., de Oliveira, E.S.: Designing an outdoor learning environment for and with a primary school community: a case study in Bangladesh. Landscape research (2019)
53. Holmes, R.M., Procaccino, J.K.: Preschool children's outdoor play area preferences. Early Child Dev. Care **179**(8), 1103–1112 (2009). https://doi.org/10.1080/03004430701770694
54. Reimers, A.K., Knapp, G.: Playground usage and physical activity levels of children based on playground spatial features. J. Public Health **25**(6), 661–669 (2017)
55. Patera, M., Draper, S., Naef, M.: Exploring magic cottage: a virtual reality environment for stimulating children's imaginative writing. Interact. Learn. Environ. **16**(3), 245–263 (2008)
56. Ünal, M.: The place and importance of playgrounds in child development. Inonu Univ. J. Fac. Educ. (INUJFE), **10**(2) (2009)
57. Nah, K.-O., Lee, S.-M.: Actualizing children's participation in the development of outdoor play areas at an early childhood institution. Action Res. **14**(3), 335–351 (2016). https://doi.org/10.1177/1476750315621610

58. Pearson, R., Howe, J.: Pupil participation and playground design: listening and responding to children's views. Educ. Psychol. Pract. **33**(4), 356–370 (2017).https://doi.org/10.1080/02667363.2017.1326375
59. Spang, A.: Space, Play, and Resistance: Reconsidering Spatial Design from the Bottom Up (Doctoral dissertation, University of Colorado at Denver) (2022)

Traditional Wisdom and Measurement: *Petangan*'s Role in Javanese Architecture as Seen in the *Serat Kawruh Griya*

Primaditya Hakim[1(✉)], V. Totok Noerwasito[2], and I. G. Ngurah Antaryama[2]

[1] Doctoral Program Department of Architecture, Institut Teknologi Sepuluh Nopember, Kampus ITS, Surabaya 60111, Indonesia
prima@its.ac.id

[2] Department of Architecture, Institut Teknologi Sepuluh Nopember, Kampus ITS, Surabaya 60111, Indonesia

Abstract. *Petangan* is Javanese logical wisdom that has been tested by time and space. As a cultural heritage, this tradition still has a strong influence on the lives of Javanese people. In the context of the *Serat Kawruh Griya* manuscript, *petangan* is a five-five subtraction calculation used to determine the length of a traditional Javanese house frame component. This research aims to complement Prijotomo's research by examining a different manuscript, namely the *Serat Kawruh Griya* manuscript, with a focus on the disclosure of ideas, concepts and purposes of using *petangan* in the construction of Javanese houses in the *Serat Kawruh Griya* manuscript. By interpreting the *petangan* in the *Serat Kawruh Griya* manuscript, an enrichment of Javanese architectural knowledge is obtained, especially regarding measurement and proportion in architectural design. The results of Prijotomo's research, measurement, and interpretation have revealed that there are no elements of belief, mysticism, or cliché. The system of calculations revealed in the *Serat Kawruh Griya* manuscript is worthy of being juxtaposed with the western proportions that are familiar to the author.

Keywords: Traditional architecture · Manuscript · Interpretation

1 Introduction

There have been many studies on Javanese architecture, but little attention has been paid to the design factor, namely, size, which is known as *petangan*. Previous research on *petangan* was conducted by Prijotomo based on *Kawruh Kambeng*, a manuscript originating from Yogyakarta, with a focus on the *Limasan* houses of village people [1].

Prijotomo's research on *petangan* aims to raise the logical and rational side and remove clichéd impressions so that *petangan* can be appreciated as a design component in the field of scientific architecture, especially Javanese Architecture. The results of Prijotomo's research, measurement, and interpretation have revealed no elements of belief, mysticism, or cliché. All *petangan* were performed using common reasoning to measure and assign values to building objects. Based on this analysis, it is evident

M. Awang (Ed.): ICCAEET 2024, PTT, pp. 209–217, 2025.
https://doi.org/10.1007/978-3-031-94347-8_18

that *petangan* can influence the shape, appearance, room pattern, and sturdiness of the studied *Limasan* houses. Almost all the structures and parts of *Limasan* houses have survived as they were built, with the ages of the buildings varying from 40 to 120 years. This proves that the use of *petangan* is a suitable guideline for building durable houses that can accommodate user activities. At the end of his article, Prijotomo suggests that there is a need to study the units of measurement of *pecak*, *asta*, *cengkal*, and *depa* so that they can be standardized as units of measurement of inch, feet, yard, and mile.

This research aims to complement Prijotomo's research by examining a another manuscript, *Serat Kawruh Griya*, with a focus on the disclosure of ideas, concepts, and purposes of using *petangan* in the construction of Javanese houses in the *Serat Kawruh Griya* manuscript.

By interpreting the *petangan* in the *Serat Kawruh Griya* manuscript, an enrichment of Javanese architectural knowledge is obtained, especially regarding measurement and proportion in architectural design.

2 Methods

This research is explorative and interprets the main source of the research, namely, the Javanese and Indonesian texts of *Serat Kawruh Griya*. Interpretation attempts to interpret and determine the meaning and relationship of existing facts and then compare one fact with another to create a harmonious and logical series. Interpretation aims to synthesize some facts obtained from various sources and, together with theories, arrange the facts into a comprehensive interpretation so that they can be considered a form of analysis.

This interpretation theory is called hermeneutics. Paul Ricoeur, the originator of methodological hermeneutics, stated that hermeneutics is a study that reveals the objective meaning of texts that are separated in time and space from the reader. Ricouer's view shows that the main objective of hermeneutics is the text. In short, Ricouer shows that interpretation is an attempt to uncover unknown and hidden meanings in texts [2].

Paul Ricoeur's hermeneutic method focuses on interpreting texts by bridging the temporal and spatial distances between the reader and the text.

Ricoeur asserts that interpretation begins with the text, which holds an objective meaning that the reader seeks to uncover [3].

1. The text is autonomous and independent from its author, and its task is to reveal the latent meanings embedded in it.
2. Ricoeur's method involves a continuous process of The Hermeneutic Circle moving between the entire text and its parts to gradually build understanding. This iterative approach helps reveal deeper, hidden meanings
3. For manuscript analysis, Ricoeur's hermeneutics involves interpreting symbols, cultural context, and the historical distance between the reader and the manuscript.

The goal is to reveal cultural insights, architectural principles, or concepts that may not be immediately clear but are encoded within the text [4].

3 Results and Discussion

3.1 *Petangan*

Local wisdom refers to the knowledge and understanding that a community or society has developed over time. This knowledge is based on their experiences, traditions, and practices. It helps people in the community live better lives and solve their problems. Every community has its own unique way of experiencing life. This means that what is considered local wisdom in one place might be different in another. Local wisdom is not merely about having knowledge; it is also about sharing and expressing that knowledge. This can happen through stories, traditions, and practices that are important to the community. As communities face new challenges or learn new things, their wisdom evolves. This development is based on experiences that have been tested over many years. Understanding and using local wisdom is crucial for a community's survival. They help them adapt to their environment, solve problems, and maintain their cultural identity.

As a world society, the Javanese have always encountered various natural phenomena. Humans have faced many natural events throughout history, and these experiences have shaped their understanding of the world. From their intelligence, humans have produced varied scientific knowledge about nature and astronomical objects. These are important for many cultures, including the Javanese. Over time, basic scientific ideas about nature have grown and changed. This growth has created specific fields of study. These concepts of scientific knowledge have evolved rapidly, leading to other, more specialized fields, such as geography, astronomy, astrology, and horoscopes. The Arabs, Chinese, Babylonians, Indians, and Greeks were the forerunners and pioneers of scientific development. These scientific ideas and practices eventually made their way to the Malay region and the Island of Java. This means that the Javanese learned from these ancient cultures and adapted their knowledge to fit their lives and environment.

The Javanese are sensitive to cosmic forces that work in a holistic, comprehensive, and integrated manner. This sensitivity helped them understand their place in the world and how to live in harmony with it. Based on this, they contemplated or learned not just about gaining knowledge but also about paying attention to the "metacosmic" aspects. to things beyond the physical world, like spiritual beliefs and values. They also think about how their knowledge connects to the larger universe and their community. This leads humans to assimilate their conditions and the surrounding environment. Therefore, innovation in science and technology must consider its values for society. Local wisdom is positive human behavior related to nature and the surrounding environment [5].

Petangan Jawa was developed in Javanese cosmology and is the logical wisdom of the Javanese community. For many years, this tradition has been significant to the lives of the Javanese. *Petangan* Jawa is a valuable part of Javanese culture that has been passed down through generations. As a cultural heritage, the tradition still exists and has a strong influence on Javanese life. Orally handed down from one generation to the next. It helps them feel connected to their culture and history.

Javanese *petangan* has many variations, forms and categories. The written source of *petangan* can be found in primbon, a Javanese ancestral heritage book that contains guidance on the relationship between human life and the universe. One of the most popular and comprehensive Primbon is the manuscript Kitab Primbon Betaljemur

Adamakna, compiled by Kanjeng Pangeran Harya Tjakraningrat and consisting of 337 chapters covering various aspects of Javanese culture, such as astrology, divination, fate, destiny-luck, rituals, prayers, mantras, knowledge of physiology, and medicine, traditional songs containing magical value, and *petangan*, a method for calculating fortune, traits, and character [6]. Based on their purposes, *petangan* in Javanese primbon can be classified into 16 petung, namely: (1) matchmaking, (2) making a house, (3) birth of a baby, (4) traveling, (5) when agung, (6) moving house, (7) choosing a village to build a house, (8) days and markets, (9) the nature of iron or kris, (10) dreams, (11) kalamudheng, (12) loss, (12) buying livestock, (14) farming, (15) rain, and (16) causes of human illness. Javanese *petangan* is an accumulation of local wisdom that is full of wise advice for future generations [7].

3.2 *Petangan* in *Serat Kawruh Griya* Manuscript

Petungan or *petangan* is a Javanese word that means two things: first, arithmetical calculation; while the second is 'weighing and deciding'. The definition of *petangan* includes the notions of measuring, calculating, counting, considering and making judgments; *petangan* is an entity that has quantitative and/or qualitative aspects in its actions and results. The word '*petangan*' in the context of this *Sêrat Kawruh Griyå* text is counting, given that it is the remainder of a five-five subtraction that is the main objective, relating to knowing the number or amount of an item or thing. Obtaining the length of a house frame component through measurement, for example, is achieved by running the *petangan*; the same goes for counting the number of timbers to be used for the *usuk* of the roof. Also known as *kaso* or rafters, the wooden structure that becomes the ribs of the roof of the house, usually installed longitudinally from the top to the edge of the roof. The method to determine the length of the frame component is performed by applying *petangan*, which consists of the *sri*, *kitri*, *gana*, *liyu*, and *pokah* calculations. Each of these designations is believed by the Javanese to have a certain meaning.

The measurements used in Petungan are based on human size, specifically a unit called pecak. This unit is about the length of an adult's foot. Using human-sized measurements makes it easier for builders to create comfortable and practical spaces for people to use. With the unit of measurement '*pecak*' for example, the measurement of the first *pecak* is called *petangan* '*sri*', the second *pecak* is called *kitri*, the third *pecak* is called *gana*, the fourth *pecak* is called *liyu*, and the fifth *pecak* is called *pokah*. The sixth *pecak* is again called *sri*, the seventh pecak is called kitri, and so on. In the application of *petangan*, for example to determine the length of the *balandar* girder of a limasan house, it falls on the *sri petangan*, so it can point to 11 *pecaks*, 16 *pecaks*, 21 *pecaks* and so on [8].

The research data source is the result of a philological study conducted by Dwi Yuniarto, who translated the *Sêrat Kawruh Griyå* manuscript stored in the Sonobudoyo Museum Library Yogyakarta with collection number PB C. 6 Rol 103, number 4. The translation done by Yuniarto is a contextual translation, which is a combination of literal translation, content or meaning translation, and free translation. Literal translation is performed by translating Javanese words into Indonesian [9].

The following excerpt from *Serat Kawruh Griya* describes the measurement guidelines or *petangan* used to measure the length of the structural components of a Javanese house:

On page 26 of chapter 5 of the manuscript, the following text has been added:

The length and shortness of the *Pamidhangan* component, which consists of *pamidhangan rêmanjang* and *pamidhangan panyêlak*, determine the shape and variety of *Joglo* houses.

> "...*Pedoman ukuran panjang pendek pamidhangan, seperti; pamidhangan rêmanjang dan pamidhangan panyêlak rumah joglo...*" (Yuniarto, 2012: 143)

Pamidhangan is a special name for the *Blandar pangêrêt* in Joglo houses. *Blandar pangêrêt* is a structural component in *limasan* houses, *kampung* houses, and *masjid houses* and serves as a guide, determinant, or reference for other structural components, because on *Blandar pangêrêt* other structural components.

> "...*Blandar pangêrêt rumah limasan, rumah kampung, serta rumah masjid yang sebagai guru atau pedoman bertempatnya kerangka dan sejenisnya, seperti di atas, tidak hanya diharapkan pantas bentuk dan kuatnya saja, tetapi juga ada tujuan lain...*" (Yuniarto, 2012:144)

This is not only based on the context of the appropriateness of the shape and strength of the structure and on other contexts or purposes mentioned in the text. This calculation is based on the context of bringing sustenance, safety, and prosperity to the people who live in the house.

> "...*Orang Jawa percaya bahwa semua itu dapat mendatangkan re- jeki, keselamatan, ...*" (Yuniarto, 2012: 144)

The context or concept of bringing sustenance, safety and prosperity to the people who inhabit the house is realized in the use of calculating the size of the structural components of the *pamidhangan* or *blandar pangêrêt* with a system of calculating the remaining multiples of 5, namely 1. *Sri, 2. Kitri, 3. gånå, 4. Liyu, 5. Pokah.*

> "...*Harapan tadi diambil dari perhitungan ukuran /26/ panjang pendeknya pamidhangan atau blandar pangêrêt dengan adanya lima istilah, yaitu 1. Sri, 2. Kitri, 3. gånå, 4. Liyu, 5. Pokah,....*" (Yuniarto, 2012: 144)

The *pendapa* house, the length of the *blandar gandhok* is 26 feet, if you subtract a multiple of 5, which is 25 feet, there will be 1 foot left. The length of the *pangêrêt* is 16 feet, if reduced by a multiple of 5 which is 15 feet, 1 foot will remain. The calculation of size minus a multiple of 5 remaining one is called '*jatuh-sri*' calculation.

> "...*panjang pamidhangan atau blandar pangêrêt ukurannya jika dikurangi lima-lima harus tersisa satu. Misalnya, jika panjang blan- dar 26 kaki, jika dikurangi lima-lima sebanyak 5 kali atau 25 kaki akan tersisa 1 kaki. Begitu juga jika panjang pangêrêt 16 kaki, jika dikurangi lima-lima sebanyak 3 kali atau 15 kaki juga akan*

> *tersisa 1 kaki. Perhitungan ukuran yang dikurangi lima-lima sisa satu di- namakan jatuh sri..."* (Yuniarto, 2012: 144)

On page 28. For a *pendapa* with a *joglo* shape, the length of the *pamidhangan* is 17 feet, and 12 feet is the length of the *panyêlak pamidhangan*. The calculated subtraction of the multiple of 5 that remains 2, is called '*jatuh kitri*'.

> *"...Biasanya pendapa harus bentuk joglo, maka jika yang sebagai pedoman di atas panjang pamidhangan rumah /28/ joglo dibuat 17 kaki, panjang pamidhangan panyêlak 12 kaki. Hal itu disebabkan jika perhitungan dikurangi lima-lima sebanyak 3 kali atau 15 kaki tinggal 2, dikurangi lima-lima sebanyak 2 kali atau 10 kaki ugi ting- gal 2 disebut jatuh kitri..."* (Yuniarto, 2012: 145)

The purpose of applying calculations in determining the size of the house components is as a hope or prayer for the good of the users who inhabit the house in question. The word *sri* means food or wealth, glory, and light. The intention in using the calculation of the house with the provisions of *sri* has a hope or prayer that the people who occupy the house can get abundant fortune. People who occupy the house can get abundant fortune, do not lack food, and can have a lot of wealth, find glory, and light in their hearts.

> *"...Ada pendapat bahwa maksudnya adalah sebagai harapan atau doa terhadap kebaikan. Seperti kata sri, berarti makanan atau kekayaan, kemuliaan, serta terang. Jadi, dalam perhitungan rumah yang dijatuh- kan sri mempunyai harapan atau doa supaya yang menempati dapat mendapat rejeki yang melimpah, dengan demikian tidak kekurangan dalam makanan, serta dapat mempunyai banyak kekayaan, menemukan kemuliaan, serta terang hatinya..."* (Yuniarto, 2012: 147)

The word *sri* also means appropriate and good, meaning that it is carried out in accordance with the instructions (*pakèn*), so that it is in place, and so that the results of its creation are good or appropriate.

> *"Karena kata sri juga bermakna pantas serta kebaikan, maka ada pen- dapat mungkin sebagai ajaran supaya dikerjakan (pakèn), supaya ber- tempat pada tempatnya, serta supaya dalam pembuatannya pantas atau diharapkan hasilnya dapat bagus atau pantas."* (Yuniarto, 2012:147)

On page 32. The word *kitri* means plant. The meaning of this word is the shade produced from plants, so that the occupants feel cold from the shade. The coldness in question is the coldness of the body, the coldness of the heart, or not being spared from fortune. This means that the people who live in the house will be happy, refreshed, and healthy.

> *"Kitri berarti tanaman. Kata tanaman mungkin diambil dari teduhnya, sedangkan /32/ maksud untuk yang menempati diharap dinginnya. Dingin dapat berarti bermacam-macam, yaitu dinginnya badan, dinginnya hati, atau tidak terhindar dari rejeki dan barang ke- punyaannya. Singkatnya supaya yang menempati dapat senang, segar, dan sehat."* (Yuniarto, 2012: 148)

In explaining *petangan* for measuring the length of the frame components and counting the number of usuk, the manuscript of *Serat Kawruh Griya* includes *pakèn* describing the use of Javanese house forms. In Kawi, *pakèn* from *pakèna* means: benefit, purpose, for, useful. *Pakèn* is a guideline for the quality of a building, and as such, it tells us about its finishing, reliability, strength, and capacity.

The manuscript *Serat Kawruh Griya* describes the applicability of *pakèn* to Javanese house forms, particularly in terms of strength and stability. Another purpose of *pakèn* is to make it only necessary for its use. Therefore, the purpose of using the *kitri* rule in the calculation of *pandhåpå, paringgitan,* and *pesanggrahan* is to make them strong and durable, similar to trees or plants that live long. And there is no need to use a lot of decorations in the house.

> "*Ada pendapat lagi mungkin maksudnya sebagai ajaran supaya dik- erjakan (pakèn). Supaya pembuatannya hanya seperlunya saja dalam kegunaannya. Jadi perhitungan pandhåpå, paringgitan, serta pesang- grahan semua dijatuhkan kitri supaya dalam pembuatannya hanya di- harap kuat dan tahan lama seperti pohon atau tanaman yang berumur panjang. Tidak perlu menggunakan banyak hiasan seperti* rumah." (Yuniarto, 2012: 148)

On page 35. The reason that all sizes use the foot, dim, or *sêtrip* guidelines is because this is the size used in the palace or leaders (*keraton*) at that time. Other parties may also use this guideline. In ancient times, size guidelines were taken from the *pêcak, kilan, têbah* or *dhêpå* of the house owner. At this time, there are also people who still use such measurements. The start of the use of foot, dim or *sêtrip* size according to new estimates is approximately 50 years. This was mentioned in the introduction of the *Serat Kawruh Griya.* Initially, this wooden house construction technology was first applied to the damaged king's palace building. After successfully building palaces with good and sturdy construction, this wooden structure technology was applied to people's homes.

> "*Adapun alasan semua ukuran menggunakan pedoman kaki, dim atau sêtrip seperti di depan tadi karena yang dipakai dalam keraton atau para pemimpin pada saat ini. Orang lain juga ada yang menggunakan pedoman seperti itu. Jika pada jaman kuna pedoman /35/ ukuran tadi diambil dari pêcak, kilan, têbah atau dhêpå dari yang punya rumah dan lain-lain. Bahkan, sekarang juga masih ada yang menggunakan ukuran seperti itu. Mulai adanya ukuran kaki, dim atau sêtrip dari perkiraan baru kurang lebih 50 tahun.*" (Yuniarto, 2012: 150)

As for the guidelines for the direction facing the house used, in order to be sustainable applied, because if not explained will cause harm and so on, ordinary people are often not believed. Ancient people may have used a method derived from causation, and the *Qibla* direction was guided by the rotation and standing of the earth.

> "*Adapun pedoman arah yang seperti itu, dari perkiraan hanya supaya lestari dilakukan, karena jika tidak dijelaskan menyebabkan celaka dan sebagainya, serta sering tidak dipercaya. Keinginan orang kuna mungkin pengambilannya berasal dari sebab dan kiblat yang sebenarnya berpedoman pada perputaran dan berdirinya bumi.*" (Yuniarto, 2012: 153)

4 Conclusions

Unlike the primbon form, *Serat Kawruh Griya* is not in verse or *tembang* form, but in prose form with a writing structure similar to the writing structure of scientific report books today, consisting of an opening section or chapter, a number of content chapters and a closing section.

Serat Kawruh Griya is a specific manuscript in the form of a guide to building houses for *undagi*, Javanese traditional house builders. It should be noted that the scope of *petangan* in the *Serat Kawruh Griya* manuscript is a guide to the rules regarding the size and location of the construction components of a Javanese house, which differs from the understanding of *petangan* in the calculation of time or good days for building a house, marriage, or a match.

The text of *Serat Kawruh Griya* explains how Javanese people value the importance of a house as a place to live. This is a prime example of how Javanese house architecture reflects the values of the Javanese way of life.

The Javanese view their homes as deeply connected to their identity and spirit. The size and design of a house are carefully considered. The house dimensions were determined by comparing the body sizes of the owner's body sizes. The unity between the house and the owner's soul suggests that the home is reflecting owner's personality and cultural values. There is a special connection between a house and the person who lives in it. Striving for harmony between the owner and the house. For the Javanese, the home is a place where family bonds are formed and traditions are passed down [10].

The *Serat Kawruh Griya* manuscript measures the size of a Javanese house's framework using *petangan* and obtains the number of *usuk* on the roof of a Javanese house. *Petangan* is a system adopted from western measurement systems whose naming and measurement systems are adapted to local Javanese measures and terms. This system was originally only used and known by the palace or high-ranking government officials of the Javanese kingdom, *Keraton*, so it had a high, hierarchical, and foreign impression for laymen. *Petangan* is deeply rooted in Javanese culture, reflecting the community's values and beliefs. It connects the architectural design to the spiritual and social aspects of life, emphasizing the relationship between humans and nature.

The calculation system revealed in the *Serat Kawruh Griya* manuscript is worthy of juxtaposing with the western proportions that we have known before, as evidenced in the study conducted by Irwan Harnoko, that the Javanese architectural calculation formula can also be used in Javanese visual communication design in this case in the proportion of Javanese puppets [11].

References

1. Prijotomo, J., Octa, L., Roosandriantini, J.: Petungan dan Proporsi di Arsitektur Jawa-Yogyakarta. Published online, 15–26 (2022)
2. Arthana, I.NN.: The knowledge building of construction process of bali arya's architecture: interpretation of the manuscript asta Kosala Kosali. J Archit. Res Educ. **1**(2) (2019)
3. Kublikowski, I.: Reflections on analysis in qualitative research: the hermeneutic circle in ricoeur. Paideia. **33** (2023). https://doi.org/10.1590/1982-4327e3319

4. Tan, H., Wilson, A., Olver, I.: Ricoeur's theory of interpretation: an instrument for data interpretation in hermeneutic phenomenology. Int. J. Qual. Methods **8**(4) (2009). https://doi.org/10.1177/160940690900800401
5. Denny Firmanto, A., Setiyaningtiyas, N.: Local wisdom as integration of the cosmic and the metacosmic perspectives in indonesian philosophy of education. In: 3rd International Conference on Education and Training (ICET 2017) (2017). https://doi.org/10.2991/icet-17.2017.29
6. Widodo, S.T., Saddhono, K.: Petangan tradition in Javanese personal naming practice: an ethnoliguistic study. GEMA Online J. Lang. Stud. **12**(4) (2012)
7. Hartono. Petung dalam Primbon Jawa. Litera. **15**(2), 256–268 (2016)
8. Prijotomo, J.: (Re)-Konstruksi Arsitektur Jawa: Griya Jawa Dalam TradisiTanpatulisan. PT. Wastu Lanas Grafika (2006)
9. Yuniarto, D.: Kajian Filologi Dan Isi Dalam Sêrat Kawruh Griyå. Published Online (2012)
10. Santosa, A., Rachmawati, M., Noerwasito, V.T.: Material and cultural values in Joglo architecture in Indonesia. ISVS e-journal. **10**(2) (2023)
11. Harnoko, I.: Petungan sebagai Sistem Ukuran dalam Desain Komunikasi Visual Jawa. J Desain. **4**(01), 38 (2016). https://doi.org/10.30998/jurnaldesain.v4i01.913

Evaluation of Enterprise Resource Planning (ERP) Systems for Contractors Using TRL and TAM Methods

Rizqi Choirul Wahdana(✉) and Jojok Widodo Soetjipto

Teknik Sipil, Fakultas Teknik, Universitas Jember, Jalan Kalimantan No. 37 – Kampus Bumi Tegalboto Kotak POS 159, Jember, Jawa Timur 68121, Indonesia
rizqidana99@gmail.com

Abstract. This study aims to evaluate user acceptance and technology readiness of an ERP system implemented in a contracting company. The research employs two primary evaluation methods: the Technology Acceptance Model (TAM) to assess user perceptions regarding the system's perceived usefulness and ease of use, and the Technology Readiness Level (TRL) to evaluate the ERP system's technological readiness in supporting the company's daily operations. Findings indicate that while the ERP system enjoys a high level of user acceptance, there are still areas requiring improvement to achieve a higher level of technological readiness. The study offers strategic recommendations to enhance the implementation and utilization of the ERP system. Furthermore, it identifies several critical factors affecting ERP adoption and success, such as management support, user training, and alignment between business processes and ERP features. By understanding these factors, companies can develop more effective strategies to address challenges and maximize the benefits of ERP. Additionally, the study emphasizes the importance of ongoing evaluation to ensure that the ERP system remains relevant and capable of meeting evolving business needs.

Keywords: Enterprise Recourse Planning (ERP) · Technology Acceptance Model (TAM) · Technology Readiness Level (TRL) · Evaluasi Sistem (system evaluation)

1 Introduction

Effective project management in construction requires the efficient management of project resources, which must be organized within a robust management system to be utilized optimally [1]. One of the increasingly prominent systems in this context is Enterprise Resource Planning (ERP).

Enterprise Resource Planning (ERP) systems are information systems designed to integrate and automate business processes within a company with transparency and high accountability, thereby facilitating automation and standardization of operational processes [2]. Often referred to as Back Office Systems, these systems do not involve

M. Awang (Ed.): ICCAEET 2024, PTT, pp. 218–231, 2025.
https://doi.org/10.1007/978-3-031-94347-8_19

direct interaction with consumers or system users. Typically modular, ERP systems handle manufacturing, logistics, distribution, inventory, shipping, payment, and accounting processes [3]. By integrating systems, each functional unit within a company can share data and information, which ultimately enhances synergy among company elements [4]. This integration enables organizations to operate more efficiently and respond more effectively to changes, as real-time information can be accessed by various departments, aiding in faster and more accurate decision-making. Furthermore, system integration reduces data duplication and errors, which frequently occur in manual processes or when using disparate systems, thereby improving overall operational quality. ERP systems also help companies automate and standardize operational processes that were previously managed manually, implementing best practices to boost productivity, reduce inefficiencies, and enhance product quality. In many companies, ERP systems are seen as a pillar of business intelligence, offering integrated processes across various functional areas with improved workflow quality, standardized business practices, and real-time data access [5].

Integrated systems used by companies to unify all corporate resources are essential for efficient planning and resource management [6]. In the digital era, ERP systems have become crucial for integrating and managing various operational aspects. PT Brantas Abipraya (Persero), a state-owned construction company, has implemented an ERP system to enhance operational efficiency and data integration across departments. This study focuses on evaluating user acceptance and technology readiness of the ERP system using TAM and TRL methodologies.

The implementation of ERP systems in the construction industry often encounters various challenges, including data integration from different departments and the need to tailor the system to specific industry practices. Research indicates that a deep understanding of these challenges and the application of appropriate solutions can improve the success of ERP implementation in construction firms, thereby enhancing project management and overall operational efficiency [7]. In ERP implementation within construction companies, project management plays a crucial role in ensuring effective integration into organizational workflows [8].

This study is essential for evaluating user acceptance and technology readiness, which are key factors in the success of ERP implementation. By employing TAM and TRL, the research aims to provide comprehensive insights into how users accept new technology and the extent to which the technology is ready for daily operations. Therefore, the study will evaluate: (1) the application of ERP business models by PT Brantas Abipraya (Persero); (2) the readiness of the ERP system for use at PT Brantas Abipraya (Persero); (3) the company's acceptance level of the ERP system; and (4) recommendations for improving the ERP system.

2 Literature Review

2.1 Enterprise Resource Planning System (ERP)

ERP is an integrated business management system that combines all aspects of a company's operations, including planning, inventory, purchasing, sales, marketing, finance, and human resources. In the construction industry, ERP is used to manage projects, integrate various departments, and enhance visibility into business processes [3].

There are several types of ERP systems based on their deployment or implementation within a company:

1. ERP On Premise
2. CLoud ERP
3. Hybrid ERP [9]

2.2 Business Model Canvas (BMC)

The Business Model Canvas (BMC) is a framework for describing, visualizing, assessing, and changing business models [10].

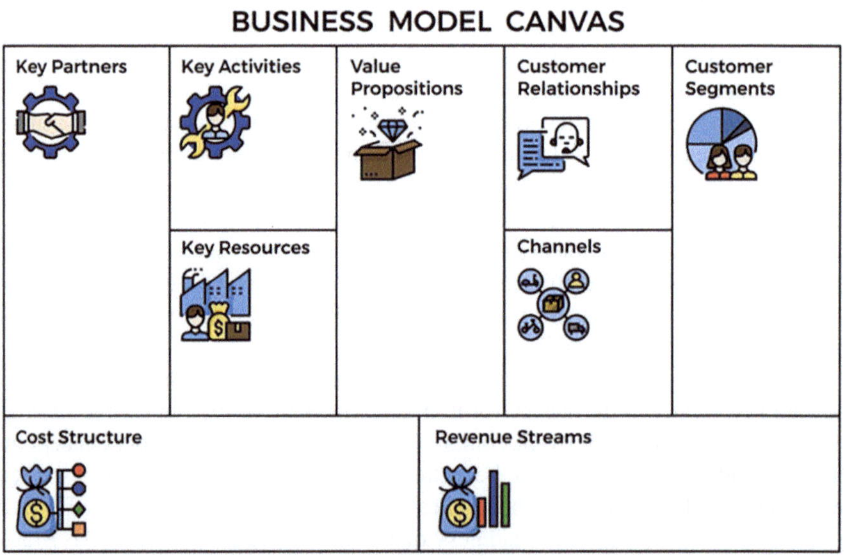

The BMC consists of nine essential elements that guide the development of a company's operational system:

1. ***Key partners***: Relationships and collaborations that facilitate smooth business operations and service organization according to company expectations.
2. ***Key activities***: The activities necessary to achieve business goals and objectives.
3. ***Key resource***: The critical assets required for a business model to function effectively (Kuhne & Bohmann 2018)
4. ***Value proposition***: The reasons customers choose one company over another.

5. ***Customer relationship***: The ways in which a company establishes and maintains connections with its customers.
6. ***Channels***: Media or channels used for marketing products and communicating with potential consumers.
7. ***Customer segment***: Identifying the target buyers, market segments, and the suitability of products for these segments.
8. ***Cost structure***: The financial structure supporting the company's operations.
9. ***Revenue stream***: The income received from different market segments [11]

2.3 Technology Acceptance Model (TAM)

The Technology Acceptance Model (TAM), developed by Davis (1989), is used to predict and explain how users accept and utilize technology related to their work [12].

TAM focuses on two main variables: Perceived Usefulness (PU), which measures the extent to which users believe that using the technology will enhance their performance, and Perceived Ease of Use (PEOU), which measures how easy users find the technology to use [12].

Using TAM to evaluate user acceptance and TRL to assess technology readiness provides a holistic view of the factors influencing ERP implementation success. This evaluation ensures that the ERP system meets technical requirements and is effectively accepted and utilized by users [13].

1. TAM is developed from the Theory of Reasoned Action.
2. It provides a basis for tracking how external variables influence beliefs, attitudes, and intentions to use new technology.
3. It predicts new IT acceptance and has been reliable in explaining acceptance behavior in various information systems contexts in Indonesia [14]

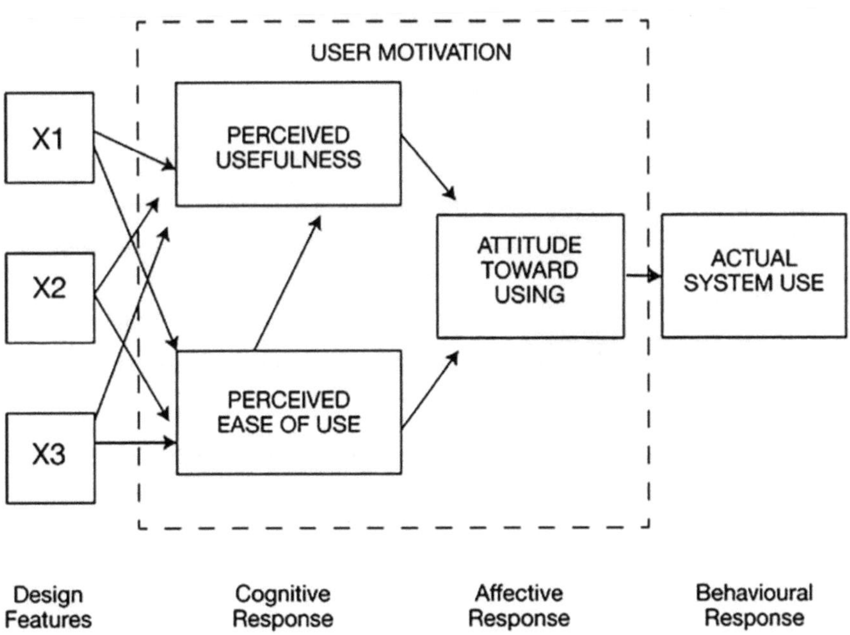

2.4 Technology Readiness Level (TRL)

TRL is an evaluation framework used to assess the maturity level of a technology for implementation in operational environments. TRL consists of 9 levels, ranging from basic research (TRL 1) to proven systems functioning in operational environments (TRL 9). TRL evaluation helps identify areas needing improvement to achieve higher technology readiness.

TRL serves as an "indicator" of how ready or mature a technology is for real-world application and user adoption [15].

Technology readiness can be assessed from two interrelated perspectives:

1. **Objective Perspective**:
2. Examines the characteristics or attributes of the technology.
3. **Subjective Perspective:**
4. Considers the views of stakeholders regarding the technology [15].

NASA's TRL framework includes nine levels, an evolution from an earlier seven-level concept [15].

According to the US GAO, TRL is crucial for aligning technology with product launch requirements. To enhance the success of technology programs, it is recommended that research institutions and development organizations achieve at least TRL 6 [15].

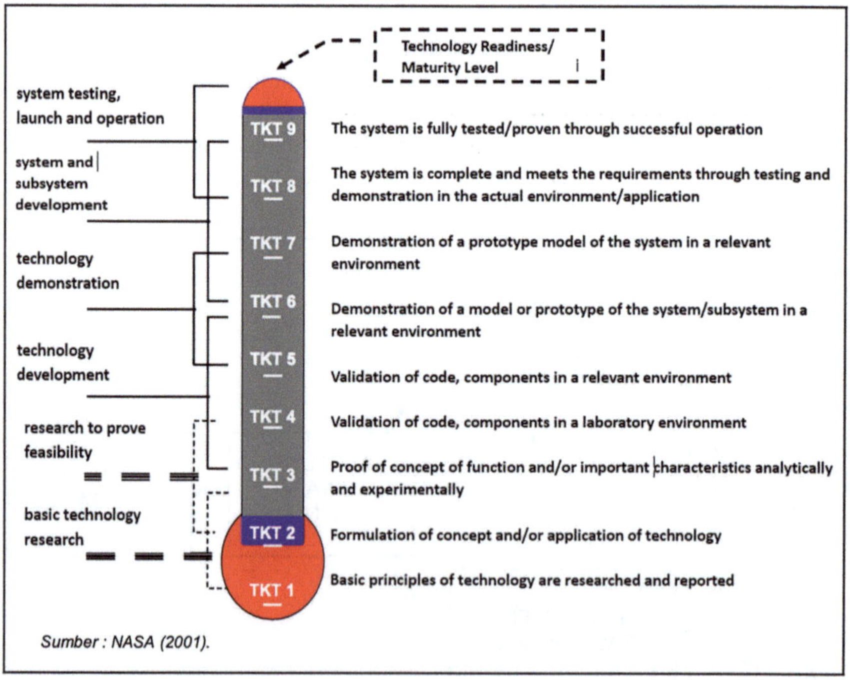

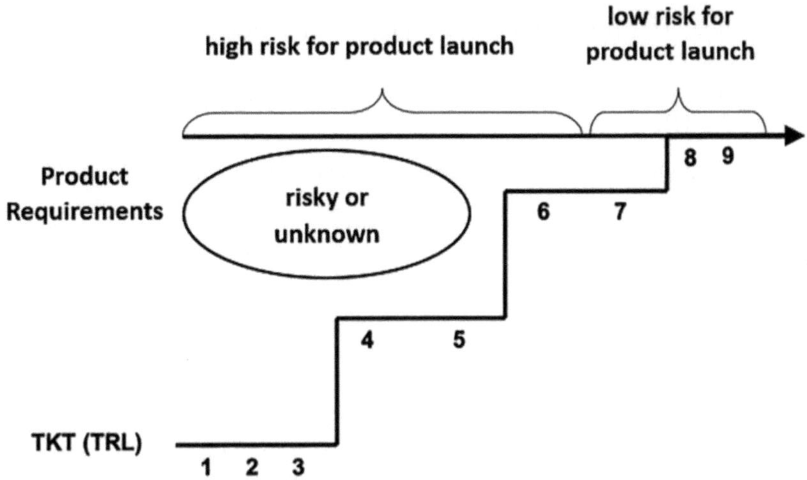

Evaluating technology readiness using the TRL model provides essential insights into how mature ERP technology is before its implementation in the construction sector. Concurrently, user acceptance measured through TAM helps identify potential adoption barriers and assess user readiness to integrate new technology into their workflows [16].

In construction companies, ERP system acceptance is often influenced by factors such as technology readiness and ease of use. Both the TAM and TRL models offer comprehensive frameworks for evaluating how users accept new technologies and how ready the technology is for deployment. This evaluation is crucial to ensure that ERP implementation meets both functional needs and technological readiness for integration into company operations. [17].

Combining TAM and TRL provides an effective approach to assessing ERP systems in the context of contractor implementation. This research reveals that integrating these models allows for a better assessment of technology readiness and user acceptance, ultimately enhancing the success of ERP implementation [18]. Comparing technology readiness and user acceptance in ERP implementations in developing countries highlights the importance of evaluating TRL and TAM for the successful deployment of systems in companies [19].

3 Research Methodology

This study employs a quantitative approach with a survey method to evaluate the ERP system at PT Brantas Abipraya (Persero). Data collection is conducted through a questionnaire designed to measure constructs from both the Technology Acceptance Model (TAM) and the Technology Readiness Level (TRL).

The questionnaire is distributed to ERP system users at PT Brantas Abipraya (Persero), including project managers, project staff, and central staff. The survey is administered online via Google Forms to randomly selected respondents. The questionnaire consists of two main sections:

1. Business Process Model

2. TAM Section: This part measures perceptions of the ease of use and usefulness of the ERP system.
3. TRL Section: This part assesses the technology readiness level of the ERP system based on the TRL scale.

The sample includes project managers, project staff, and central staff at PT Brantas Abipraya (Persero).

$$n = \frac{N}{1 + N(e)^2} \tag{1}$$

n = Number of Samples
N = Population Size
E = Error Rate

Based on Eq. (1), the study requires 21 respondents for the Technology Readiness Level (TRL) and 31 respondents for the Technology Acceptance Model (TAM).

The data analysis process for this research is outlined as follows:

1. Business Process Model:
   - Identification: Identify the Business Process Model at PT Brantas Abipraya (Persero) using the Business Model Canvas (BMC) tool.
   - Analysis: Conduct an analysis based on the nine elements of the BMC.
2. TAM Section:
   - Feedback Analysis: Analyze feedback from the questionnaire distributed to respondents regarding the use of the ERP application.
   - TAM Analysis: Apply TAM analysis to assess acceptance of the technology based on the collected feedback.
3. TRL Section:
   - Questionnaire Design: Develop a questionnaire based on the ERP technology readiness framework.
   - Questionnaire Distribution: Distribute the questionnaire randomly to respondents who are directly involved with the technology.
   - Data Analysis: Analyze the data to determine the technology readiness level of the ERP system.

4 Discussion and Result

4.1 Business Model Canvas (BMC)

The application of the Enterprise Resource Planning (ERP) business model by PT Brantas Abipraya (Persero) using the Business Model Canvas (BMC) method is as follows:

Customer Segments
The ERP system at PT Brantas Abipraya supports various customer segments by providing integrated solutions for project management, financial tracking, and resource

allocation. This integration enhances the company's ability to effectively serve diverse market segments.

Value Proposition

The ERP system contributes to the company's value proposition by improving operational efficiency, accuracy, and decision-making capabilities. It enables timely access to comprehensive data, supporting better project management and customer service.

Channels

The ERP system facilitates internal channels for communication and data sharing, enhancing coordination between departments. It also supports external channels by providing accurate information to clients and partners.

Customer Relationships

The ERP system enhances customer relationships by offering a centralized platform for managing client interactions and tracking project progress. This integration ensures consistent and transparent communication processes.

Revenue Streams

The ERP system supports various revenue streams by streamlining financial processes such as billing, invoicing, and cost management. It enables accurate financial reporting and analysis, contributing to improved revenue management.

Key Resources

Key resources supported by the ERP system include human resources, financial assets, and technological infrastructure. The system provides tools for efficient and effective management of these resources.

Key Activities

The ERP system automates and integrates key activities such as procurement, inventory management, and project tracking. This automation reduces manual efforts and enhances process efficiency.

Key Partnerships

The ERP system supports key partnerships by providing integrated data and communication tools. This integration facilitates collaboration with suppliers, contractors, and other external partners.

Cost Structure

The ERP system impacts the cost structure by optimizing resource utilization and reducing operational inefficiencies. It allows for better cost tracking and management, contributing to overall cost savings [20] (Fig. 1).

Fig. 1. Business Model Canvas (BMC) Results

The results of the Business Model Canvas (BMC) analysis, based on qualitative data gathered through interviews, are as follows.

Technology Acceptance Model (TAM)

The results of the survey conducted among ERP system users at PT Brantas Abipraya (Persero)—including project managers, project staff, and central staff—are presented in the following diagram:

The results of the questionnaire indicate that the majority of respondents hold the position of Project Manager. These individuals have significant experience, typically ranging from 5 to 10 years, and possess a Bachelor's degree (S1). Additionally, the respondent group is predominantly composed of employees aged between 20 and 30 years.

The findings from the Technology Acceptance Model (TAM) questionnaire are presented in the table below (Table 1).

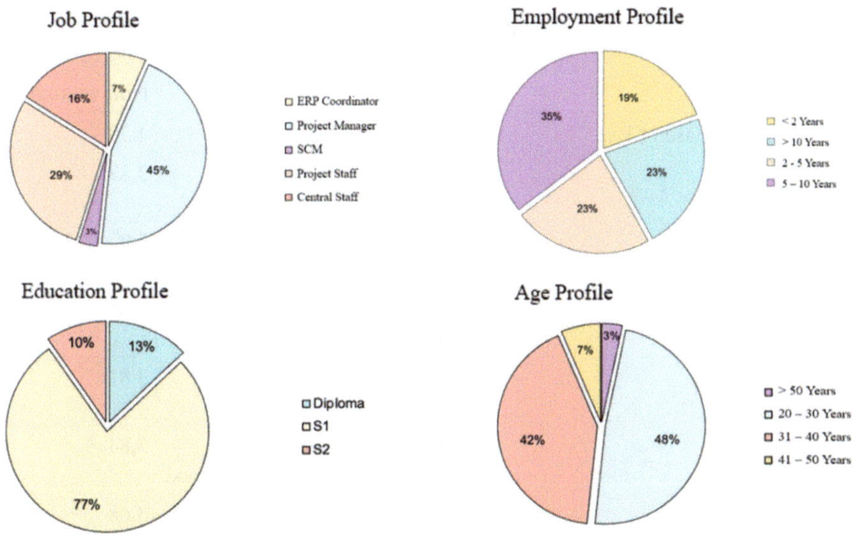

Table 1. Technology Acceptance Model (TAM) Respondent Questionnaire Results

No	Variable	Item	Nilai rata-rata
1.	Perceived Usefulness of ERP	ERP systems can increase overall productivity	3,77
		ERP systems can improve work effectiveness	3,74
		ERP system has accurate data	3,61
		ERP system allows integration with other information technology systems	4,06
		ERP system meets the needs of users	3,65
		ERP system improves customer service	3,74
		ERP system can reduce organizational costs	3,84
		ERP system can help in running business processes faster	3,94

(continued)

Table 1. (*continued*)

No	Variable	Item	Nilai rata-rata
2.	Perceived Ease of Use	I know what ERP system is	4,29
		ERP system is clear and understandable	3,87
		I understand how to operate ERP system	3,97
		ERP system is easy to learn and operate	3,65
		I understand the features in ERP system	3,81
		Information generated by ERP system can be used and understood	3,84
		Information generated by ERP system is concise, relevant, and available	3,68
3.	Attitude Toward Using	I am motivated to use ERP system	4,10
		I feel the advantages and benefits in implementing ERP system	3,90
		My understanding has increased through training program to use ERP system	4,03
		I feel helped by ERP system	3,94
		Using ERP system makes work under control	3,90
4.	Level of Acceptance of ERP Technology	ERP system is easy to accept in Company system	3,71
		Is the output result of ERP system according to what is desired	3,52
		ERP system is reliable	3,58
		I feel ERP system features are complete	3,29
		Satisfaction of using ERP system for work	3,65
		Total 3,80	

4.2 Technology Readiness Level (TRL)

The results of the survey, conducted via questionnaires distributed to ERP system users at PT Brantas Abipraya (Persero)—including Project Managers, Project Staff, and Central Staff—are presented in the following diagram (Fig. 2).

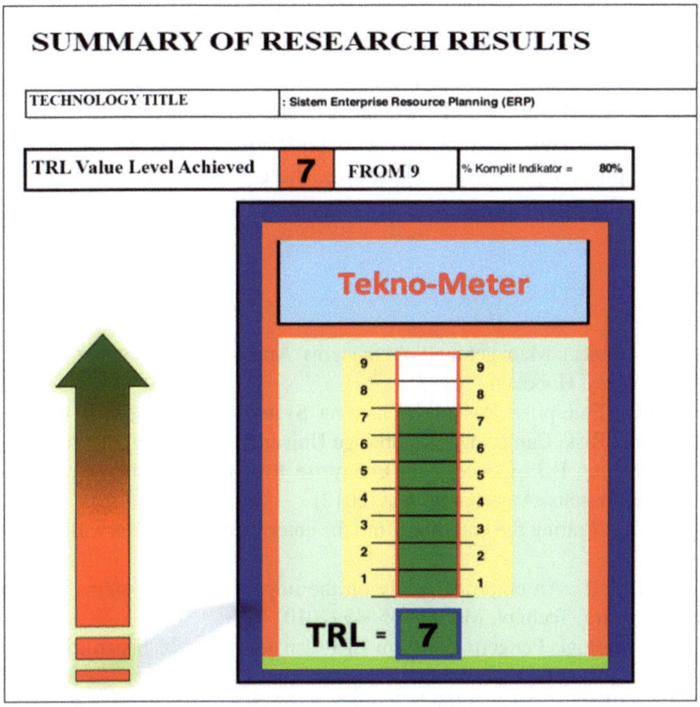

Fig. 2. Results of Technology Readiness Level (TRL) Measurement

Technology Readiness Level (TRL) = 7, This level indicates that the prototype is approaching or aligned with its operational system plan. This state reflects an advancement from TRL 6 and requires the demonstration of a functional prototype within an operational environment.

5 Conclusion and Recommendation

PT Brantas Abipraya (Persero), a national contractor, has successfully developed an ERP system categorized as "excellent." This assessment is based on the analysis of the Business Model Canvas (BMC) designed by the company. The "excellent" category is evident from the ERP system's capability to accommodate all business processes of the company. Specifically, the system has effectively streamlined operations, enhanced procurement processes, increased profitability, and provided early warnings of potential project losses.

The evaluation of the Technology Readiness Level (TRL) revealed a TRL of 7, indicating that the system is close to or aligned with its planned operational system. This level suggests that while the prototype is advanced, it still requires demonstration within an operational environment. The TRL score was derived from both qualitative and quantitative assessments of the technology's capabilities, implying that further testing and refinement of the prototype are necessary. In terms of technology acceptance, the

ERP system has achieved a level 4 on a scale of 1 to 5, as measured by the Technology Acceptance Model (TAM). This level indicates that the staff at PT Brantas Abipraya (Persero) are well-prepared to adopt the Enterprise Resource Planning (ERP) system.

However, the ERP implementation faces some challenges, particularly from the IT and Supply Chain Management (SCM) departments, due to incomplete integration and ongoing adaptation processes. Therefore, the company should continue to make improvements and refinements to enhance the system's effectiveness and integration.

References

1. Kerzner, H.: Project Management: A Systems Approach to Planning, Scheduling, and Controlling. Wiley, Hoboken (2017)
2. O'Leary, D.E.: Enterprise Resource Planning Systems: Systems, Life Cycle, Electronic Commerce, and Risk, Cambridge. Cambridge University Press, Inggris (2000)
3. Monk, E., Wagner, B.J.: Concepts in Enterprise Resource Planning. Course Technology, Boston, Massachusetts, Amerika Serikat (2012)
4. Davenport, T.H.: Putting the enterprise into the enterprise system. Harv. Bus. Rev., 121–131 (1998)
5. H., Mottaghi, A.H.: An empirical study on the impact of ERP systems on organizational performance. J. Inf. Technol. Manag., 35–45 (2010)
6. D. S. I. d. Teknologi: Pengertian sistem ERP dan fungsinya bagi perusahaan, Jawa Barat (2021)
7. C., Liu, W.Y.: ERP implementation in construction industry: challenges and solutions. Int. J. Proj. Manag., **33**(7), 1527–1538 (2015)
8. El-Sabaa, S.: The role of project management in the success of ERP implementation: a case study in the construction industry. J. Constr. Eng. Manag., 135(12), 1225–1233 (2009)
9. A., Maulana, P.: Enterprise Resource Planning (ERP), December 2020. https://www.researchgate.net/publication/346969919_ENTERPRISE_RESOURCE_PLANNING_ERP
10. Osterwalder, A., Pigneur, Y.: Business Model Generation: A Handbook for Visionaries, Game Changers, and Challengers. Wiley, Hoboken (2012)
11. Keane, S.F.,Cormican, K.T., Sheahan, J.N.: Comparing how entrepreneurs and managers represent the elements of the business model canvas. J. Bus. Venturing Insights, 9, 65–74 (2018)
12. Davis, F.D.: A Technology Acceptance Model for Empirically Testing New End User Information Systems. Sloan School of Management, Massachusetts Institute of Technology (MIT), Cambridge (1986)
13. S., Nugroho, R.F.: Penggunaan model TAM dan TRL dalam evaluasi sistem ERP pada Kontraktor. Jurnal Sistem Informas **18**(1), 35–45 (2020)
14. Wu, L., Li, J.Y., Fu, C.Y.: The adoption of mobile healthcare by hospital's professionals. Decis. Support Syst., 587–596 (2011)
15. Taufik, T.A.: Konsep dan Metode Pengukuran Tingkat Kesiapan Teknologi/TKT (Technology Readiness Level/TRL). ResearchGate, Indonesia (2005)
16. T., Hidayat, S.N.: Kesiapan teknologi dan penerimaan pengguna dalam implementasi ERP pada industri konstruksi. Jurnal Manajemen dan Teknologi **14**(2), 142–156 (2021)
17. R., Aditya, P.R.: Valuasi penerimaan sistem ERP pada perusahaan konstruksi menggunakan model TAM dan TRL. Jurnal Teknologi Informasi **12**(4), 567–578 (2019)
18. I., Jahnke, H.G.: Technology acceptance model and ERP implementation: a case study. Eur. J. Inf. Syst. **21**(6), 643–656 (2012)

19. V., Kumar, V.H.J.: RP implementations in developing countries: a comparative study of technological readiness and user acceptance. Inf. Syst. J. **17**(3), 223–244 (2007)
20. Osterwalder, A., Pigneur, Y.: Business Model Generation: A Handbook for Visionaries, Game Changers, and Challengers. Wiley, Hoboken (2010)

The Development of a Pavement Damage Assessment Model for Runway Surfaces Using Low-Cost Drone LiDAR: A Review Paper

Adhitya Surya Manggala[1], Ervina Ahyudanari[1(✉)], Mokhamad Nur Cahyadi[2], and Catur Arif Prastyanto[1]

[1] Department of Civil Engineering, Institut Teknologi Sepuluh Nopember, Surabaya, Indonesia
ervina@ce.its.ac.id

[2] Department of Geomatics Engineering, Institut Teknologi Sepuluh Nopember, Surabaya, Indonesia

Abstract. Maintenance and assessment of runway pavement conditions are vital to ensure airport safety. Traditional methods of runway inspection, which often rely on manual surveys and visual assessments, can be time-consuming, labor-intensive, and prone to errors. In recent years, the integration of unmanned aerial vehicles (UAVs) equipped with LiDAR technology has emerged as a promising alternative for improving the accuracy and efficiency of pavement damage detection. This review paper provides a comprehensive analysis of the current state of research on the use of low-cost drone-based LiDAR systems for runway surface assessment. The review explores key advancements in the development of pavement damage assessment models, including methods for detecting surface irregularities, quantifying damage severity, and classifying various types of distress such as cracking, rutting, and surface deformation. Additionally, the paper discusses the advantages of this approach over traditional methods, such as faster data collection, cost-effectiveness, and enhanced spatial accuracy. Challenges related to data processing, system integration, and regulatory compliance are also highlighted. By synthesizing recent findings and identifying research gaps, this review aims to provide insights into the potential of low-cost drone LiDAR technology to revolutionize runway maintenance practices. The paper concludes by outlining future research directions, particularly in the areas of automated data processing, real-time damage detection, and system scalability for larger airport infrastructure.

Keywords: Pavement Damage · Assessment Model · Runway Surfaces · Low-Cost Drone LiDAR

1 Introduction

The assessment and maintenance of infrastructure, particularly roadways and runways, are essential to ensuring safety and operational sustainability in transportation systems [1]. Runways, as critical components of airports, require frequent inspections to identify and address any pavement distress that could compromise the safety of aircraft operations

M. Awang (Ed.): ICCAEET 2024, PTT, pp. 232–241, 2025.
https://doi.org/10.1007/978-3-031-94347-8_20

[2]. Current methods for runway damage assessment typically rely on manual visual inspections and conventional measurement tools, both of which are time-consuming, resource-intensive, and prone to human error [3]. These traditional approaches are particularly inadequate for large-scale infrastructure like runways, where the vast area and the need for accuracy present significant challenges. Consequently, there is an urgent need for more efficient and reliable assessment techniques that can deliver accurate results while minimizing operational disruptions [4].

In recent years, Light Detection and Ranging (LiDAR) technology has gained significant attention for its potential to revolutionize infrastructure assessment [5]. LiDAR uses laser-based sensors to generate high-resolution, three-dimensional representations of surface conditions, allowing for the precise detection of pavement irregularities [6]. This technology has proven effective in identifying cracks, rutting, and other forms of surface damage that may be difficult to detect through traditional methods [7]. Despite its advantages, LiDAR systems are often prohibitively expensive and require skilled personnel to operate and analyze the data [5], which limits their accessibility, particularly for smaller airports or those with limited maintenance budgets. The high cost and complexity of traditional LiDAR solutions have thus created a barrier to their widespread adoption in routine runway inspections [8].

To address these challenges, the development of low-cost drone-based LiDAR systems has emerged as a promising alternative. By mounting lightweight LiDAR sensors on unmanned aerial vehicles (UAVs) [9], it is possible to capture detailed surface data with greater efficiency and at a fraction of the cost of traditional systems [10]. Low-cost drones equipped with LiDAR technology offer several advantages, including the ability to cover large areas quickly, perform assessments without interrupting runway operations, and reduce the manpower required for inspections [11–13]. Moreover, drones provide flexibility in accessing hard-to-reach or hazardous areas, enhancing the safety and comprehensiveness of inspections [14]. These benefits make drone LiDAR systems an attractive option for airports seeking more cost-effective solutions for runway maintenance.

Despite the growing interest in low-cost drone LiDAR technology, there remain several challenges that need to be addressed before widespread implementation can occur. These include the technical limitations of low-cost LiDAR sensors, such as lower accuracy compared to high-end systems, as well as challenges related to data processing and analysis [15]. Additionally, regulatory issues regarding the use of drones in controlled airspace, such as around airports, must be carefully managed to ensure compliance with aviation safety standards. Addressing these challenges will be key to unlocking the full potential of low-cost drone LiDAR for runway pavement damage assessment.

This review paper provides a comprehensive examination of the current state of research on the use of low-cost drone LiDAR for assessing runway pavement conditions. It explores the advantages and limitations of this technology compared to traditional assessment methods, with a focus on its potential to improve the accuracy, efficiency, and cost-effectiveness of runway inspections. Additionally, this paper discusses the key challenges that must be overcome to facilitate the broader adoption of drone LiDAR systems in airport infrastructure maintenance. By synthesizing recent advancements and identifying gaps in the literature, this review aims to provide insights into the future of drone-based LiDAR in enhancing runway pavement management practices.

2 Runway Inspections and Technological Progress

2.1 Surface Condition Evaluation

The included pavements of Airport were evaluated using the PCI method (Fig. 1). The runways, taxiways, taxilinks, and rapid exits were assessed using the ASTM D5340–12 "Standard Test Method for Airport Pavement Condition Index Surveys" methodology [16]. The GSE roads encircling the interior airport facilities were assessed using ASTM D6433–18, "Standard Test Methods for Roads and Parking Lots Pavement Condition Index Surveys." Furthermore, the interlocking concrete GSE roads were evaluated in accordance with ASTM E2840–11 (2015), "Standard Test Methods for Pavement Condition Index Surveys for Interlocking Concrete Roads and Parking Lots."

Fig. 1. PCI Assessment of the GSE roadways

2.2 Functional Evaluation

Using the Australian Road Research Board (ARRB) laser profiler (Fig. 2), the runway, taxiway, and rapid exit roughness was measured. The system is a mobile device for gathering data on roughness measurement that combines a high-resolution camera with a precise laser profiler. With two precision laser sensors and accelerometers to account for vehicle body movement, the laser profiler is a World Bank Class 1 profiler. The IRI measurement lines were restricted to the middle strip of the facility under evaluation, meaning that each facility had two lines, spaced six meters apart from each other, in the most popular traffic direction [17]. In order to analyze the roughness of the data, the average IRI values for each sensor's 25 m, 100 m, and 200 m lengths were determined. For every test run, the laser profiler Hawkeye analysis program generated ERD output files in addition to the IRI computation. The computer application ProVAL was used to examine the generated ERD files and determine the Rolling Straight Edge (RSE) values for each section (Fig. 3).

Fig. 2. Utilized a laser profiler to assess the roughness of pavement

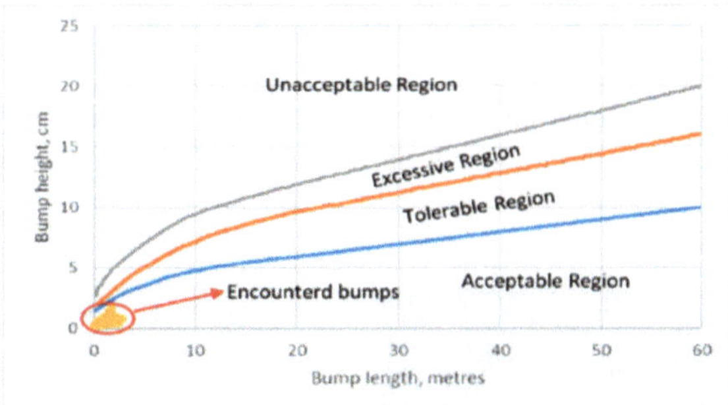

Fig. 3. Observed RSE bump lengths and heights on surveyed taxiways overlaid on ICAO roughness standards

2.3 Light Detection and Ranging (LiDAR)

Significant technological progress has been made in the field of runway inspections. One notable advancement is the use of remote sensing technologies, such as Light Detection and Ranging (LiDAR). LiDAR provides high-resolution, three-dimensional surface data, allowing for more detailed analysis of runway conditions. In addition to LiDAR, Unmanned Aerial Vehicles (UAVs), or drones, have emerged as a promising tool for runway inspections. The integration of UAVs with advanced sensors (Fig. 4), such as LiDAR and photogrammetry equipment, allows for rapid data collection and processing. Drones can survey vast runway surfaces in a fraction of the time it would take using traditional methods, all while minimizing disruption to airport operations [18].

The Livox Mid-40 sensor, a low-cost LiDAR, has emerged as a highly efficient tool for various applications, including infrastructure inspection and environmental mapping. With its unique non-repetitive scanning pattern, the Livox Mid-40 provides a dense point cloud that enhances data resolution, making it suitable for detecting surface irregularities such as cracks, deformations, and other pavement distress. Despite being more affordable than traditional LiDAR sensors, the Mid-40 offers competitive performance in terms of accuracy and range, covering distances of up to 260 m [19]. Its compact design and

Fig. 4. Low-Cost LiDAR Sensor (Livox Mid-40)

integration with UAV systems further increase its versatility, enabling efficient, high-resolution data collection in large-scale infrastructure assessments, including runway surface inspections.

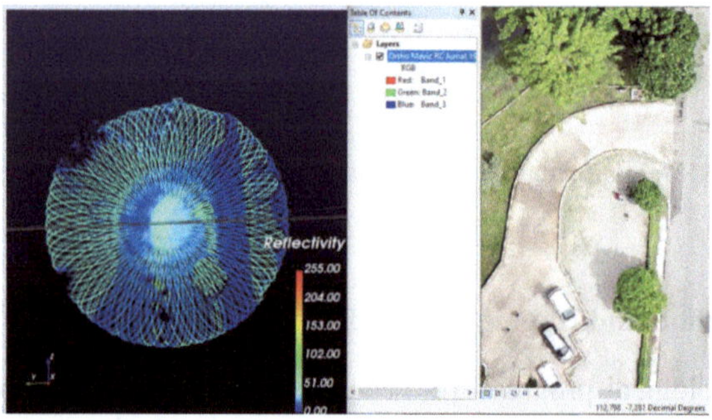

Fig. 5. RAW LiDAR Data from the Livox Mid-40 sensor [15]

Specific parameters are included in the raw data from the Livox Mid-40 LiDAR sensor for data processing and analysis (Fig. 5). These parameters include ground height from a Digital Elevation Model (DEM), mask resolution, height threshold, flat height threshold, normal distance threshold, shared points for coincident planes, and angle between parallel planes [20]. The raw LiDAR data set for the Livox Mid-40 is a three-dimensional array with dimensions for measurement time, channel, and range bin. It includes the measured raw LiDAR data as well as analog and photon-counting signals [21].

3 Methodology

This review paper follows a structured approach to analyze and synthesize the existing literature on the development of pavement damage assessment models for runway surfaces using low-cost drone LiDAR technology. The methodology is divided into several key stages: literature selection, data extraction, analysis, and synthesis of findings. Each stage was designed to ensure a comprehensive and critical review of the current state of research, focusing on both the technical capabilities and limitations of drone-based LiDAR systems for runway assessments (Fig. 6).

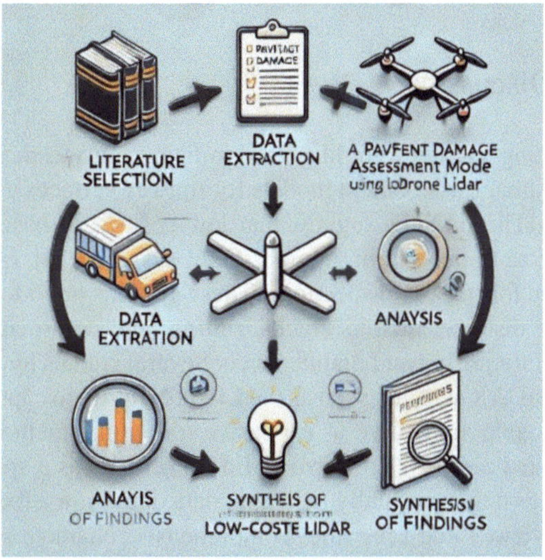

Fig. 6. The methodology diagram in this review paper

In the first stage, literature selection, a systematic search was conducted across major academic databases, including IEEE Xplore, ScienceDirect, and Google Scholar. The search terms used included combinations of "drone LiDAR," "runway pavement assessment," "UAV-based LiDAR," "low-cost LiDAR," "pavement distress detection," and "runway surface monitoring." Peer-reviewed articles, conference papers, and technical reports published in the last 10 years were prioritized to capture recent advancements in the field. Studies were included based on their relevance to the topic, with a particular focus on those that explored the use of UAVs, LiDAR technology, and their integration for pavement assessment. Exclusion criteria were applied to filter out studies that did not directly address runway surfaces or low-cost LiDAR systems.

The second stage, data extraction, involved collecting detailed information from the selected studies. Key data points included the type of LiDAR sensor used, drone specifications, data collection methods, and the algorithms or models employed for damage detection. The accuracy of LiDAR in detecting specific types of pavement distress, such as cracks, ruts, and surface deformations, was also noted. In addition, the cost and operational efficiency of drone LiDAR systems compared to traditional methods were recorded. These parameters provided the foundation for evaluating the practical applications and scalability of low-cost drone LiDAR for runway inspections.

In the analysis stage, the extracted data were organized into thematic categories, such as the technical capabilities of low-cost drone LiDAR systems, accuracy in detecting pavement damage, operational challenges, and potential cost savings. A comparative analysis was conducted to identify trends, common findings, and discrepancies across the reviewed studies. Particular attention was paid to studies that reported field trials or real-world applications of drone LiDAR for runway pavement assessment. The results of this analysis provided insights into the strengths and weaknesses of current models and highlighted gaps in the research that need to be addressed for broader adoption of the technology.

4 Result and Discussion

The review of existing literature highlights significant advancements in the development of pavement damage assessment models for runway surfaces using low-cost drone LiDAR technology. The findings suggest that low-cost drone-based LiDAR systems, while not as precise as traditional high-end LiDAR systems [13], offer a cost-effective and scalable solution for runway inspections. These systems provide sufficient accuracy in detecting surface distress, such as cracks, rutting, and deformations, making them suitable for routine monitoring and maintenance. Several studies have demonstrated the ability of low-cost LiDAR sensors, such as the Livox Mid-40, to capture high-resolution data, offering a valuable alternative to expensive traditional methods, particularly for small to medium-sized airports. The Livox Mid-40 sensor has a special non-repetitive rosette scanning pattern that gradually increases data density in a fixed direction [22]. It has a 38.4° field of view. Table 1 provides comprehensive characteristics for the Mid-40 sensor. The sensor is rated IP67, costs $599 USD, and weighs less than one kilogram. Its volume is less than ten centiliters.

Table 1. Table Livox Mid-40 Manufacturer Instrument Specifications

Parameter	Specification
Field of View	38.4° circular
Beam Divergence	0.28° (vert.) by 0.03° (horz.)
Range Precision	2 cm (1σ @ 20 m)
Angular Accuracy	$<0.1^{\circ}$
Laser Wavelength	905 nm (Class 1)
Detection Range	90 m @ 10% reflectivity
Measurement Rate	100 kHz

One of the key findings from this review is the accuracy of low-cost LiDAR technology in identifying different types of pavement damage [23]. Although some limitations exist, particularly in detecting smaller cracks or subtle deformations, the overall performance of these systems is promising. The non-repetitive scanning pattern of sensors like the Livox Mid-40 ensures a dense point cloud, enabling better surface analysis than conventional visual inspections. Additionally, the integration of advanced algorithms

Fig. 7. Low-Cost LiDAR Sensor integration with DJI Matrice 300

and machine learning techniques to process LiDAR data further enhances the accuracy and reliability of damage detection, as identified in multiple studies (Fig. 7).

The cost-efficiency of low-cost drone LiDAR systems was another important aspect discussed in the literature [8]. Traditional runway inspections are resource-intensive, requiring significant labor and equipment. By contrast, low-cost drones equipped with LiDAR technology can survey large areas in a fraction of the time, reducing both inspection costs and downtime [13]. This benefit is particularly relevant for airports with limited budgets, as the lower operational costs of drones make frequent inspections more feasible. Several case studies confirmed that low-cost systems provide considerable savings while maintaining a level of accuracy suitable for early damage detection and preventive maintenance.

The review also identifies several challenges associated with the adoption of low-cost drone LiDAR for runway inspections. One key issue is the variability in data quality depending on environmental conditions, such as weather, lighting, and surface reflectivity. In addition, the learning curve associated with operating drones and interpreting LiDAR data presents a barrier for some airport operators. These challenges suggest the need for ongoing research and development to improve the robustness and user-friendliness of these systems. Moreover, the integration of AI and machine learning techniques for automating damage detection processes requires further refinement to ensure consistent and accurate results across different airport environments [24].

The results indicate that low-cost drone LiDAR technology presents a viable solution for runway pavement damage assessment, offering a balance between accuracy, cost-effectiveness, and operational efficiency (Fig. 8). While there are still hurdles to overcome, particularly in terms of data quality and ease of use, the potential benefits of this technology make it a promising tool for future runway inspection practices. Further research is recommended to address the identified challenges and to improve the scalability and adoption of these systems in diverse airport settings.

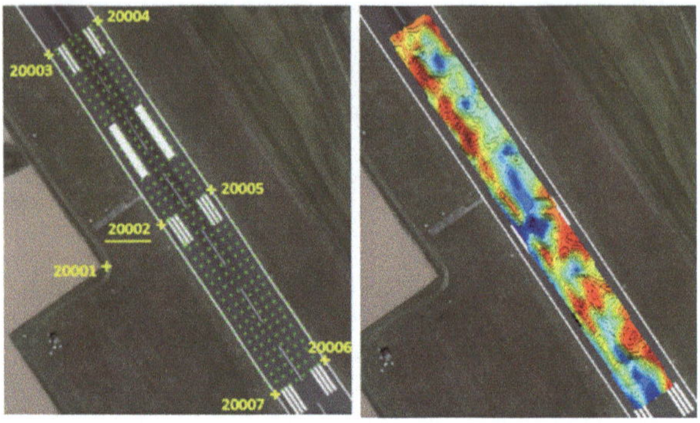

Fig. 8. Analysis of Changes in the Condition of the Runway (deviation surface)

5 Conclusion

The development of a pavement damage assessment model for runway surfaces using low-cost drone LiDAR technology presents a viable and cost-effective alternative to traditional inspection methods. This review has demonstrated that while low-cost LiDAR systems may not match the precision of high-end models, they provide sufficient accuracy for detecting critical surface distresses such as cracks and deformations. The integration of UAVs with LiDAR technology significantly enhances the efficiency and frequency of inspections, making it particularly beneficial for airports with limited resources. Despite challenges related to environmental factors and data interpretation, the potential for improving runway maintenance practices through this approach is clear, offering a scalable and practical solution for enhancing airport safety and operational sustainability.

References

1. Solla, M., Perez-gracia, V., Fontul, S.: Trends in GPR and other NDTs for transport infrastructure assessment (2021). https://doi.org/10.3390/books978-3-0365-2124-4
2. Sholichin, I., Ahyudanari, E., Muntini, M.S.: Review of runway pavement monitoring. IOP Conf. Ser. Earth Environ. Sci. **971**(1) (2022). https://doi.org/10.1088/1755-1315/971/1/012017
3. Asi, I.M., Al-asi, A.I.: Airports And Highways Pavement Performance Evaluation For Maintenance Need- Case Study, pp. 24–38
4. Mouratidis, A.: The 7 challenges of road management towards sustainability and development. J. Infrastructure Policy Dev. **4**(2), 249 (2020). https://doi.org/10.24294/jipd.v4i2.1174
5. Gargoum, S., El-Basyouny, K.: Transportation infrastructure asset management using LiDAR remote sensing technology. Int. J. Trend Res. Dev. ISSN, 143 (2018). www.ijtrd.com
6. Elaksher, A., Ali, T., Alharthy, A.: A quantitative assessment of LIDAR data accuracy. Remote Sens. **15**(2) (2023). https://doi.org/10.3390/rs15020442
7. Gao, Q., Fan, L., Wei, S., Li, Y., Du, Y., Liu, C.: Differences evaluation of pavement roughness distribution based on light detection and ranging data. Appl. Sci. **13**(14) (2023). https://doi.org/10.3390/app13148080
8. Sanaei, P., Salman, B.: Use of photogrammetry and laser scanning technologies in runway inspections. Constr. Res. Congr. CRC 2024, 1, 386–396 (2024). https://doi.org/10.1061/9780784485262.040

9. Astor, Y., Nabesima, Y., Utami, R., Sihombing, A.V.R., Adli, M., Firdaus, M.R.: Unmanned aerial vehicle implementation for pavement condition survey. Transp. Eng. **12**, 100168 (2023). https://doi.org/10.1016/j.treng.2023.100168
10. Imani, D.W., Cahyadi, M.N., Farid, I.W., Mardianto, R.: Integration analysis of drone multi sensor-GNSS-Lidar-camera for 3d mapping (case study: Pt Garam, Pamekasan, Madura). J. Mar. Sci. Technol. **4**(1), 23–32 (2023). https://doi.org/10.12962/j27745449.v4i1.699
11. Tan, Y., Li, Y.: UAV photogrammetry-based 3D road distress detection. ISPRS Int. J. Geo-Inf. **8**(9) (2019). https://doi.org/10.3390/ijgi8090409
12. Waliulu, Y.E.P.R., Suprobo, P., Adi, T.J.W.: Volume calculation accuracy and 3D visualization of flexible pavement damage based on low-cost LiDAR. In: 2023 IEEE Asia-Pacific Conf. Geosci. Electron. Remote Sens. Technol. Glob. Challenges Geosci. Electron. Remote Sens. Futur. Dir. City, Land, Ocean Sustain. Dev. AGERS 2023, vol. C, pp. 109–113 (2023). https://doi.org/10.1109/AGERS61027.2023.10490615
13. Tamimi, R., Toth, C.: Accuracy assessment of UAV LiDAR compared to traditional total station for geospatial data collection in land surveying contexts. Int. Arch. Photogramm. Remote Sens. Spat. Inf. Sci. - ISPRS Arch., 48–2–2024, 421–426 (2024). https://doi.org/10.5194/isprs-archives-XLVIII-2-2024-421-2024
14. Mehta, S., Peach, J., Weinert, A.: To expedite roadway identification and damage assessment in LiDAR 3D imagery for disaster relief public assistance. Infrastructures **7**(3) (2022). https://doi.org/10.3390/infrastructures7030039
15. Manggala, A.S., Ahyudanari, E., Cahyadi, M.N., Raharjo, A.B., Harun, Z., Lebuna, M.L.E.: Optimization of land area mapping and volume calculations using drone lidar livox Mid-40 data with the downsampling method. Bio Web Conf. **89**, 1–14 (2024). https://doi.org/10.1051/bioconf/20248901007
16. Heathrow, L.: Runway maintenance strategy : a summar (2023)
17. Golov, E., Evtyukov, S., Protsuto, M., Evtyukov, S., Sorokina, E.: Influence of the road surface roughness (according to the international roughness index) on road safety. Transp. Res. Procedia **63**, 999–1006 (2022). https://doi.org/10.1016/j.trpro.2022.06.099
18. Di Benedetto, A., Fiani, M., Petti, L., Repetto, E.: Road surface modelling and characterization from terrestrial lidar data. Int. Arch. Photogramm. Remote Sens. Spat. Inf. Sci. - ISPRS Arch. **48**(1/W1–2023), 113–120 (2023). https://doi.org/10.5194/isprs-archives-XLVIII-1-W1-2023-113-2023
19. Kelly, C., Wilkinson, B., Abd-Elrahman, A., Cordero, O., Lassiter, H.A.: Accuracy assessment of low-cost lidar scanners: an analysis of the velodyne HDL–32E and Livox Mid–40's temporal stability. Remote Sens. **14**(17) (2022). https://doi.org/10.3390/rs14174220
20. Awrangjeb, M., Fraser, C.S.: Automatic segmentation of raw LIDAR data for extraction of building roofs. Remote Sens. **6**(5), 3716–3751 (2014). https://doi.org/10.3390/rs6053716
21. D'Amico, G., Amodeo, A., Mattis, I., Freudenthaler, V., Pappalardo, G.: EARLINET single calculus chain-technical and n dash; Part 1: Pre-processing of raw lidar data. Atmos. Meas. Tech. **9**(2), 491–507 (2016). https://doi.org/10.5194/amt-9-491-2016
22. Glennie, C.L., Hartzell, P.J.: Accuracy assessment and calibration of low-cost autonomous LiDAR sensors. Int. Arch. Photogramm. Remote Sens. Spat. Inf. Sci. - ISPRS Arch. **43**(B1), 371–376 (2020). https://doi.org/10.5194/isprs-archives-XLIII-B1-2020-371-2020
23. Mihić, M., Sigmund, Z., Završki, I., Butković, L.L.:An analysis of potential uses, limitations and barriers to implementation of 3D scan data for construction management-related use—are the industry and the technical solutions mature enough for adoption?. Buildings **13**(5) (2023). https://doi.org/10.3390/buildings13051184
24. Kovacic, B., Doler, D., Sever, D.: Innovative business model for the management of airports in purpose to identify runway damage in time. Sustain. **13**(2), 1–20 (2021). https://doi.org/10.3390/su13020613

Removal of Microplastic Pollutant in Surface Water Using Electrocoagulation Process with Fe Electrode

Fitria Ummuzzahra, Arlini Dyah Radityaningrum(✉), and Nadia Rana Abiyya Kholish

Department of Environmental Engineering, Faculty of Civil Engineering and Planning, Institut Teknologi Adhi Tama Surabaya, Surabaya, Indonesia
dyah@itats.ac.id

Abstract. Microplastics (MPs) were considered as a problem due to their small size. Previous research has proved that MPs resulted in some negative impacts in a long term for human, biota, ecosystems, and environments including aquatic environment. Electrocoagulation (EC) is a potential technology to reduce number of MPs in surface water. This study aimed to determine the removal efficiency of MP abundance in surface water using EC method. This research employed a 2 L sample which was treated using EC with iron (Fe) electrodes for 30 min. The EC processes used voltage variation of 15 and 30 Volts which were batch operated. The number of electrodes were 2 and 4 plates (Fe-Fe and Fe-Fe-Fe-Fe). The MP abundance was observed using digital microscope. Characteristics of MPs was analyzed using Fourier Transformed Infra-Red (FT-IR) Spectrophotometer. The results showed that the highest removal efficiency (95.65%) was achieved in the variation of 30 V EC with 4 Fe plates. Meanwhile, the lowest removal efficiency (91.30%) had occurred in the 15 V with 2 Fe plates. The range of removal efficiencies showed that the EC process enabled to achieve high MP removal efficiency in surface water.

Keywords: Electrocoagulation · Fe · Microplastics · Removal

1 Introduction

Microplastic (MP) is a small plastic particle with size of <5 mm which might has ecotoxicological effect and tends to accumulate in a food chain [1, 2]. MP is classified into primary and secondary MP [3]. Primary MPs are defined as plastics produced in microscopic sizes, for example scrubber in household products, while secondary MPs are the result of large plastic decomposition [4]. MPs have several characteristics, such as high absorbency, very small size, low weight, large surface area, and low density [5]. MPs is ubiquitous, including in surface water which generally come from daily human life, such as agriculture, industry, and household activities [6].

MP pollutant in surface water might carry toxic substances that potentially causes blockage of the digestive tract, reduces fitness, and causes death when consumed by aquatic biota [2]. In addition, ingestion of MPs are also harmful to human health [7].

M. Awang (Ed.): ICCAEET 2024, PTT, pp. 242–249, 2025.
https://doi.org/10.1007/978-3-031-94347-8_21

Moreover, MPs are hydrophobic materials which are insoluble in water [8, 9]. This causes MPs to become a carrier of an existing toxic contaminants material in the environment [8]. Therefore, MP abundance in surface water needs to be eliminated to avoid the MP negative impact.

Previous studies developed many methods to eliminate MP contaminants in surface water. Electrocoagulation (EC) suggested to be an alternative of water treatment to remove MPs. EC technology is a coagulation process that uses direct conduction of electricity with a simple electrochemical process which is a cost-effective treatment [10, 11]. EC technology is known to be widely used in laboratory scale water treatment to reduce several parameters. Previous study revealed that EC was effective in removing MPs, achieving 86% removal efficiency [8]. In comparison, this result was slightly higher than another previous research which used a coagulation-flocculation process (35–85%) [12]. EC can be an alternative for controlling MPs in waters. This study focused on MP removal in water using EC process. The study aim was to determine the MP removal efficiency in surface water using EC with iron (Fe) electrodes.

2 Material and Methods

2.1 Material

This study used samples from the Jagir Surabaya River, DC power supply, Fe electrode plate, and an EC reactor with a capacity of 3.5 L (Fig. 1). The EC reactor employed 2 variations of Fe electrode plate numbers (2 and 4 Fe plates). The distance between Fe plate was 3 cm. Besides, the EC reactor was supplied with 2 voltage variations (15 and 30 V). To separate the MP, hydrogen peroxide (H_2O_2) 30% and ferrous sulfate solution ($FeSO_4$) were used for the oxidation process. H_2O_2 was responsible for oxidizing natural organic substances in the sample, whereas $FeSO_4$ function was to catalyze reactions [6]. PTFE membrane paper with pores of 0.2 μm and a diameter of 47 mm was used for MP sorting process and was cleaned with 70% alcohol. MP particles were identified with the digital Dino-Lite Premier microscope. MP polymers were characterized using Fourier Transformed Infra-Red (FT-IR) Spectrophotometer.

Fig. 1. EC reactor

Description:

1. EC reactor
2. DC power supply

2.2 Methods

This study was conducted from March to June 2024 at the Environmental Engineering Laboratory, Institut Teknologi Adhi Tama Surabaya. The research began with sampling in the Jagir River Surabaya at a depth of about 0.3 m from the river surface. The abundance of MP is most commonly found at a depth of 0.16–0.3 m from the river surface [4]. The sampling method was grabbing sampling, in accordance with SNI 6989.57:2008 Water and Wastewater – Part 57: Surface Water Sampling Methods. Furthermore, an initial oxidation process was carried out. After that, a 2 L water sample was filtrated, and the MP contaminants were sorted to determine the abundance of the initial MP. The water sample was then poured into the EC reactor. Then, the sample was treated using batch EC reactor for 30 min. This process resulted in flocs that floated on the water surface and settled after 60 min. Effluent water samples were then collected to identify MP abundance and removal efficiency. This process was conducted in two replications. The abundance of MP was expressed in units of particles/L with the following equation [13]:

$$\text{MP abundance (particles/L)} = \frac{\textit{number of MP particles (particles)}}{\textit{water volume (L)}} \quad (1)$$

The removal efficiency was used to determine the percentage of MP which was diminished in the EC process (Eq. 2) [14]:

$$\text{Removal efficiency} = \frac{\textit{Initial MPs (in influent)} - \textit{Final MPs(in effluent)}}{\textit{Initial MPs(in influent)}} \times 100\% \quad (2)$$

3 Results and Discussion

3.1 Microplastic Abundance

MPs were identified twice, before and after the EC process. Before the EC process, it was found that the initial MP abundance was 92 particles/L which had different shapes. The sample was then given EC processing with the results of MP abundance after processing as in Fig. 2.

Based on Fig. 2, the least abundance of MP with EC process is in the 30 V Fe-Fe-Fe-Fe variation which is 8.5 particles/L, while the most abundance of MP is in the 15 V Fe-Fe variation, which is 13.5 particles/L. The abundance value can determine the efficiency of the EC process in removing MP in surface water.

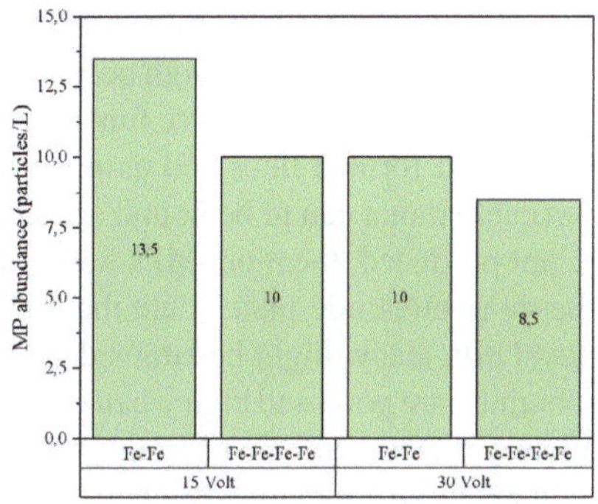

Fig. 2. MP abundance after the EC process

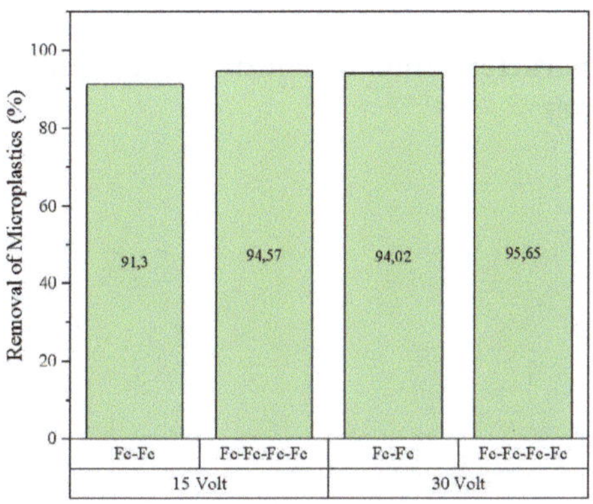

Fig. 3. MP removal efficiencies of EC processes

3.2 Removal of Microplastics

The MP removal efficiencies of EC process with two variations of voltage and plate number were depicted in Fig. 3.

Based on data in Fig. 3, the MP removal efficiencies in the EC processes reached a value above 90%. The highest and lowest efficiencies were 95.65% and 91.3%. Other efficiencies were 94.57% and 94.02%. The difference in efficiency values was caused by the conditions, characteristics, and content of the samples used [15].

Figure 3 showed that the highest efficiency was found in the 30 V EC process. This condition might be caused by the higher voltage (30 V) than another (15 V), which was estimated to remove more MP particle number. Meanwhile, the EC process with 4 Fe electrode plates resulted in more efficient MP removal than 2 Fe electrode plates. This was caused by the process of a reduction-oxidation (redox) reaction. The redox led to the release of ions in the cathode and anode, where the reaction was affected by the current in the electrodes [9]. The oxidized metal ions then bond to OH^- produced by the cathode, leading to form an interbond reaction as follows [15]:

$$Fe^{2+} + 2H \qquad Fe(OH)_2 \; [15,16,17] \qquad (3)$$

The bonded ions then moved freely in the water and served as coagulants which bonded pollutants into flocs [16]. Coagulants with high polymerization enable to capture more pollutants [17]. Coagulants in EC process were functioned to bind pollutants that have a negative charge which then formed flocs and assisted by hydrogen gas to float [8]. MPs is a very small particles so that can to be bound and adsorbed to the flocs. This meant that the more coagulant produced, the more MPs will be bound [8]. Accordingly, this was concluded that the more electrode plates then the more MPs might be bound. Therefore, the higher removal efficiency might be attained.

The efficiency results obtained are good and high when compared to other studies that use the coagulation-flocculation method with an efficiency of 35–85% [12]. Although it has good efficiency, EC requires quite a lot of costs when compared to the coagulation-flocculation process, especially if it is used for large scale treatment. Therefore, EC can be used as a good technology option for removing MP in surface water at a smaller scale.

3.3 Microplastics Characterization by Fourier Transformed Infra-Red (FT-IR) Spectrophotometer

(Table 1)

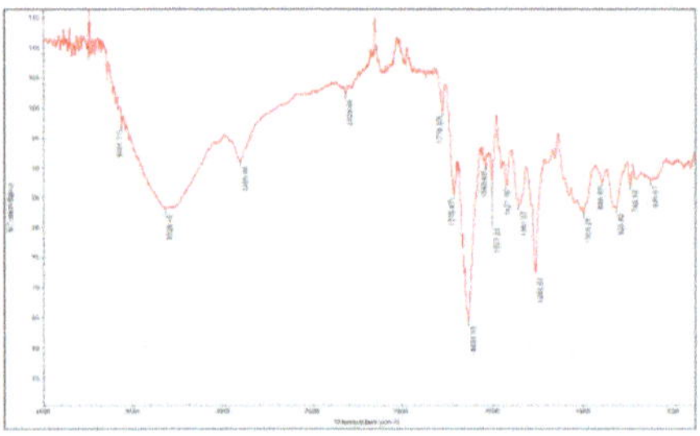

Fig. 4. FT-IR analysis of MPs on 15Volt Fe-Fe sample

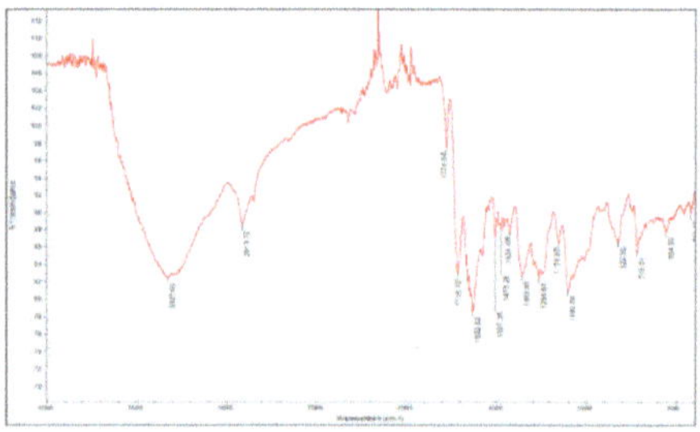

Fig. 5. FT-IR analysis of MPs on 15Volt Fe-Fe-Fe-Fe sample

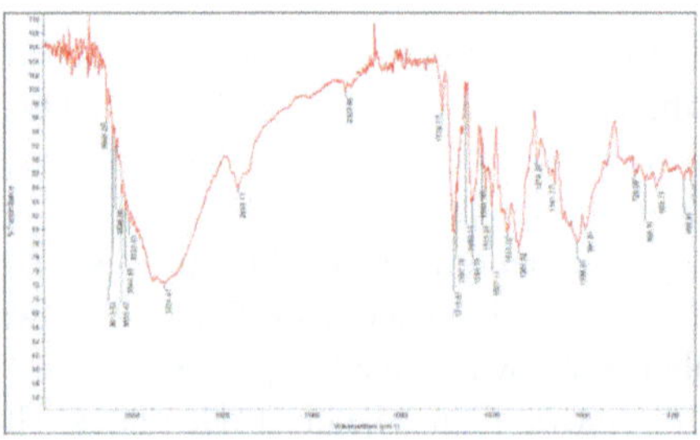

Fig. 6. FT-IR analysis of MPs on 30Volt Fe-Fe sample

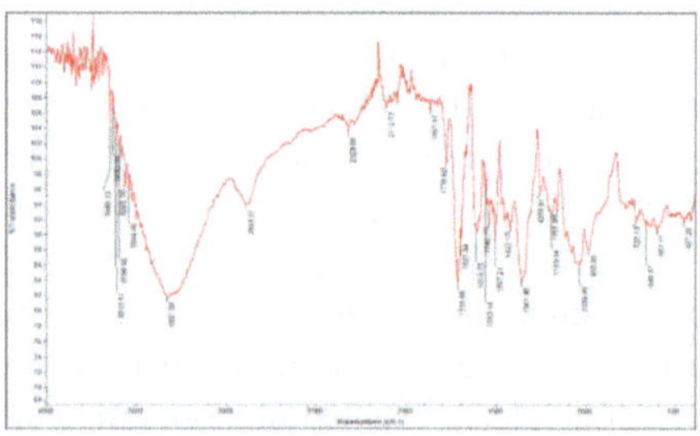

Fig. 7. FT-IR analysis of MPs on 30Volt Fe-Fe-Fe-Fe sample

Figure 4. Indicated the presence of nylon 6/6 polyamide (PA) polymer. Nylon 6/6 PA was known as an artificial or synthetic PA [18]. PA nylon 6/6 was widely used in

Table 1. Analysis of Microplastic

Sample	Polymer analysis	Chemical structure
Figure 4	Polyamide nylon 6/6	$(C_{12}H_{22}N_2O_2)_n$
Figure 5	Polymide	(-CO-NR-CO-)
Figure 6	Polystyrene	$(C_8H_8)_n$
Figure 7	Polystyrene	$(C_8H_8)_n$

the manufacture of toothbrush bristles, textile base materials, fishing line materials, and others [19]. Meanwhile, data in Fig. 5. Indicated the presence of polyimide polymers. Polyimide was widely used in plastic materials, adhesives, nonlinear optics, and membranes [19]. Polyimide polymer was one of the polymers with good heat resistance (high temperature). Besides, polymide had the characteristics of easy to process [19].

Figure 6 and Fig. 7. Showed the presence of polystyrene (PS) polymer. PS is known to be one of the primary microplastic particles that has a diameter between 1 and 5 μ [20]. PS had the property of being easily soluble in organic solvents (ketones, esters, and aromatic hydrocarbons); resistant to acids, bases, salts, mineral oils, organic acids, and alcohols; and have good thermal insulation properties [21]. As a hard plastic, PS was widely used for food packaging made of styrene, single-use plastic cups, plastic tableware, compact disk and the container, and other [22].

4 Conclusion

This research had revealed that EC enabled to remove MP pollutant in surface water of Jagir Surabaya up to 95.65% at the 30 V Fe-Fe-Fe-Fe variation. This result proved that the EC method was suggested to be a potential technology for MP removal in raw waters. Further research is highly recommended to be developed with different variables. The EC method can also be suggested as one of the units in drinking water treatment plant by considering the electrical power and cost.

References

1. Barkmann-Metaj, L., et al.: Quantification of microplastics in wastewater systems of german industrial parks and their wastewater treatment plants. Sci. Total. Environ. **881**(April), 163349 (2023). https://doi.org/10.1016/j.scitotenv.2023.163349
2. Sandra, S.W., Radityaningrum, A.D.: Study on the abundance of microplastics in aquatic biota. J. Environ. Sci. **19**(3), 638–648 (2021). https://doi.org/10.14710/jil.19.3.638-648. (In Indonesian)
3. Liu, Y., Wang, B., Pileggi, V., Chang, S.: Methods to recover and characterize microplastics in wastewater treatment plants. Case Stud. Chem. Environ. Eng. **5**(January), 100183 (2022). https://doi.org/10.1016/j.cscee.2022.100183
4. Lestari, P., Trihadiningrum, Y., Wijaya, B.A., Yunus, K.A., Firdaus, M.: Distribution of microplastics in Surabaya river Indonesia. Sci. Total. Environ. **726**, 138560 (2020). https://doi.org/10.1016/j.scitotenv.2020.138560

5. Zhang, B., et al.: Distribution and removal mechanism of microplastics in urban wastewater plants systems via different processes. Environ. Pollut. **320**, 121076 (2023). https://doi.org/10.1016/j.envpol.2023.121076
6. Sutanhaji, A.T., Rahadi, B., Firdausi, N.T.: Analysis of microplastic abundance in surface water in metro river, Malang. J. Natl. Resour. Environ. **8**(2), 74–84 (2021). https://doi.org/10.21776/ub.jsal.2021.008.02.3. (in Indonesian)
7. Raharjo, M.A.: Identification of Microplastic Content in Fish in the Gajah Wong River. Thesis: Islamic University of Indonesia (2022). https://dspace.uii.ac.id/handle/123456789/40635%0Ahttps://dspace.uii.ac.id/bitstream/handle/123456789/40635/18513076.pdf?sequence=1&isAllowed=y. (in Indonesian)
8. Rosariawari, F., Rachmanto, A., Mirwan, M., Rahmayanti, D.: Electrocoagulation process to reduce microplastic in wonokromo surface water. In: 2nd International Conference Eco-Innovation in Science, Engineering, and Technology, pp. 142–147 (2021). https://doi.org/10.11594/nstp.2021.1423 https://doi.org/10.11594/nstp.2021.142
9. Rizkia, P.N., Hendrasarie, N.: Reduction of fiber microplastic levels in laundry waste by electrocoagulation method. J. Serambi Eng. **7**(3), 3516–3524 (2022). https://ojs.serambimekkah.ac.id/jse/article/view/4536. (In Indonesian)
10. Ananda, E.R., Irawan, D., Wahyuni, S.D., Kusuma, A.D., Buadiarto, J., Hidayat, R.: Manufacture of liquid waste processing equipment with electrocoagulation method for the tofu industry of Samarinda City. JTT (J. Integr. Technol.) **6**(1), 54 (2018). https://doi.org/10.32487/jtt.v6i1.439. (in Indonesian)
11. Lee, J., Kim, Y.-T., Choi, J.: Recycling microplastics to fabricate anodes for lithium-ion batteries: from removal of environmental troubles via electrocoagulation to useful resources. Adv. Sci. **10**, 1–11 (2023)
12. Wiśniowska, E., Moraczewska-Majkut, K., Nocoń, W., Popenda, A.: Microplastics removal from natural surface water by coagulation process. Desalin. Water Treat. **319**(March), 100462 (2024). https://doi.org/10.1016/j.dwt.2024.100462
13. Nurdiana, M., Trivantira, N.S.: Identification of types and abundance of microplastics in the shipping river of the brantas river tributary, Sidoarjo regency, East Java. Environ. Pollution J. **1**(3), 245–254 (2022). https://doi.org/10.58954/epj.v1i3.71. (In Indonesian)
14. Mar'atusholihah, M., Trihadiningrum, Y., Radityaningrum, A.D.: Abundance and characteristics of microplastics in IPAM Karangpilang III, Surabaya City. ITS Eng. J. **9**(2) (2021). https://doi.org/10.12962/j23373539.v9i2.55473. (In Indonesian)
15. Putri, A.S., Nurhalimah, L., Azzahra, M.F.: Identification of characteristics and abundance of microplastics in Surabaya river water samples. ECOTON **2**(2), 1–10 (2022). (In Indonesian)
16. Othman, N.A., Isa, N., Amri, N., Abdul Wahab, N.A., Bashirom, N.: The effect of voltage on polypropylene microplastics removal by electrocoagulation process using Fe electrode. ESTEEM Acad. J. **20**(March), 54–64 (2024). https://doi.org/10.24191/esteem.v20imarch.610.g535
17. Shen, M., et al.: Efficient removal of microplastics from wastewater by an electrocoagulation process. Chem. Eng. J. **428**(July 2021), 131161 (2022). https://doi.org/10.1016/j.cej.2021.131161
18. Yaranal, N.A., Kuchibhotla, S.A., Subbiah, S., Mohanty, K.: Identification, removal of microplastics and surfactants from laundry wastewater using electrocoagulation method. Water Emerg. Contam. Nanoplastics, **3**(1), 1–15 (2023). https://doi.org/10.20517/wecn.2023.46
19. Pramadhani, P., Dewata, I.: Optimization of extraction methods with wet peroxide oxidation (WPO) for the identification of polyamide (PA) microplastics. J. Tambusai Educ. **7**(3), 22907–22915 (2023). (In Indonesian)

20. Liaw, D.J., Wang, K.L., Huang, Y.C., Lee, K.R., Lai, J.Y., Ha, C.S.: Advanced polyimide materials: syntheses, physical properties and applications. Prog. Polym. Sci. **37**(7), 907–974 (2012). https://doi.org/10.1016/j.progpolymsci.2012.02.005
21. Hwang, J., Choi, D., Han, S., Jung, S.Y., Choi, J., Hong, J.: Potential toxicity of polystyrene microplastic particles. Sci. Rep. **10**(1), 1–12 (2020). https://doi.org/10.1038/s41598-020-64464-9
22. Li, W.C.: The Occurrence, Fate, and Effects of Microplastics in the Marine Environment. Microplastic Contamination in Aquatic Environments: An Emerging Matter of Environmental Urgency. Elsevier Inc (2018). https://doi.org/10.1016/B978-0-12-813747-5.00005-9

Accidents Analysis in Sugar Factories: The Impact and Causes Using Government Regulation and Human Factor and Classification System (HFACS)

Edi Santoso(✉), Dewi Kurniasih, and Priyambodo Nur Ardi Nugroho

Master Program -Safety and Risk Management Engineering, Shipbuilding Institute of Polytechnic Surabaya, Surabaya 60111, Indonesia
edisan261180@gmail.com

Abstract. The Sugar Factory uses more than 150 pieces of equipment in production process. With so many tools, the possibility of accidents also very large, especially because of the age of machine and equipment. On July 22 2022, anety A accident occurred which resulted 1 (one) person dying as a result of being hit by a lifting beam and sugarcane scale due to a broken wire rope. Investigation and analysis of the causes of accidents as well as follow-up actions must be carried out, because the production process must continue. This research aims to analyze accidents that occur based on human factors using Government Regulation and Human Factor Analysis and Classification System (HFACS) approach. Government regulations are used to identify the types of accidents with a major impact on lost work hours and fatalities. Government regulations are used to determine the types of accidents that have a major impact on lost work hours and deaths. Using data from sugar factories and interview results, this study shows that accidents with the greatest impact are caused by lifting equipment. Using the Human Factor And Classification System (HFACS) method to analyze the 6 accidents, the results show that at level 1, the biggest causes of accidents are at the sub-level decision error, skill-based error, perceptual error and routine violation. At level 2, the biggest causes of accidents are at the sub-level technological environment and personal readiness. While at levels 3 and 4, Failed to correct problem and supervisory violation have the largest percentage at level 3 and resource management at level 4. The conclusion shows that there are still various problems from the human factor to implementate occupational safety which could potentially lead to accidents.

Keywords: Accident analysis · government regulation · HFACS · Sugar Factories

M. Awang (Ed.): ICCAEET 2024, PTT, pp. 250–263, 2025.
https://doi.org/10.1007/978-3-031-94347-8_22

1 Introduction

Occupational Safety and Health are all activities to guarantee and protect the safety and health of workers through efforts to prevent accidents and occupational diseases, in accordance with Government Regulation Number 5 of 2012 concerning the Occupational Safety and Health Management System. Basically, the implementation of K3 aims to protect workers from potential hazards in the workplace, namely work accidents and work-related diseases. Occupational Safety and Health program really needs to be implemented in the workplace, because every day workers spend their time at work and will always facing potential hazards in the workplace. Most companies have implemented occupational safety and health program in the workplace. However, the number of accidents is still relatively high. According to global data released by the International Labor Organization (ILO), the number of work accident cases in the world reaches 430 million per year. This number consists of 270 million (62.8%) cases of work-related accidents and 160 million (37.2%) cases of work-related diseases, with 2.78 million workers dying each year. (Profil Keselamatan dan Kesehatan Kerja Nasional Indonesia Tahun 2022, 2023).

Accidents in Indonesia are still quite high. The Ministry of Manpower released data, in 2019, the number of work accident cases was 15,486 with 13,519 victims. When compared to 2019, the number of accidents in 2020 decreased slightly with the number of cases being 6,037 with the number of victims being 4,287 workers. However, accident cases increased again in 2021, reaching 7,298 cases with the number of victims being 9,224 workers. Based on the BPJS Ketenagakerjaan annual report, the number of work accident cases in 2019 was 210,789 cases, then became 221,740 cases in 2020 and 234,370 cases in 2021 as seen in Fig. 1.1 below. (Profil Keselamatan dan Kesehatan Kerja Nasional Indonesia Tahun 2022, 2023).

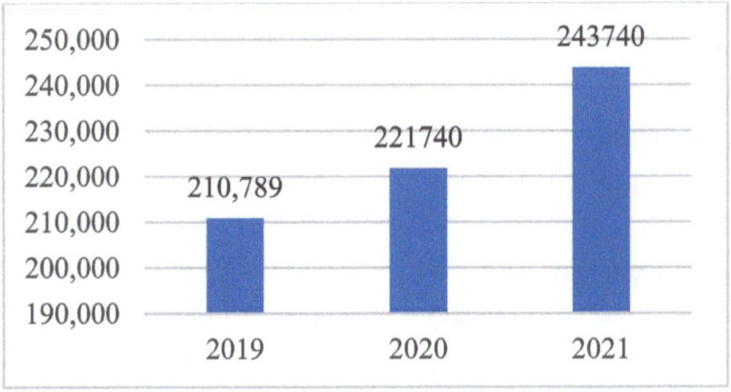

Fig. 1. Accident in Indonesia (Source: Profil Keselamatan dan Kesehatan Kerja Nasional Indonesia Tahun 2022, 2023)

Almost 20% of national work accident cases occur in East Java Province. Based data from BP Jamsostek East Java Regional Office, the number of work accidents in East Java in 2021 was 40,332 cases. This number increased to 47,524 cases in 2022 and 58,709 cases in 2023. Sidoarjo, Gresik and Mojokerto contributed the largest number of accidents with 27,818 cases for Sidoarjo, 18,182 cases for Gresik and 17,570 cases

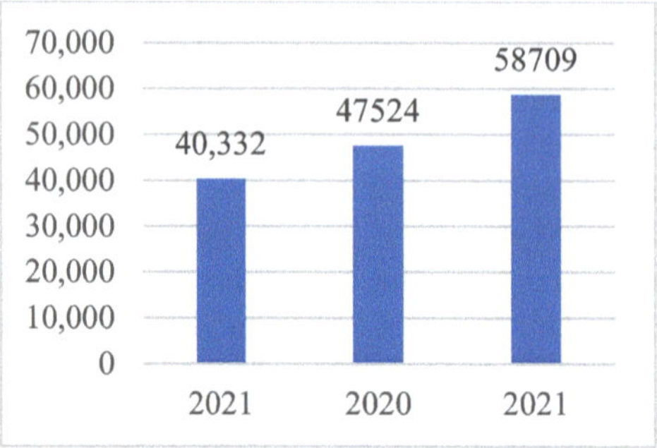

Fig. 2. Accident in East Java (Source: BP Jamsostek East Java Regional Office 2023)

for Mojokerto. More complete cases of work accidents that occurred in East Java in the period 2021 - 2023 can be seen in Fig. 2 below. (Data from BP Jamsostek East Java Regional Office, 2023).

Based on the causes, accidents in Indonesia from 2019 to 2021 occurred in several business sectors. The largest percentage occurring in the industrial sectors with 22.3%, followed by the trade and services sector (21.4%), and agriculture, fisheries, plantations, forestry with 17.3%. More complete data on accident cases based on business sectors can be seen in Table 1.below:

Table 1. Percentage of accident by sector.

No	Sector	2019	2020	2021	Total	%
1	Industrial	49.524	50.898	48.195	148.617	22,3
2	Energy, telecomunication, transportation	8.917	9.163	10.214	28.294	4.2
3	Consumable industry	31.361	33.483	38.879	103.723	15.5
4	Chemical and basic industry	27.381	26.039	27.187	80.607	12.1
5	Finance and investment	3.408	3,492	3.651	10.551	1,6
6	Trading	43.502	46.426	52.224	142.152	21.4
7	Mining	2.494	3.131	6.565	12.190	1.8
8	Agriculture, fisheries, plantations, forestry	36.301	40.947	38,476	115.724	17,3
9	Properti dan Real estate	3.739	3.551	3,402	10.692	1,6
10	Others	4.162	4.610	5.577	14.349	2.2

The sugar industry is one of the industrial sector. This is became one of the national strategic industries to supports the fulfillment of the basic needs of the Indonesian people. So far, national sugar production has not been able to meet national needs. In 2021, Central Bureau of Statistics (BPS) reported that national sugar production was 2.35 million tons, supplied by state-owned sugar factories of 1.06 million tons and private sugar factories of 1.29 million tons. Meanwhile, sugar needs in 2022 reached 6.48 million tons (BPS 2022). From this data, we can see that the national sugar industry is still unable to meet national sugar needs.

Increasing effectiveness and efficiency in the production process is the one of effort to increase the amount of sugar production. However, this is not easy, because the sugar production process goes through quite a long stage. The age of machine and equipment is too old. Fixed-term work agreement who are only recruited during the milling process, also contributed the number of accidents because the awareness of accidents in the workplace is still minimal.

In the last 3 years, there have been more than 100 accidents resulting in dying or loss of working hours, usually referred to as downtime, which disrupts the production process. The data of lost working hours in the last 3 years can be seen in Table 2 below (Based on sugar company X).

Table 2. Percentage of downtime machine.

Year	Machine downtime	Production hours	Percentage
2020	162,08	1695	10%
2021	229,43	1132,5	20%
2022	205,33	2170,5	9%

Table 3. Accident at sugar factories.

No	Sector	2019	2020	2021	Total
1	Nov 09,2020	Boiler station	The heel (ankle) fell off the iron plate	Injuries	1
2	Oct 12,2020	Milling station	Eyes exposed to iron particles from grinding	Injuries	1
3	July 05,2021	Besali station	Fingers caught in lathe	Injuries	1
4	Aug 16,2021	Street plemahan - pagu	Traffic accident	Injuries	2
5	Aug 23,2021	Boiler station	Slipped while going through the stairs	Injuries	1
6	July 27,2022	DCS Scale station	Hit by DCS scales due to broken sling	Dying	6000
7	Dec 12,2022	Boiler station	Exposure to iron splashes during grinding	Injuries	1

From Table 2 above, we can see that over the past 3 years, the average annual time lost due to accidents more than 10% of the total production hours in the company which usually takes between 3 to 6 months in 1 year. In 2020, the percentage of time lost was 10% which increased to 20% in 2021 and decreased to 9% by making several improvements in 2022. Beside machine downtime, accidents also resulted 7 injuries, which one of the victims is dying.

To solve the problem, the first step is to investigate accidents using toolkit in Surat Keputusan Direktur Jenderal Pembinaan Pengawasan Ketenagakerjaan dan Hubungan Industrial Departemen Tenaga Kerja Republik Indonesia Nomor Kep.84/BW/1998 tentang Cara Pengisian Formulir Laporan dan Analisa Statistik Kecelakaan attachment 2. Second step is to determine the type of accident that has the greatest impact. The parameter is number of days lost due to workers being unable to work due to accidents and the number of hours of machine downtime, with reference to the Regulation of the Minister of Manpower No. Per.03/MEN/1998 concerning Procedures for Reporting Work Accidents. To determine the days lost due to accidents, for accidents that cannot be calculated as days lost, the conversion table contained in the Decree of the Director General of Manpower Supervision and Industrial Relations of the Ministry of Manpower of the Republic of Indonesia Number Kep.84/BW/1998 concerning How to Fill Out the Accident Report Form and Statistical Analysis is used. This stage is what distinguishes this study from previous studies. The convertion of days lost is:

To analyze the causes of accidents, the Human Factor Analysis and Classification System method or abbreviated as HFACS. This method, developed by Weigmann and Shapell in 2003, is used to analyze the causes of accidents from the human factor aspect, which divided into 4 level, including unsafe action, precondition of unsafe action, unsafe supervision, and organizational influences (Giu Fu et al. 2020). The human factor aspect greatly influences the occurrence of accidents in the milling process in sugar factories. This is include human error, safety capabilities and safety culture (Giu Fu et al. 2020). The company's policy in recruiting employees during the milling process prioritizes those who have participated in milling in previous periods and the surrounding community, not based on competence. Also the number of Occupational Safety and Health personel is minimal. In general, the description of the accident analysis framework using the HFACS method is as follows. The stages in conducting an accident analysis are:

- Identifying the factors causing the accident
- Creating an HFACS table
- Categorizing the factors causing the accident into the HFACS sublevel
- Entering the accident category into the HFACS table

2 Research Methodology

Research methodology is a description of the entire series of activities carried out during the research process, from the beginning of the activity to the end of the research. Research methodology is used to direct and facilitate the process of solving problems and analyzing the results of processing through good research management so that the research carried out can be of higher quality. Each stage in the research methodology is an important part so that it must be carried out properly and carefully. The description of the research steps is as follows.

Table 4. The Conversion of days lost.

No	Amputation of all or part of a bone	Thumb	Index finger	Middle finger	Ringer finger	Little finger
a	Hand and Finger					
	End segment	300	100	75	60	50
	Middle segment	-	200	150	120	100
	Base section	600	400	300	240	200
	Palm between fingers and wrist	900	600	500	450	-
	Hand to wrists	3000				
b	Hand and Finger					
	End segment	150	35			
	Middle segment	-	75			
	Base section	300	150			
	Palm between fingers and wrist	600	350			
	Hand to wrists	2400				
c	Elbow					
	Every part from the wrist to the elbow	3600				

(continued)

1. Identifying accident for the last 3 years

The definition of accident is an unwanted and unexpected event cause human casualties and/or property (Minister of Manpower Regulation Number Per.03/MEN/1998,1998). Based on this definition, the category of accidents are downtime due to machine breakdown and loss of work days due to injuries.

2. Determine the type of accident has the greatest impact

To determine the days lost due to accidents, that cannot be calculated as days lost, the conversion table contained in the Decree of the Director General of Manpower Supervision and Industrial Relations of the Ministry of Manpower of the Republic of Indonesia Number Kep.84/BW/1998 concerning How to Fill Out the Accident Report Form and Statistical Analysis.

3. Accident analysis using HFACS method

Table 4. (*continued*)

No	Amputation of all or part of a bone	Thumb	Index finger	Middle finger	Ringer finger	Little finger
	Each part from the elbow to the shoulder joint	4500				
d	Legs					
	Every part above the ankle to the knee	3000				
	Every part above the knee to the groin	4500				
e	Lost of function					
	One eyes	1800				
	Both of eyes in one accidents	6000				
	One ears	600				
	Both of ear in one accidents	3000				
f	Total Paralysis and Death					
	Total paralysis	6000				
	Death	6000				

Accident analysis using the Human Factor Analysis and Classification System (HFACS) method. HFACS was developed by Dr. Scott Shappel and Dr. Doug Wiegmann which is used to measure the level of human error in depth. The output of the analysis using the HFACS method is the cause of the accident in terms of human factors. The stages in conducting an accident analysis are:

- Identifying the factors causing the accident
- Creating an HFACS table
- Categorizing the factors causing the accident into the HFACS sublevel
- Entering the accident category into the HFACS table

3 Result and Discussion

3.1 Downtime Due to Breakdown Machine and Lost day Because of Accident

There are 12 categories of equipment or machines become the source of accidents according to Surat Keputusan Direktur Jenderal Pembinaan Hubungan Industrial Dan Pengawasan Ketenagakerjaan RI Nomor Kep.84/BW/1998 concerning the Method of Filling in Accident Forms and Analysis of Accident Statistics. During 2020, the number of machine downtime was 99.39 h with a milling time of 4 months, with the largest percentage occurring in production machines was 61.91 h. Compared to the previous period, machine downtime in 2021 increased to 229.4 h for the 4-month milling period. Meanwhile, in 2022, with a milling period of 5 months from June to October, there was an increase in machine downtime. to 377.54 h. Recapitulation of downtime machine and the calculation of lost days because of injury in Table 5 and Table 6.

Table 5. Recapitulation of downtime machine.

Source	2020	Years	2022	Total
		2021		
Production machine	61,91	63,94	47,07	172,92
Prime mover	0,83	0	4	4,83
Lift	0	0	0	0
Lifting equip,mnet	4,67	0	139	143,67
Conveyor	0,58	1,83	9,66	12,07
Transport equipment	0	0	5,83	5,83
Transmission	4,49	2,09	6,41	12,99
Hand tool	0	0	12,92	12,92
Boiler and pressure vessel	22,16	147,42	25,25	194,83
Electrical	4,75	14,16	23,5	42,41
Chemical	0	0	0	0
Environment	0	0	103,9	103,9
Total	99,39	229,44	377,54	706,37

Based on Table 5 and 6 the worst impact due to accidents was caused by lifting equipment. The equipment caused an accident that resulted in 1 (one) person dying which, when converted, resulted in 6000 lost days. And 143.67 h losing during the period 2020, 2021 and 2022, or 5.9 days. So the type of accident that has the worst impact is an accident caused by lifting equipment with a total of 6005.9 days lost.

Table 6. Calculation of lost days because of injury.

Source	Years			Total	Lost days
	2020	2021	2022		
Production machine	1	1	1	3	3
Prime mover				0	
Lift				0	
Lifting equip,mnet			1	1	6000
Conveyor				0	
Transport equipment				0	
Transmission				0	
Hand tool				0	
Boiler and pressure vessel	1	1		2	2
Electrical				0	
Chemical				0	
Environment				0	

3.2 Study of Accident Caused by Lifting Equipment

In the 2020, 2021 and 2022, there are 6 accidents caused by lifting equipment with the 1 victim died and 143.67 h losing. The detail of accident by liftin g equipment can be seen in Table 7.

Table 7. Accident caused by lifting equipment.

No	Date	Type	Injuries /dying	Lost time (hours)
1	28 Juli 2020	Sling broken	0	0,17
2	30 Juli 2020	Sling broken	0	0,08
3	05 Aug 2021	Crane trouble	0	1,17
4	06 Aug 2021	Crane trouble	0	2,5
5	08 oct 2022	Bucket elevator trouble	0	8,42
6	27 Juli 2022	Wire rope broken	1	120

The first step to analyze the accident is translate into an accident flow to find out the chronology of the accident, like in Fig. 4 and 5, and also same with another incident.

(a) (b)

Fig. 3. (a) Flow accident in 27 July,2022, (b) Flow accident in 28 July,2020.

3.3 Human Factor Analysis and Classification System (HFACS)Please Note that the First

Human Factor Analysis and Classification System (HFACS) is a tool used to analyze an accident case systematically and organized. The results of the analysis using the HFACS method by Focus Group Discussion with expert and have been validated by the company are as shown in Fig. 8 for incident in July 27,2022. This figure explain the level, and sublevel and the activity for each sub level.

For another 5 incident, the process is same. At level 1, workers often doing activities of decision error, perceptual error and routine violation with the highest percentage. This condition is indicated by the frequent of unsafe lifting or unsafe binding by workers. Workers also do not check the equipment used or the lifting aids before using it. Workers often seen not using personal protective equipment and do not have an OHS license. At level 2, the condition of the equipment that is not suitable indicates that the technological environment level also occurs. While the dusty working environment conditions indicate the presence of physical environment factors. The lack of willingness of workers to learn new things, so they only work based on experience in the milling process last year. The foreman's lack of supervision ability, assignment of operators who do not have an OHS license and the condition of the equipment whose suitability has not been confirmed are conditions that reflect the occurrence of unsafe supervision levels in the field. Unsafe supervision here, if not controlled, will trigger the occurrence of preconditions for unsafe acts. At the organizational influence level, it shows that management has a very large influence on accidents. Wrong policies in the recruitment process, determining recruitment schedules that are too tight, or wrong budgeting can trigger unsafe actions carried out by workers. The recapitulation of HFACS of 6 incidents is like in Table 8.

At level 1, decision error and perceptual error as well as skill-based error, are quite large contributors, namely 22%. The workers often make mistakes in operations due to lack of knowledge and information. Also they work only based on previous experience, because workers are the same every year. Furthermore, at the sub-level violation, it can be seen that the sub-level routine is often carried out with a percentage of 22%. This condition is indicated by the many violations, ranging from operations that are not in accordance with procedure, do not have an OHS license, meaning that the Company violates the provisions of the Minister of Manpower Regulation Number 8 of 2020 concerning OHS for Lifting and Transport Equipment. Certification of operators, tie-ups and technicians can help the Company reduce these incidents. Periodic socialization and

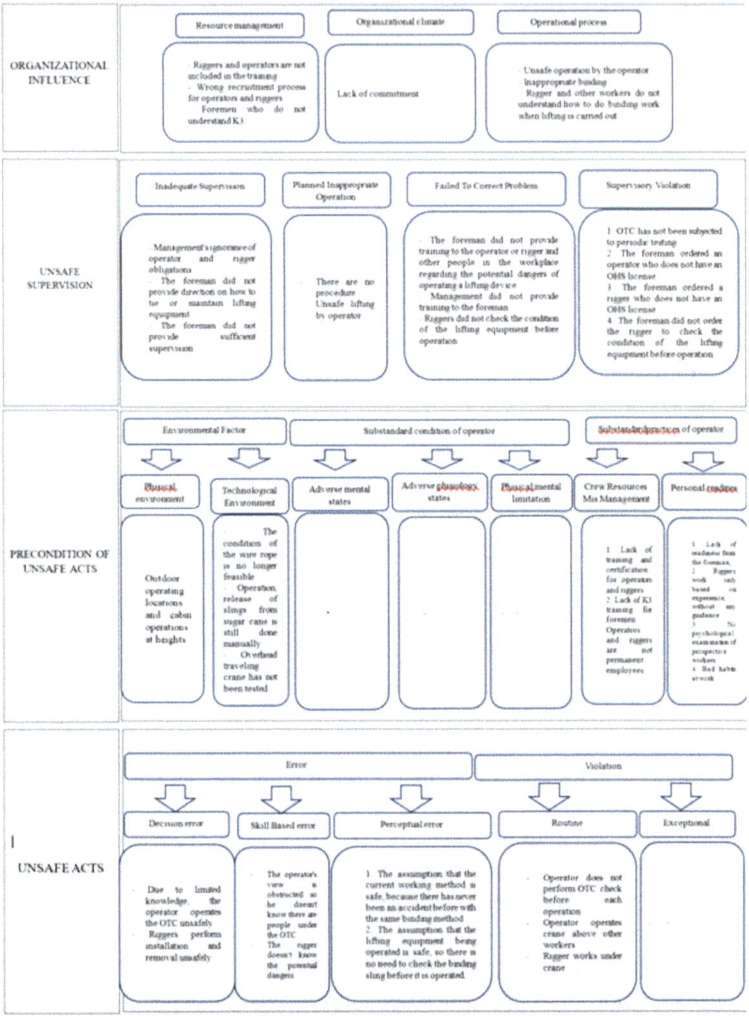

Fig. 4. HFACS of accident on 27 July, 2022

coaching as well as increased supervision of worker performance will also reduce the level of errors made by workers. Technological environment, crew resources mismanagement and personal readiness contribute the largest percentage at the precondition of unsafe action level or level 2 with a percentage of 27%. This condition is indicated by the large number of equipment that is more than 5 years old, which reflects that the technological environment plays a fairly important role as a contributor to the cause of accidents. Supported by the lack of knowledge from the operator so that it operates unsafely, it will ultimately contribute to accidents. This condition can be controlled by conducting regular checks and tests on equipment that is vulnerable or often causes accidents, as an effort to ensure that the equipment is safe to operate. Crew resources mismanagement and personal readiness are indicated by the lack of certification and

Table 8. Recapitulation of each level for 6 accident.

HFACS Category			N	Percentage
Organizational Influence (Level) 4			**8**	
-	Operational Process		2	25%
-	Organizational Climate		0	0%
-	Resource management		6	75%
Unsafe Supervision (Level 3			**19**	
-	Inadequate Supervision		6	32%
-	Planned Inappropriate Supervision		-	11%
-	Failed To Correct Problem		6	32%
-	Supervisory violation		6	32%
Precondition of unsafe action (level 2)			**22**	
-	**Environmental Factor**			
	-	Phisical Environment	2	9%
	-	Technological Environment	6	27%
-	**Substandard Condition Of Operator**			
	-	Adverse mental states	-	
	-	Adverse physiological states	-	
	-	Phisical / Mental Limitation	-	
-	**Substandard Practices Of Operator**			
	-	Crew Resources Mis management	6	27%
	-	Personal readiness	6	27%
Unsave Action (Level 1)			**27**	
-	**Error**			
	-	Decision error	6	22%
	-	Skill Based Error	6	22%
	-	Perceptual Error	6	22%
-	**Violation**			
	-	Routine	6	22%
	-	Exceptional	0	-

authority as indicated by the operator or the binder and technician not having an OHS license. This can happen because there is indeed no budget from the Company to prepare workers who will work in terms of operational safety and maintenance as well as binding goods. In addition, the lack of knowledge from the foreman regarding the implementation of OHS. Level 3 has a smaller percentage than levels 1 and 2. At this level, the largest sub-levels that cause accidents are failed to correct problems and supervisory violations and inadequate supervision. Lack of supervision from the foreman and the assumption that the current operation is safe are some of the behaviors that cause this level to have a fairly large percentage. This can happen because of the lack of knowledge of the foreman, as well as the reluctance of the workforce to learn new things. Budget allocation from the Company for foreman training as a form of knowledge upgrading is one solution that can be done by the Company.

In general, the Company has committed to implementing occupational safety and health. However, the lack of facilities and infrastructure or personnel who understand OHS has triggered the emergence of sub-level resource management with the largest percentage at this level. This is indicated by the lack of certification for personnel operating dangerous equipment. Coaching is only carried out by OHS experts in the Company in the form of internal coaching. Of the many sub-levels at each level, those that need special attention because they have the largest percentage in causing accidents are decision errors and perceptual errors at level 1 and technological environment also at level 2. Furthermore, at level 3, failed to correct problems and supervisory violations must also be considered. As well as improvements in terms of resource management, to control the causes of accidents at level 1.

4 Conclusions

1. The results of the calculation of severity of the accident at the Sugar Factory using the method in the Decree of the Director General of Industrial Relations and Manpower Supervision of the Republic of Indonesia Number Kep.84/BW/1998 concerning the Method of Filling in Accident Forms and Analysis of Accident Statistics for accidents during the last 3 years, starting in 2020, 2021 and 2022, showed that the accident with the greatest impact was caused by equipment included in the category of lifting equipment with a total machine downtime of 143.67 h and the number of lost working days of 6000 days with a total of 6 incidents.
2. By using the Human Factor and Classification System (HFACS) method to analyze the 6 accidents, the results showed that at level 1, the biggest cause was at the sub-level decision error, skill-based error, perceptual error and routine violation. At level 2, the biggest cause was the sub-level technological environment and personal readiness. Meanwhile, at level 3 and level 4, Failed to correct problem and supervisory violation have the largest percentage at level 3 and resource management at level 4.
3. Applicable improvement recommendations were obtained as follows:

- Provide guidance to foremen and workers in accordance with the Company's obligations in Article 9 of Law No. 1 of 1970
- Conduct health checks on workers before acceptance in accordance with the provisions in Article 8 of Law No. 1 of 1970
- Include a budget for the implementation of K3 in the Company in the preparation of the budget in accordance with the provisions of Government Regulation No. 50 of 2012 concerning SMK3
- Conduct inspections and tests on lifting equipment in accordance with the Company's obligations in Minister of Man Power Regulation Number 08 of 2020 concerning Lifting Equipment and Transport Equipment

- Add personnel responsible for OHS, such as operators, rigger, technicians, firefighters, first aid officers, general K3 experts.
- Include operators, binders and technicians in related training so that they can obtain a OHS license in accordance with the Regulation of the Minister of Manpower Number 08 of 2020 concerning Lifting Equipment and Transport Equipment

References

1. Law Number 1 of 1970 concerning occupational safety
2. Government Regulation Number 50 of 2012 concerning the Occupational Safety and Health Management System
3. Decree of the Director General of Manpower Supervision and Industrial Relations of the Ministry of Manpower of the Republic of Indonesia Number Kep.84/BW/1998 concerning How to Fill Out the Accident Report Form and Statistical Analysis
4. Indonesian Ministry of Man Power. Regulation of the Minister of Manpower Number 08 of 2020 concerning K3 for Lifting and Transporting Equipment (2020)
5. Tanjung, A.A., M. Si, Muliyani, S.E., M. SI, MA.: Metodologi Penelitian Sederhana, Ringkas, Padat dan Mudah Dipahami. Scopindo Media Pustaka (2021)
6. BP Jamsostek Kanwil Jatim. Peran BP Jamsostek dalam Inmplementasi K3 (2024)
7. Kementerian Ketenagakerjaan Republik Indonesia. Profil Keselamatan dan Kesehatan Kerja Nasional Indonesia Tahun 2022 (2023)
8. Fahd, M., Puspitasari, N.B., Rumita, R.: Analisis Kecelakaan Kerja denga Menggunakan Metode Human Factor Analysis and Classification System (HFACS) dan 5 Whys di Divisi Stamping PT. Mekar Armada Jaya (2019)
9. Widyanti A, Reyhannisa A.: (2019) Human factor analysis & classification system (HFACS) in the outpatient medication errors. Int. J. Technol. **11**(1), 167–179 (2020)
10. Joe Asare, T., Stemn, E., Amegbey, N.: Applicability and usefullness of the HFACS-GMI. Ghana Mining J. **21**(2), 33–45 (2022)
11. Shapell, S.A., Wiegman, D.A.: The Human Factor Analysis and Classification System – HFACS, US Department of Transportation, Virginia (2000)
12. Fu, G., Xie, X., Jia, Q., Li, Z., Chen, P., Ge, Y.: The development history of accident causation models in the past 100 years: 24 model, a more modern accident causation model. Process. Saf. Environ. Prot. **134**(2020), 47–82 (2019)
13. Fu, G., Xie, X., Jia, Q., Li, Z., Chen, P., Ge, Y.: Accident analysis and prevention of coal and gas outburst : understanding human error in accident. Process. Saf. Environ. Prot. **134**(2020), 1–23 (2019)
14. Buksuhazy, K., Matuchova, E., Zuvala, R., Moracvova, P., Kostikova, M., Mikulec, R.: Human factor contributing to the road traffic accident occurance. Transp. Res. Procedia **45**(2020), 555–561 (2019)

Critical Points Analysis of Potential Fire Risk in the Graving Dock XYZ Area Using FMECA Method

Sudi Utomo(✉), Yuning Widiarti, Priyambodo Nur Ardi Nugroho, Intan Eka Puspitasari, and Nourmanita Naja Hanifa

Shipbuilding Institute of Polytechnic Surabaya, Surabaya 60111, Indonesia
suditomo@gmail.com

Abstract. Graving dock XYZ is one of the work areas in the shipyard owned by PT. ABC has a high fire risk due to flammable materials and hot work processes such as cutting steel plates and pipes, grinding, and welding. The hot work is carried out simultaneously in the same work area. Fires are one of the most common operational obstacles in shipyards. The study aims to determine the cause of the fire in Graving Dock XYZ. This study uses the Failure Modes, Effect, and Criticality Analysis (FMECA) method to find priority risks, analyze equipment or production processes, identify failure types, and determine the consequences of these failures, including identifying critical points. The Fishbone Diagram is used to find the root cause of these priority risks. In this study, seven potential fire risks can occur in Graving Dock XYZ, and there was one risk at the high-risk level, namely the burning of existing plate coating during hot work.

Keywords: Critical Point · FMECA · Hot Work · Graving Dock · Shipyard

1 Introduction

Along with the increasing market demand for new shipbuilding and ship maintenance, the shipbuilding industry in Indonesia is experiencing rapid development. However, behind this progress is a considerable threat of risk, especially the risk of fires that are often overlooked. In complex work environments such as shipyards, where hot work such as metal cutting, welding, and grinding are routinely performed, fire hazards are a real threat that must be appropriately managed. Fires not only endanger worker's lives but can also damage the company's infrastructure and assets, ultimately causing significant losses and disrupting operations. The impact of work accidents can result in enormous losses for the company (Zaman et al. 2021). In addition, shipyard fires can affect the company's reputation domestically and internationally in the maritime industry. A fire incident that must be managed appropriately can create a positive perception among customers and business partners, ultimately impacting business trust and opportunities. Therefore, fire risk reduction is essential not only to guarantee the safety of employees but also to ensure the company's business continuity. This requires a systematic approach to identify,

M. Awang (Ed.): ICCAEET 2024, PTT, pp. 264–273, 2025.
https://doi.org/10.1007/978-3-031-94347-8_23

assess, and manage fire risk at every stage of the shipyard's production process. FMECA greatly emphasizes reliability factors in systems such as marine equipment, which may be relevant when effectively analyzing risks in shipyards (Goo et al. 2019). The method was chosen for this study because it combines the FMEA method with critical point analysis using a critical matrix to accurately assess potential risks that may occur in tools, processes, and systems (Rahman 2021). FMECA is a preventive effort to avoid tool or process failures, and applying risk assessment using FMECA can significantly reduce risks (Goo et al. 2019). FMECA is a quantitative version of the FMEA method, in which critical analysis is added (Ahmed & Gu 2020). The primary purpose of the FMECA method is to identify and analyze the severity and criticality that can cause unintended impacts on the system. In addition, to recognize system risks that can result in equipment damage and financial losses. This technique effectively improves the system and provides recommendations for maintenance planning (Suwanasri et al. 2021). Using appropriate risk analysis methods, such as Failure Mode, Effects, and Severity Analysis (FMECA), allows companies to identify potential fire risks and determine appropriate mitigation measures. By implementing effective mitigation strategies, businesses can reduce the likelihood of fires and minimize the impact if fires do occur.

Based on the above background, this study aims to determine the risks that can cause fire in shipbuilding projects and the MRO process, risk probability, and high-level risk, as well as determine the proper steps to mitigate the risks that are the top priority.

2 Method

The quantitative method Failure Mode Effect and Criticality Analysis (FMECA) method implemented in the graving dock XYZ area is an extension of the FMEA (qualitative) method for identifying potential failures by incorporating critical point analysis through the use of a criticality matrix. The advantage of the quantitative method is its higher accuracy than the qualitative method.

The application process of the FMECA method in the graving dock XYZ area is broadly divided into two steps:

1. Conduct Failure Modes and Effect Analysis (FMEA), which consists of several stages:
   a. Analyse the failure points of the process and assign their identification numbers.
   b. Evaluate the failure points by calculating the Risk Priority Number (RPN), determined by multiplying Severity (S), Occurrence (O), and Detection (D).

Severity refers to the level of seriousness resulting from a failed process in production. The scale used to assess severity ranges from 1 to 10 (Table 1).

Occurrence refers to determining the disturbance value that can lead to production process failure. The ranking used to assess occurrence is between 1–10, as shown in Table 2. Below:

Detection refers to the ability to detect process failures before they occur. The probability of failure is rated on a scale of 1–10, shown in Table 3 below (Table 4):

2. Collected data related to the problem from various sources and divided into two sections: primary and secondary sources of data.

a. Primary Data

Table 1. Severity Rating (Nugraha & Sari 2019)

No	Effect	Rating	Criteria
1	No Effect	1	No Effect
2	Very Slight	2	A very slight effect on Effect
3	Slight Effect	3	A slight effect on product or system performance
4	Minor Effect	4	Minor effect on product or system performance
5	Moderate Effect	5	Moderate effect on product or system performance
6	Significant Effect	6	Product performance degrade but operable and safe
7	Mayor Effect	7	Product performance was severely affected but driveable and safe. System function impaired
8	Extreme Effect	8	Product inoperable, but safe system inoperable
9	Serious Effect	9	Potential hazardous effect. Able to stop product without mishap-gradual failure
10	Hazardous Effect	10	Hazardous effect. Safety-related, sudden failure

Primary data was obtained from observed sources through direct measurement, worker interviews, or questionnaires. The primary data used in this research were interviews with the graving dock XYZ workshop team and expert judgment. These data will be crucial for the researchers to visualize work process failure modes in the form of small fire outbreaks causing fires. Expert judgment refers to the Senior Leader Workshop and the HSE Team who meet the following criteria:

1) Have experience and knowledge in their field.
2) Have a good reputation in the Company.
3) They have the availability of time to be a resource person.
4) Be neutral and honest (not biased two)

b. Secondary Data

Secondary data are data obtained from the company documents to support primary data. The secondary data in this research include:

1) Manual Management System PT ABC, 2024.
2) Standard Operating Procedure (SOP) of the HSE Department-PT ABC Regarding Hazard Identification and Risk Assessment (Fig. 1).

The FMECA method process flow starts by explaining the work process, creating a block diagram, completing the header on the FMEA from the work process sheet, identifying failure modes, analyzing the effects/causes of failure programs, determining the Severity (S) rating, determining the Occurrence factor (O), determining Detection methods (O), determining the Risk Priority Number (RPN), performing criticality analysis, determining recommended actions to address potential failures with high RPN values, and completing the FMECA.

Table 2. Probability of Failure (Nugraha & Sari 2019)

No	Probability of Failure	Possible Failure Rates	Rank
1	Very High: Failure is almost inevitable	≥1 per 10 or ≥10%	10
2	Very High: Failure is almost inevitable	≥1 per 20 <1 per 10 or ≥5% <10%	9
3	High: Repeated failures	≥1 per 40 <1 per 20 or > 2.5% <5%	8
4	High: Repeated failures	≥1 per 100 <1 per 40 or ≥1% <2.5%	7
5	Moderate: Occasional failures	≥1 per 200 <1 per 100 or > 0.5% <1%	6
6	Moderate: Occasional failures	≥1 per 1000< 1 per 200 or≥ 0.1% <0.5%	5
7	Moderate: Occasional failures	≥1 per 1000 or 0,1%	4
8	Low: Relatively few failures	0,5 per 1000 or 0,05%	3
9	Low: Relatively few failures	≥0,1 per 1000 or 0,01% <0,5 per 1000 or 0,05%	2
10	Remote: Failure is unlikely	<0,01 per 1000 or 0.0001%	1

3 Result and Discussion

3.1 Criticallity Analysis

The risk analysis approach has utilized a risk matrix as the foundation for mapping criticality based on the results of the probability and consequence analysis (Siswantoro et al. 2020), which involves several stages:

a. Determine the level of Severity and Occurrence.
b. Determine the Risk Priority Number (RPN) for each failure point.
c. Place each failure point on the criticality matrix considering the severity and occurrence values. If the failure point is positioned further to the left of the column, it means the failure point is higher, and if the failure point is positioned further up the row, the failure rate is also higher. Conversely, the failure rate is lower if its position is further to the right of the column and below the row (Fig. 3).

Table 3. Probability of Failure (Nugraha & Sari 2019)

No	Detection Ability	Rank	Criteria
1	Very High	1	Remote likelihood that the product or service will be delivered. The defect is functionally obvious and readily detected. Detection reliability at least 99,99%
2	High	2–5	Low likelihood that the product would be delivered with defect. The defect is obvious. Detection reliability at least 99,80%
3	Moderate	6–8	Moderate likelihood that the product will be delivered with defect. The defect is easily identified. Detection reliability at least 98,00%
4	Low	9	High likelihood that the product would be delivered with defect. The defect is subtle. Detection reliability at greater than 90%
5	Very Low	10	Very likelihood that the product and/or service will be delivered with defect. Item is usually not checked or not checkable. Quite often the defect is latent and would not appear during the process or service. Detection reliability 90% or less

Table 4. Risk Priority Number (RPN) Values of Failure Modes

Process	No ID	Failure Mode	Severity (S)	Occurrence (O)	Detection (D)	RPN
Cutting and Welding	1.1		7	8	5	280
	1.2		7	4	5	140
	1,3		6	6	5	180
Confined Space	2.1		7	6	5	210
	2.2		5	4	6	120
	2.3		4	3	6	72
Docking & Undocking	3.3		6	5	5	150

3.2 Discussion

Data on safety factors, occupational health, and hazardous areas collected are the essential components in addition to secondary data from the company. The interview and site visit process with HSE personnel and the graving dock XYZ area leader was conducted to clarify and understand the function of each work area and its potential hazards. During these observations, a list of hazards and risk values for each activity in the work area and the failure points were evaluated. The determination of the Risk Priority Number (RPN) is as follows:

$$RPN = S \times O \times D \tag{1}$$

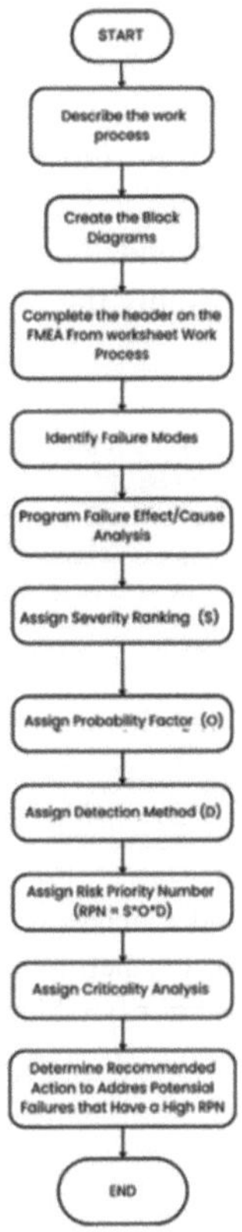

Fig. 1. FMECA Process Flowchart

where:
S = Severity
O = Occurance
D = Detection
The following table contains several failure points in the production process of ship maintenance, repair, and overhaul (MRO) in the graving dock XYZ area, along with the details of severity, occurrence, and detection values, which will produce the RPN value.

The ranking is done by sorting sub-risks based on the Risk Priority Number (RPN) value from the largest to the smallest. The ranking results are based on the calculation in the table. Table 4 shows the results of risk ratings from the highest to the lowest RPN. The risk of burning the existing plate coating during hot work (1.1) obtained the first rank with RPN 280, and the risk of Insufficient lighting (2.3) obtained the lowest rank with RPN of 72. This RPN will then be followed up and used to determine the root cause of critical risks and mitigation strategies. Efforts in identification and mitigation are expected to be able to prevent risk events in the Graving Dock.

3.3 Critically Matrix

The criticality matrix ranks failure modes based on their severity and occurrence levels. Following the Pareto diagram, nine failure modes were identified. The top priority is small fire outbreaks engendered by burning existing plate coatings during hot work, with exposure to hazardous gases ranking second. The subsequent step involves reevaluating the severity and occurrence values, utilizing the criticality graph (Fig. 2).

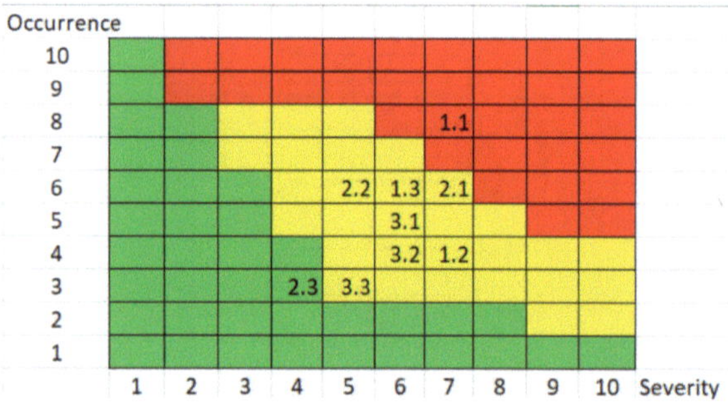

Fig. 2. The criticality graph of failure modes makes it easier to identify the ranking of the nine failure modes that need corrective action. Furthermore, these nine failure modes are arranged sequentially in the following failure mode priority table (Table 6.)

3.4 Cause and Effect Diagram

The fishbone diagram, also known as the cause-and-effect diagram, is utilized to identify the root cause of failure modes. The priority failure mode discussed in the fishbone diagram is small fire outbreaks caused by the burning of existing plate coatings during hot work (Fig. 3).

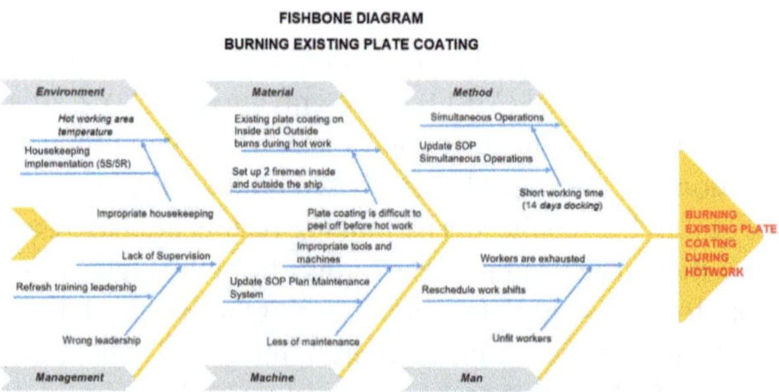

Fig. 3. Fishbone Diagram

The fishbone diagram analysis revealed that the cause of the failure mode of small fire outbreaks during hot work is due to the burning of the existing plate coating during welding and cutting because the inner and outer ship area coating layers are difficult to clean/peel off before hot work is carried out. Based on the root cause analysis of the risk of delayed material arrival using the Fishbone Diagram, it can be described as follows:

a. Man: Workers who are not physically or mentally fit can increase the risk of work errors that can cause fires during hot work. Worker fatigue decreases concentration and alertness, increasing the likelihood of accidents. In addition, the lack of supervision from the management can lead to a disregard for safety standards, which ultimately increases the risk of fire.
b. Method: Operations performed simultaneously over a short period, such as during a 14-day docking period, often result in the neglect of essential safety measures, thereby increasing the risk of fire. Improper rescheduling of work shifts can stress workers, increasing the chances of errors. When SOPs (Standard Operating Procedures) for maintenance systems are not updated regularly, the procedures followed may not be relevant to current conditions, increasing operational risk.
c. Machine: Using improper or non-standard tools and machines can cause malfunctions, Triggering fires. In addition, the high temperature of the work area during hot work can increase the likelihood of burning the remaining coating, increasing the risk of fire.
d. Material: Coatings that are difficult to peel off before hot work can be a source of harmful fuel when exposed to heat or sparks during the work process, thus increasing the risk of fire.
e. Environment: The lack of application of the 5S principle (Sort/Seiri, Set in Order/Seiton, Shine/Seiso, Standardize/Seiketsu, Sustain/Self-discipline/Shitsuke) in managing the cleanliness and regularity of the work area can cause flammable materials not to be handled properly, which increases the potential for fire hazards during hot work.
f. Management: Weak leadership in the work team can lead to the implementation of SOPs and safety procedures needing to be more optimal, which increases the risk of accidents. In addition, the lack of training or the absence of retraining for leadership and occupational safety makes workers unprepared to deal with emergencies such as fires, so the risk of incidents increases.

4 Conclusions and Suggestions

4.1 Conclusions

Risk assessment was chosen for the three main work processes in the graving dock XYZ: Cutting and welding work, confined space work, and ship docking and undocking work, as they have high potential failure modes that require priority in risk management. Based on the discussion above, the researchers conclude:

1. Two possible failure modes need to be prioritized:

The burning of the existing plate coating on the inside and outside of the ship during hot work, RPN= $S \times O \times D$ ($7 \times 8 \times 5=280$)

Exposure to hazardous gases during confined space work. RPN= SxOxD ($7 \times 6 \times 5=210$)

2 Based on the analysis using the Fishbone Diagram, there are several root causes of the risk of Burning existing plate coating during hot work:

a. Man: Workers who are not physically or mentally fit can increase the risk of work errors that can cause fires during hot work. Worker fatigue decreases concentration and alertness, increasing the likelihood of accidents. In addition, the lack of supervision from the management can lead to a disregard for safety standards, which ultimately increases the risk of fire.
b. Method: Operations performed simultaneously over a short period, such as during a 14-day docking period, often result in the neglect of essential safety measures, thereby increasing the risk of fire. Improper rescheduling of work shifts can stress workers, increasing the chances of errors. When SOPs (Standard Operating Procedures) for maintenance systems are not updated regularly, the procedures followed may not be relevant to current conditions, increasing operational risk.
c. Machine: Using improper or non-standard tools and machines can cause malfunctions, triggering fires. In addition, the high temperature of the work area during hot work can increase the likelihood of burning the remaining coating, increasing the risk of fire.
d. Material: Coatings that are difficult to peel off before hot work can be a source of harmful fuel when exposed to heat or sparks during the work process, thus increasing the risk of fire.
e. Environment: The lack of application of the 5S principle (Sort/Seiri, Set in Order/Seiton, Shine/Seiso, Standardize/Seiketsu, Sustain/Self-discipline/Shitsuke) in managing the cleanliness and regularity of the work area can cause flammable materials not to be handled properly, which increases the potential for fire hazards during hot work.
f. Management: Weak leadership in the work team can lead to ineffective SOPs and unsafe safety procedures, which increase the risk of accidents. In addition, the lack of training or the absence of retraining for leadership and occupational safety makes workers unprepared to deal with emergency situations such as fires, increasing the risk of incidents.
g. Mitigation strategies for existing coating plate fire risks during hot work (1.1) involve improving risk management and handling and safety inspections to minimize the potential for fire by reducing the probability and impact of risk occurrence.
h. Please note that the first paragraph of a section or subsection is not indented. The first paragraphs that follows a table, figure, equation etc. does not have an indent, either.

4.2 Suggestions

The leader of the graving dock XYZ area should strive to prevent work accidents, especially fires, with all parties by:

1. Always conduct risk assessments before starting work. The researcher recommends using the Failure Modes Effect and Criticality Analysis (FMECA) method.
2. Before conducting high-risk hot work such as cutting, welding, and grinding, hot work equipment must be inspected and strictly supervised by firefighters inside and outside the ship. Ensure the number of firefighters monitoring inside and outside the vessel is sufficient before starting hot work.

3. A monitoring program has been developed as a continuous improvement mitigation measure. It upgrades the existing CCTV cameras in the graving dock XYZ area into a fire early warning system based on the Internet of Things (IoT) for real-time notifications, which has proven effective in reducing the initially high fire risk to a lower level.

References

Akyuz, E., Celik, E.: A quantitative risk analysis by using interval type-2 fuzzy FMEA approach: the case of oil spill. Marit. Policy Manage. **116**,(2018). https://doi.org/10.1080/03088839.2018.1520401

Ahmed, S., Gu, X.C.: Accident-based FMECA study of Marine boiler for risk prioritization using fuzzy expert system. Results Eng. **6**(November 2019), 100123 (2020). https://doi.org/10.1016/j.rineng.2020.100123

Goo, B., Lee, J., Seo, S., Chang, D., Chung, H.: Design of reliability critical system using axiomatic design with FMECA. Int. J. Naval Archit. Ocean Eng. **11**(1), 11–21 (2019). https://doi.org/10.1016/j.ijnaoe.2017.11.004

Nugraha, E., Sari, R.M.: Analisis defect dengan metode fault tree analysis dan failure mode effect analysis. Organum: Jurnal Saintifik Manajemen Dan Akuntansi, **2**(2),62–72 (2019). https://doi.org/10.35138/organum.v2i2.58

Rahman, A.: Penggunaan Metode Fmeca (failure modes effects criticality analysis) Dalam Identifikasi Titik Kritis Di Industri Kemasan. Jurnal Teknologi Industri Pertanian, **31**(1), 110–119 (2021). https://doi.org/10.24961/j.tek.ind.pert.2021.31.1.110

Siswantoro, N., Shaneza, F., Pitana, T.: Criticality assessment for marine diesel engine using failure mode and effect criticality analysis (FMECA) approach: case study on lubricating oil system. Int. Rev. Mech. Eng. (IREME) (2020)

Standard Operating Procedure K3LH PT XX Tentang Identifikasi Bahaya dan Penilaian Risiko (2021)

Suwanasri, C., Saribut, S., Suwanasri, T., Phadungthin, R.: Risk analysis using failure modes, effects, and criticality analysis for transmission network assets. Energies **14**(4) (2021) . https://doi.org/10.3390/en14040977

Zaman, M.B., Baheramsyah, A., Ashari, I.: Analysis of Work Accident Factors in the Shipyard. In: IOP Conference Series: Earth and Environmental Science, vol. 698, no. 1 (2021). https://doi.org/10.1088/1755-1315/698/1/012016

Relationship Between the Meaning of Symbolic Messages and Spatial Order in the Penti Ceremony of the Todo Indigenous Community in Manggarai (A Semiotic Approach)

Eleonora Vilgia Putri Beyan(✉)

Prodi Teknik Sipil Fakultas Teknik Universitas Katolik Indonesia St. Paulus Ruteng, Jln. Ahmad Yani 10, Tenda, Ruteng District, Manggarai Regency, East Nusa, Tenggara 86511, Indonesia
vhyrrabeyan@gmail.com

Abstract. Indonesia has thousands of traditional villages. One of the traditional villages is in Todo Village, West Satar Mese District, Manggarai Regency, East Nusa Tenggara Province. The traditional village known as the Todo Tribe Traditional Village has a settlement pattern dominated by ritual practices. In this context, culture is a determining factor in the process of forming settlements. The people of the Todo Tribe Traditional Village have maintained this tradition for generations. One of the traditions that is the focus of this research is the annual penti ritual. This ritual is a practice of thanksgiving for the crops carried out by the community. This series of rituals has influenced the uniqueness of this village settlement pattern. This study was conducted through qualitative methods with the type of field research. The analysis uses semiotics, namely the science of signs, so that it can explore and relate the elements of form, function and meaning to the Todo Tribe Traditional Village (interviews and field observations). Graphical analysis is used to describe and understand the visual aspects of the Todo Tribe Traditional Village which are appropriately presented graphically (charts or pictures). In the end, the existence of the settlement of the Todo Traditional Village is spatially determined by form, function and meaning (verbal and non-verbal symbolic messages).

Keywords: Spatial arrangement · Semiotics · Penti ceremony · Todo Tribe Traditional Village

1 Introduction

In traditional villages, located far from urban centers, there are customary settlements. These customary settlements represent the community's adaptation to their environment, grounded in their beliefs and worldview, manifested in traditional settings (customary environment). The process of settlement occurs within a structured housing unit following a settlement pattern, which is then continued through each generation. Settling is closely related to places and spatial patterns created by humans as a means for their activities, such as working, recreating, residing, and aspiring, or realizing a life perspective that includes symbolic aspects of space (Samadhi 2004). Customary settlements are

M. Awang (Ed.): ICCAEET 2024, PTT, pp. 274–286, 2025.
https://doi.org/10.1007/978-3-031-94347-8_24

cultural legacies passed down from generation to generation that maintain identity and dignity, and the ability to sustain balance in all aspects of life. Knowledge and experience acquired by the community through ancestral heritage are formed naturally and passed down through generations. Today, this heritage is preserved amidst modernization, which increasingly erodes local cultures in Indonesia.

Culture and space shape the spatial organization of settlements. Human backgrounds, including life perspectives, beliefs, values, and norms, impact an individual's behavior. The way one lives and the roles they choose in society are examples of this impact. Rapoport's (1977) study of behavior demonstrates this. This sociocultural paradigm underlies human systems and activities. Lifestyles and activity systems underpin the types and media for related activities. Form refers to spaces that are interconnected over a period. The emphasis on cultural identity in communal spaces is a manifestation of cultural identity (cultural identity), which in this case is local wisdom (genius loci). The concept of "place" through the theory of genius loci is used to identify a space in a traditional village as accurately as possible. Space in a traditional village has meanings derived from human experiences. By using the theory of genius loci, we identify how the local spirit or essence of a public space can function as a social space.

Customary ceremonies (ritual activities) are related to life events and specific locations within a settlement. Douglas, as cited in Sasongko (2005), explains that a ritual has two aspects: the zone for purification and order, and the zone beyond that which is chaos, based on the ritual's purpose of protecting the body from harm. Various types of ritual activities are associated with architectural boundaries (space). Certain events and ritual activities are linked to specific locations and have particular settings or arrangements. "Place" is interpreted with various expansive meanings. A place is identified through several relationships, including economic, cultural, psychological, social, and emotional aspects. A space becomes a place when the space is better understood and assigned value (Tuan 1977). Norberg-Schulz (1979) describes a place as a space with unique character. Both Tuan and Norberg-Schulz explain that human experience is a crucial element in the formation of a place. East Nusa Tenggara is one of several regions with cultural wealth embraced by its people from various areas scattered throughout the region. The meaning of a culture can be defined not only based on traditions and customs passed down in each area but also on the social interactions between individuals and groups with different backgrounds and religious beliefs.

Local wisdom (genius loci) is an integral part of local culture and a cultural heritage from ancestors that must be preserved and maintained, especially for the current generation. Traditional villages that endure are communities that consistently uphold their customs. A traditional village is characterized by elements that define it as a group with customary law, with several characteristics that distinguish it from other social groups. These distinguishing features include having clearly defined territorial boundaries, a majority of the population residing there, and the existence of sacred buildings belonging to the traditional village itself. One such traditional village is the Todo Tribe Traditional Village, which is administratively located in Todo Village, North Satar Mese District, Manggarai Regency, East Nusa Tenggara Province.

Todo Village is located at the foot of Mount Anak Ranaka, which was the center of the Manggarai Kingdom's administration in ancient times. Manggarai was the largest kingdom that controlled Flores Island before moving to Ruteng, which is now the capital of Manggarai Regency. One of the pieces of evidence of King Todo's grandeur and his impact on the Manggarai Kingdom is the existence of the traditional house known as niang or mbaru niang, with the central structure being the Niang Mbowang, or the main house. Niang Todo is a traditional house resembling a round stilt house with a conical thatched roof, known as the palace of the former King Todo. The Todo traditional house is the oldest traditional house in Manggarai Regency, making it crucial for study. The lack of specific research on Todo Tribe Traditional Village has prompted this study to enhance knowledge. The unique Todo traditional house has gained global attention and was nominated for the Aga Khan Award for Architecture in 2013 by the United Nations Educational, Scientific and Cultural Organization (UNESCO).

The Todo Tribe Traditional Village has a circular village layout, which includes several main components: 9 Mbaru Niang (traditional houses), Natas (village courtyard), Pa'ang (village entrance gate), and Compang (offering altar) for ceremonial rituals. The community maintains its customs and has a clear understanding of the symbolic meanings embedded in the objects and language used in their rituals. The people of Todo Tribe Traditional Village are agrarian, as evidenced by their inseparable connection to farming and gardening activities. Consequently, the area is well-known for its agricultural and horticultural products such as rice, corn, ginger, and candlenuts. In addition to farming, they also engage in animal husbandry, raising livestock including chickens, pigs, goats, and cattle. Furthermore, they practice weaving traditional Todo songke cloth.

There are many ways that people express their gratitude for the blessings given by God, ranging from prayers during worship to performing local religious, customary, and cultural traditions. One such customary ceremony practiced by the community is the Adat Penti Latang Mori Jari agu Dedek (a thanksgiving ceremony to God the Creator and the spirits of ancestors). The Penti ceremony is a traditional ritual conducted annually, either at the end of the year or the beginning of the year, depending on the agreement among the residents of the Mbaru Niang in the Todo Tribe Traditional Village. The purpose of this ceremony is to give thanks for all the results obtained from rice fields, gardens, livestock, and weaving throughout the year to God Almighty and to the ancestors of Todo who have passed away.

There is also a Penti Mese event that takes place every five years. The rituals used are the same, except for the sacrificial animal. While the annual Penti uses pigs as sacrificial animals, the five- year Penti uses buffalo. The meaning of the Penti Mese is to give thanks for the results obtained from rice fields, gardens, livestock, and weaving over the past five years to God Almighty and to the ancestors of Todo who have passed away.

The religious aspects of the Todo Tribe Traditional Village community remain closely tied to the norms and types of customary ceremonies and the values they embody. The customary ceremonies commonly performed are integral to the community's life or parts of it. Therefore, these ceremonies not only serve as traditional rituals but also function as a repository of local wisdom for the community, as they are intended to be carried out through generations.

As people live and engage in activities in a place, their attachment to that place increases. The formation of a place is a process where human relationships with the place are established through social interactions and activities that imbue meaning to the place. According to Canter (1997), studies on this topic connect physical (tangible) and non-physical (intangible) aspects. Physical attributes relate to spatial configuration, activities, and concepts. The sustainability of the Todo Tribe Traditional Village depends on the meaning and perspective of its community, as these elements significantly impact the formation of the place and are closely tied to the village's sustainability. According to Ferdinand de Saussure (Benny 2014:17), semiotics is a field of study that examines the role of signs as part of social life. Meanwhile, Charles Sanders Peirce (Benny 2014:16) defines semiotics as a field of study that investigates the relationship between signs, objects, and meanings.

In relation to this, the researcher aims to specifically discuss and describe the meaning of both physical (tangible) and non-physical (intangible) aspects found in the Penti ceremonial ritual. The goal is to provide broad knowledge and inspiration regarding the culture that is particularly significant to the Todo Tribe Traditional Village community. This background motivates the researcher to investigate how the ritual processes in the Todo Tribe Traditional Village transform traditional houses (mbaru-mbaru niang) into spaces for ritual activities from both tangible and intangible perspectives, thereby enlivening the concept of place.

2 Method

This research was conducted in Todo Village, Satar Mese Barat District, Manggarai Regency, East Nusa Tenggara Province, according to the location of the Todo Indigenous Village. The research was carried out from February to April 2024. The type of research used was qualitative, employing field research methods. Field research is a form of study aimed at uncovering the meanings attributed by community members to their behaviors and the surrounding reality. The purpose of the research is to explore meaning through symbols in tangible forms, such as batu lekak, kepan, kapen koor, and lampek poor putes found in Niang Mbowang, as well as in intangible forms, specifically the distinct functions of each space and the five levels of the Niang Todo house, which are philosophically interconnected in the Penti customary ceremony. Data collection in the field was conducted through observation and interviews. Field observation was carried out by visiting the Todo Indigenous.

Village, which is the focus of this research. The type of interviews conducted was semi-structured interviews with the residents of the Todo Indigenous Village, who are the subjects of the study.

Data analysis was conducted qualitatively through semiotics. Qualitative research can reveal individual experiences in interacting with society in daily life and study groups and experiences that might not be previously known (Bogdan and Taylor, in New York: John Wiley & Sons, 1975:4–5). Semiotic (comparative) analysis can delve into and connect elements of form, function, and meaning related to the Todo Indigenous Village (interviews and field observation). Graphic analysis is used to outline and understand visual aspects of the Todo Indigenous Village that are best presented graphically (charts or images). This is then organized into comparison tables of form, function, and meaning.

3 Result and Discussion

Customary ceremonies or rituals are routine activities or practices often carried out by specific communities or in particular regions. These ceremonies can take various forms according to local customs, such as weddings, harvest celebrations, and others. Such ceremonies are intended to preserve traditional customs, which are part of the cultural heritage that must be maintained, and to continue the legacy of ancestors established long ago. One example is the Penti ritual in the Todo Indigenous Village (Fig. 1).

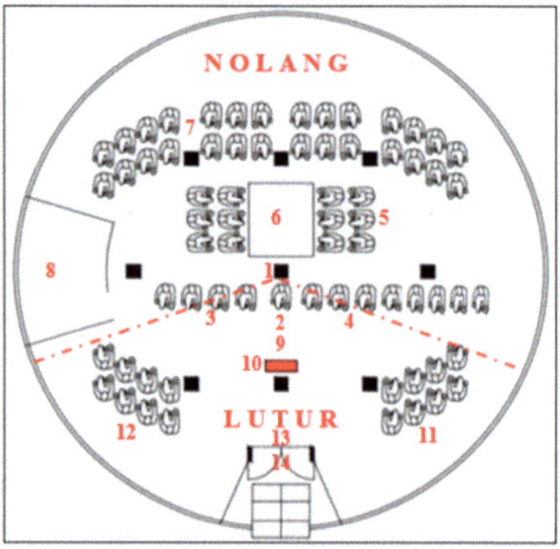

Fig. 1. Layout of the Penti ritual at Mbaru Niang Mbowang

Caption:

1. Main pillar (siri bongkok).
2. Person performing the torok (prayer and ritual leader).
3. Representative of wae tu'a(elders) from the direct descendants of Niang Mbowang.
4. Representative of wae tu'a (elders), 8 (alo) people from each niang.
5. Women (mothers and teenagers) cooking.
6. fireplace (hapo)
7. Core family (women and children).
8. Room (loang)
9. Place for tying and slaughtering sacrificial animals.
10. Place for offerings (watu lekek) and wooden plate (dulang)
11. Guests (wo'e), including anak rona and anak wina.
12. Household head (core family from 9 niang).
13. Stone/box (kapen koor) and decorative patterns (lampek poor potes).
14. Door (para).

3.1 The Penti Ritual

The Penti ritual is a traditional ceremony conducted annually, either at the end of the year or the beginning of the year, depending on the agreement of the residents of the Todo

Indigenous Village. The purpose of this event is to give thanks for the results obtained from rice fields, farms, livestock, and weaving over the year to God Almighty and to the Todo ancestors who have passed away. There is also a Penti Mese event held every five years. The rituals used are the same, but the sacrificial animal differs. While the annual Penti uses pigs as sacrificial animals, the five-year Penti uses buffaloes. The meaning of the Penti Mese is to express gratitude for the results obtained over the past five years from rice fields, farms, livestock, and weaving to God Almighty and the Todo ancestors who have passed away. The Penti ritual involves all members of the community who are bound in a genealogical unity, actively participating without exception. The starting point of the Penti ritual is a communal meeting of all village residents at the traditional house. Traditional leaders such as tu'a golo, tu'a tembong, tu'a teno, and tu'a-tu'a panga invite all residents to sit together and discuss the plans for the Penti ritual.

Pre-Penti

Before the main event, around 7:00 PM local time, each family performs a preparatory ritual called lugit kilo in the morning at their respective niang houses. This ritual is conducted to avoid any bad occurrences, ensuring the main event proceeds smoothly. If any technical errors occur during the main ceremony, the lugit kilo is meant to protect each family from adverse effects. During the lugit kilo, materials used include manuk cepang (a type of rooster), cepa (areca nut), and tuak (palm wine). Prayers are offered to the Creator and to the Todo ancestors who have passed away. Following this, the main event, torok manuk, begins, where the family presents their wishes and intentions. The manuk cepang or rooster is held by the head of the family or the ritual leader, and after the ritual, the rooster's neck is cut to collect its blood in a container. This is followed by toto urat di'a, examining the rooster's internal organs to determine the success or failure of future plans. Finally, helang (offerings) are prepared as a tribute to the ancestors, with the roasted rooster served with warm rice to the Creator and the Todo ancestors who have passed away.

Barong Wae

In the afternoon at 3:00 PM, the Barong Wae ceremony is conducted at a spring located about 400 m from Niang Mbowang. Representatives from each family of the Niang walk together to the spring. The purpose of this ceremony is to symbolically cleanse themselves so that the Creator and the ancestors who have passed away are prepared to be invited to the main event later. The sequence of the Barong Wae ritual is as follows: (1) teing sepa, offering betel and areca nuts to empo and naga; (2) teing tuak, presenting village chicken eggs as a drink for the naga (spirits of nature); (3) tura manuk, reciting traditional prayers while holding the sacrificial animal (chicken); (4) mbele manuk, ritualistically cutting the chicken with specific intentions for each different color of chicken; (5) toto urat, interpreting signs of approval or disapproval of the requests to God, empo, and naga, based on the chicken's intestines (and also pig liver); (6) hao helang, mixing rice and meat on a plate with hands while reciting traditional prayers; (7) hesak helang, offering food to empo and naga, consisting of chicken meat and warm rice.

Roi Boa

The ritual of summoning empo (ancestors) at the village cemetery is performed to invite them to attend the main Penti ritual, where sacrificial animals will be offered to the ancestors at Mbaru Mbowang. Each descendant from the 9 niang sacrifices a manuk rae (red hen). The sequence of this ritual is the same as that performed during the Barong Wae ceremony, except that the teing tuak (offering of palm wine) is not included.

Barong Compang

After the Roi Boa ritual, the participants gather at the pa'ang (village gate) and walk together to the natas. The participants proceed to the compang to perform teing cepa, teing tuak, tura manuk, and helang.

The Main Penti Event

At 7:00 PM local time, the gong is struck from Niang Mbowang or the main niang, and all members of the Todo community from each Niang proceed to Niang Mbowang and sit on the lutur inside the house, as do any guests (woe). The materials used for this event include a white chicken or manuk bakok, cepa or betel, one pig, and tuak or palm wine. Cepa or betel is placed on a dulang called kepan. The white chicken or manuk bakok is held by the traditional leader conducting the ritual, while the pig is tied to the lutur or sitting area inside the house by several men who will later slaughter it during the main event.

The ritual continues with teing mangko mese latang ise empo, where prayers are offered in the Manggarai language, concluding with the chicken's neck being cut to collect its blood on a stone called watu lekak. This is followed by toto urat, where the chicken's internal organs are examined to predict the success or failure of future plans. The final part of the ritual is helang (offerings), which are presented to the ancestors. The chicken is roasted and served with warm rice to the Creator and the Todo ancestors who have passed away, and placed on the watu lekak. After the ancestors are believed to have eaten the offerings, each attendee takes turns eating a small portion of the helang, which is believed to bring prosperity and health from the ancestors.

The climax of the Penti event is mbele ela penti. The pig tied to the lutur or sitting area inside the house is slaughtered. After the slaughter, it is taken to Niang Rato, the kitchen, where the women begin preparing it for dinner (Fig. 2).

Caption (Figs. 3, 4 and 5):

1. Niang houses (mbaru-mbaru niang)
2. Spring (barong wae)
3. Village gate (pa'ang)
4. Public cemetery (boa)
5. Offering altar (compang)

3.2 Mbaru Niang Mbowang

The interior of the Mbaru Niang resembles a large circular hall. The roof framework, made up of bamboo arranged in a circular pattern, also forms the walls of the niang.

Fig. 2. Map of the Penti ritual location in the Todo Indigenous Village.

Fig. 3. Gong and drum at Mbaru Niang Mbowang.

The largest circle is at the bottom with the most bamboo, gradually tapering towards the center of the roof at the top of the niang. This represents parliamentary democracy, according to Mr. Walter Mohon in an interview on March 27, 2023, who manages the site and is also a traditional leader. The bamboo at the bottom appears like millions of people from the Manggarai Kingdom sitting around their king who is perched at the top of the niang, in a grand assembly forum. "The highest point is the leader. Starting with the deity, followed by mori, adak, and then the king," he said. From top to bottom, the

Fig. 4. Gong and drum at Mbaru Niang Mbowang.

Fig. 5. Place for offerings (watu lekak) and wooden plate (dulang) at Mbaru Niang Mbowang.

framework of the niang roof symbolizes the hierarchy of leadership. At the top level is the king, followed by dalu, equivalent to a governor. It is not surprising that nearly all districts in Manggarai were formerly referred to as kedaluan. Below this is gelarang, equivalent to a regent, overseeing komo kaba, equivalent to a sub-district head, and finally at the bottom level are ro'eng or the common people.

The nine niang are arranged in a semi-circular formation like a council. Each niang represents the position of the king and his ministers. At the center stands Niang Mbowang, the residence of the king (Fig. 6).

Fig. 6. Mbaru Niang Mbowang.

In addition to its larger size compared to other traditional houses, this one features a distinctive ornament at its peak called a "periuk," which sets it apart from the others. The entrance door is small and short, requiring one to bow upon entering, symbolizing respect for the homeowner. Furthermore, upon entry, one steps onto a stone that was once a chest called "kapen koor." Anyone who enters the house with ill intentions will face consequences and will not return safely.

There are about five carvings on the top of the entrance door of a house named Lampek Poor Potes, which signify a man (father) waiting outside the house while his wife is giving birth. He would strike the wall of the room from outside, and on the fifth strike, the cutting of the newborn's umbilical cord would occur. This represents a vow or promise that a daughter (ata pe'ang) is considered an outsider who does not have inheritance rights, unlike a son (ata one).

Niang ini was the residence of the king in ancient times. Whenever there was an issue, King Todo would gather his eight ministers for a deliberation in Niang Mbowang. Naturally, not every discussion reached a consensus. To this day, Niang Mbowang remains the main niang, where all important discussions, customary rituals, and other significant matters of the Todo Indigenous Village are held (Figs. 7 and 8).

Fig. 7. The kapen koor in Mbaru Niang Mbowang.

Fig. 8. Lampek poor putes in Mbaru Niang Mbowang.

The Niang House Has Rooms With Different Functions:

1. The first floor is a resting area for the residents, divided into three sections: nolang, loang, and lutur. Nolang and loang (private areas) are used for activities, including cooking, while lutur (public area) is where guests engage and rest. This arrangement

reflects the local culture of mutual respect between residents and visitors. Despite the spatial separation, they live on the same floor with full tolerance.

2. The second floor, known as wase mese, serves as storage for food and various daily necessities of the Todo people.
3. The third floor, called wase koe, is used for storing seeds for staple crops such as corn, rice, and legumes.
4. The fourth floor, known as lemparea, is designated for storing food supplies in case of drought.
5. The fifth floor, called lengkiang, is used for offering sacrifices to ancestors (Fig. 9).

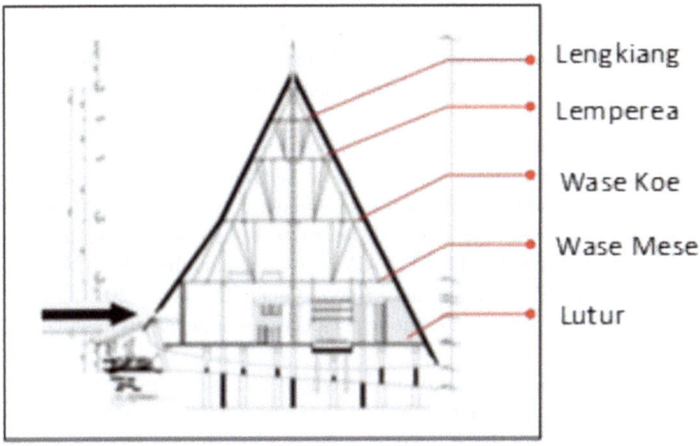

Fig. 9. The floor levels in Mbaru Niang Mbowang.

The Five Levels of the Todo Niang House Each Have Their Own Philosophy:

1. First level (lutur): Ro'eng
2. Second level (wase mese): Komo Kaba
3. Third level (wase koe): Gelarang
4. Fourth level (lemperae): Dalu
5. Fifth level (lengkiang): Raja

4 Conclusion

In the Penti ritual at Kampung Adat Todo, the boundary between the sacred and social spaces is marked by the seating arrangement of the customary elders. The ritual leader, Torok, sits in front of the siri bongkok, with the Wae Tu'a from the direct lineage of Niang Mbowang to his right and a representative of Wae Tu'a from the eight Niang to his left. Nolang is the area where women and children are active, as they are considered crucial for the continuity of the family and protection from external threats. The traditional residential space in Kampung Adat Todo is shaped by their culture, including the Penti ritual, which reflects societal values such as belief in Empo (ancestors), Naga (spirits of nature), and Mori (the Supreme Being), as well as the patrilineal social system of Wa'u or Ase Ka'e. Penti also serves to strengthen relationships within the clan, family, and relatives, and acts as a medium for artistic expression and reconciliation, emphasizing the importance of social bonds and gratitude towards God and ancestors. This ritual maintains the connection between humans, God, and ancestors, and is preserved by future generations.

References

Berger, A.A.: Pengantar Semiotika. Yogyakarta: Tiara Wacana (2010)
Budiman, K.: Semiotika Visual: Konsep, Isu, dan Problem Ikonsitas. Yogyakarta: Jalasutra (2011)
Fiske, J.: Pengantar Ilmu Komunikasi.TerjemahanolehHapsari Dwiningtyas 2014. Jakarta: Rajawali Pers (2012)
Geertz, C.: The Interpretation of Cultures. Basic Books, Inc., New York (1973)
Hoed, H.B.: Semiotika dan Dinamika Sosial Budaya. Jakarta: Komunitas Bambu (2011)
Ida, R.: Metode Penelitian Studi Media dan Kajian Budaya. Jakarta: Prenada Media Group (2014)
Riskiyanto.: Simbol-simbol Budaya dalam Upacara Adat Mogundam manuk Totolu, Kajian Semiotik. Skripsi sarjana pada FKIP UNTAD Palu: tidak dipublikasikan (2014)
Vera, N.: Semiotika Dalam Riset Komunikasi.Bogor: Ghalia Indonesia (2014)
Budi, S.: Cultural symbols in rituals: a semiotic perspective on indigenous practices. J. Cult. Stud. **12**(3), 45–60 (2020)
Dewi, R.: The role of space in traditional ceremonies: a study of Penti in Manggarai. J. Ethnographic Res. **8**(2), 112–130 (2021)
Sari, M.: Symbolic meanings in rituals: analyzing the penti ceremony of the todo community. Int. J. Anthropol. **15**(1), 27–42 (2022)
Prasetyo, A.: Rituals and their spaces: exploring the penti ceremony in Manggarai. J. Southeast Asian Stud. **29**(4), 78–95 (2023)
Wulandari, N.: Semiotics of indigenous rituals: a case study of the penti ceremony. J. Indigenous Stud. **6**(3), 50–67 (2022)
Suharto, J.: Cultural spaces and symbolism in Manggarai ceremonies. Asian J. Cult. Anthropol. **19**(1), 34–50 (2023)
Fitriani, E.: The intersection of symbolism and space in indigenous ceremonies. J. Ritual Stud. **11**(2), 90–105 (2021)
Kusuma, D.: Understanding rituals through a semiotic lens: the case of Penti. J. Cult. Anthropol. **14**(4), 67–82 (2020)
Rahayu, T.: Symbolic communication in traditional ceremonies: insights from the todo community. J. Commun. Stud. **18**(1), 22–39 (2024)
Yulianto, A.: Examining the spatial dynamics of rituals in Manggarai. J. Southeast Asian Cult. Stud. **25**(2), 112–126 (2023)

Correction to: Mitigation of Pier Structure Failure Due to Sediment

Muhammad Hawari Ikhwan, Jojok Widodo Soetjipto,
and Retno Utami Agung Wiyono

Correction to:
Chapter 1 in: M. Awang (Ed.): *1st International Conference on Civil, Architecture, Environmental Engineering, and Technology*, PTT, https://doi.org/10.1007/978-3-031-94347-8_1

In the original version of the book, the following correction has been incorporated in chapter "Mitigation of Pier Structure Failure Due to Sediment": The Author name has been corrected from Jokok Widodo Soetjipto to Jojok Widodo Soetjipto.

The updated version of this chapter can be found at
https://doi.org/10.1007/978-3-031-94347-8_1

M. Awang (Ed.): ICCAEET 2024, PTT, p. C1, 2025.
https://doi.org/10.1007/978-3-031-94347-8_25

Author Index

M. Awang (Ed.): ICCAEET 2024, PTT, pp. 287–288, 2025.
https://doi.org/10.1007/978-3-031-94347-8

MIX
Papier aus verantwortungsvollen Quellen
Paper from responsible sources
FSC® C105338

If you have any concerns about our products,
you can contact us on
ProductSafety@springernature.com

In case Publisher is established outside the EU,
the EU authorized representative is:
Springer Nature Customer Service Center GmbH
Europaplatz 3, 69115 Heidelberg, Germany

Printed by Libri Plureos GmbH
in Hamburg, Germany